AMERICAN
LUCIFERS

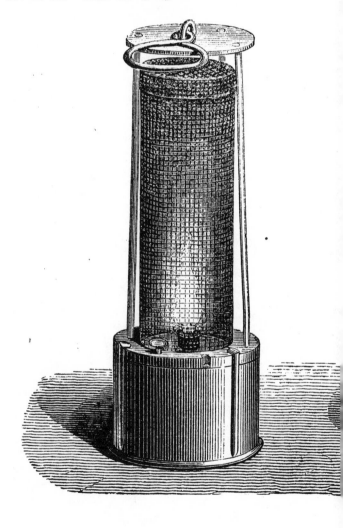

AMERICAN LUCIFERS

LUCIFERS

THE DARK HISTORY OF

ARTIFICIAL LIGHT

1750~1865

Jeremy Zallen

THE UNIVERSITY OF NORTH CAROLINA PRESS

Chapel Hill

This book was published with the assistance of the Anniversary Fund
of the University of North Carolina Press.

Designed by Jamison Cockerham
Set in Arno, Scala Sans, Rudyard, and Trattatello
by Tseng Information Systems, Inc.

Manufactured in the United States of America

The University of North Carolina Press has been a member
of the Green Press Initiative since 2003.

Portions of the epilogue appeared previously in Jeremy Zallen, "'Dead Work,' Electric
Futures, and the Hidden History of the Gilded Age," *Montana: The Magazine of
Western History* 66, no. 2 (Summer 2016): 39–65. Reprinted with permission.

Cover images: Top: Various designs of student kerosene and oil lamps. [1881?].
Charles F. A. Hinrichs, engraver. Prints and Photographs Division, Library of Congress,
LC-DIG-pga-07088. Bottom: Udo J. Keppler. *Lights and Shadows* (detail). 1909.
Prints and Photographs Division, Library of Congress, LC-DIG-ppmsca-26412.

LIBRARY OF CONGRESS CATALOGING-IN-PUBLICATION DATA
Names: Zallen, Jeremy, author.
Title: American lucifers : the dark history of artificial light, 1750–1865 / by Jeremy Zallen.
Description: Chapel Hill : University of North Carolina Press, [2019] |
    Includes bibliographical references and index.
Identifiers: LCCN 2019002918 | ISBN 9781469653327 (cloth : alk. paper) |
    ISBN 9781469653334 (ebook)
Subjects: LCSH: Lighting—United States—History—18th century. |
    Lighting—United States—History—19th century. | Lighting—
    Economic aspects. | Lighting—Political aspects. | Labor—United
    States—History—18th century. | Slave labor—History—18th century.
Classification: LCC HD9684.U62 Z34 2019 | DDC 338.4/762132097309034—dc23
    LC record available at https://lccn.loc.gov/2019002918

For Jenny

Contents

Illustrations

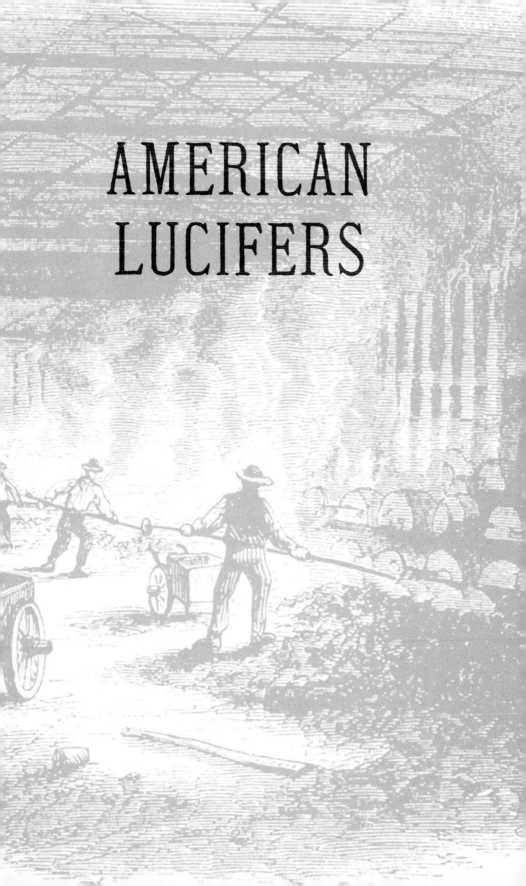

AMERICAN
LUCIFERS

PROLOGUE

Candles again.

At least this time her daughter was there to help. Martha Ballard had been making candles for over a decade, likely for most of her sixty-three years. She had probably repeated the hot, messy task each winter since she was a girl helping her own mother in Massachusetts. She was hardly unusual. Women all across the North Atlantic had, for generations, been slaughtering cows and trimming fat and guts from the carcasses. They had twisted wicks and boiled tallow. They had spent days and sometimes weeks every year dipping those wicks in the searing hot fat. Like their mothers and like their daughters, countless women had hung the dips until they hardened, then dipped again and again until the candles reached the proper size. For most of her adult life, Ballard had made nearly all of the lights consumed by her Maine household, and she'd often had to do it alone. It was no small thing that someone had come to the house to share the work with her.[1]

On February 21, 1797, Ballard recorded in her diary that her daughter Dolly Lambart "Spun me yarn for Candle wicks." If Ballard, who was a midwife, hadn't been called away to deliver a child, the candle making would likely have begun that same evening. But the delivery took all night, and Ballard returned home only at nine the following snowy morning, undoubtedly exhausted. Worse, she knew she had no time for rest. Ballard willed herself through the fatigue, and on February 22, "Dagt Lambart and I made 43 Dzn Candles." Ballard and Lambart began by twisting the spun cotton yarn into wicks, cutting them to the right length, and then doubling the wicks over specially made rods. While one of them stoked a fire under a kettle, the other cut in slices of the tallow from the cow killed in December until the kettle was full. When all the tallow was melted, they dipped the rods of wick-pairs into the boiling fat—which had to be kept continually boiling for hours—set the dips to cool, and repeated the dipping over and over until they had 516 candles, each containing slightly less than an ounce of tallow. The process took hundreds of dippings and hours of continuous labor.[2]

The work of candle making drew together not only mother and daughter, but the whole family. As Ballard noted, while they made candles, Lambart's "husband Came and Sleeps here." And the work drew the whole family into a local community of energy joining sun to land to cow to flame to people. The wicks, spun by Ballard's daughter from raw cotton likely grown by enslaved women and men on U.S. or West Indian plantations, drew further-flung connections. Most of the time, Ballard purchased, picked, carded, and spun the raw cotton into skeins herself. But neighbors often gave her their own spun cotton as payment for her midwife services. As for the tallow, a few months earlier, two of Ballard's sons had helped extract the candle fuel when they "Came and assisted us to Butcher a beef Cow." Beginning the slaughter at noon with her sons, Ballard had "Cleand the Tripe & feet and Tried the Tallow" by rendering the fat from all the other flesh and bone in a wood-fired kettle. Once the tallow was rendered, Ballard had "Straind it to cool. I Closed my Evngs work at 11h 30m." The day's labor required to fully unmake this cow—after, of course, raising and feeding it for a year—yielded enough fat to make ninety dozen candles.[3]

It was tedious work. But it could also be dangerous. "My Tallow Cought fire and alarmed us much," Ballard wrote after she and Lambart finished making the last of the candles. By then a seasoned hand, Ballard got the situation under control, putting out the flames and salvaging the remaining tallow. Lambart was not yet so capable. Panicked and breathing in smoke, "my Dagt fainted by reason of her Surprise."[4]

Unsurprisingly, rural women held candle making in a special kind of contempt. Writing over a century later, the Kansas settler Adelea Orpen recalled that besides dish washing, which she "loathed, there was another domestic job which went very hard with me—that was candle-making.... Maybe it was because in my awkwardness I scalded my fingers in the boiling fat, or scratched them when putting in the wicks into the moulds. Anyhow I hated it, and let Auntie know the fact." Orpen's story is a familiar one, marking history as a climb up from labor, a transcendence of work through new technologies of light. Orpen's "Auntie" marked her own life in much the same way, telling her ungrateful niece how well she had it in the age of matches, when at least she didn't have to hike miles through the snow to "borrow fire" from a neighbor in order to light hearth and home.[5]

Beginning in the late nineteenth century, a similar, if differently gendered, narrative came to redefine lighting as a thing of progress, history, and pleasure, rather than work, routine, and pain. In 1891, a few decades before Adelea Orpen published her *Memories of Old Emigrant Days in Kansas*,

Senator Orville Platt stood before the Congress of the United States and announced the electrical apotheosis of man. "Formerly we ascribed creative faculty or force to the Divine Being alone," Platt noted modestly. But "now when we look upon the wondrous contrivances and inventions everywhere contributing to our life wants and adding to our life enjoyments, we are forced to exclaim: 'Behold the expressed thought of the creator — man!'" Just in case his meaning was not clear enough, Platt assured his fellow legislators, "If you will think as you come to this place this evening how the thought of man has transformed black coal and viewless electricity into the agents which light your pathway, you will feel it scarcely irreverent to exclaim: 'And man said, "Let there be light," and there was light.'"[6]

Few symbols have better given form to the elusive concept of progress than the before-and-after image at the core of the myth of electric light. First we had campfires and dark nights; now we have electricity and are free of the cycles of sun and moon, of the dangers of fire, of the gravity and labor of living in nature. We've heard the story, or some version of it, our whole lives. It is celebrated in museums, schools, and even car commercials. It was also the story many people who lived through these changes told. To be fair, like all good myths, this one was at least partly true, and its truths can reveal important insights about those who told it. It was true, as Adelea Orpen recounted, that buying lights at the store was markedly easier and more pleasant than making her own candles. When possible, women in both country and city eagerly abandoned homemade candles for illuminants available in local markets, even when manufacturing candles at home would have been cheaper.[7] It was true, as Senator Platt declared, that electric lighting was, in nearly every way (except, at the time, cost), an improvement over gaslight. But such progressive narratives were, at best, half-truths. Between the bright lines of the myth of electric light is a dark history in need of telling, one that can change how Americans think not only of the eighteenth and nineteenth centuries but of their ever-more-automated present. *American Lucifers* tells such a history, exploring the struggles among the makers and owners of the overlooked but world-changing lights that illuminated the transformative century between Martha Ballard's tallow dips and Senator Platt's incandescent filaments.

Forget "the" candle, or "the" electric light. There was never just one. Instead, imagine two tallow candles, both made in 1797. The first sat in Martha Ballard's home; the second, in the hands of a Peruvian silver miner. Examined by a chemist or a candle collector, the two candles would have been indistinguishable. Chemists would say they were the same, and so would most historians of technology. They'd point out that both were made of rendered beef fat, that the wicks were cotton, and that they were nearly identical to candles from hundreds of years before, save for the small "innovation" of the cotton wicks. But the two candles were not the same. Not even close.

If, rather than trying to unlock the secrets of the candles by dissecting them, we instead try to see these candles as historically embedded energy and matter in process, then the means of light would become the question, and that would make the story deeply human, and more-than-human too.[8] Out of historical context, the candles still looked basically identical. In context, the two candles were radically, categorically different. The fact that Martha Ballard *could* have made candles with linen wicks is important, but more so was that she actually used cotton, tying herself through her lights to a political economy of slavery. Miners in the pits of Potosí *could* have used other light sources, but they didn't, and by burning historically specific tallow to illuminate historically specific treasure-producing mines, they made and remade historically specific social worlds of cattle and slaughter and colonialism.

The biography of the Ballard tallow candle flame traced ecological and social relations that were at once intensely local and unavoidably long-distance. The flame consumed the work and energy of the multiple generations of Ballards who had come together to slaughter an individual cow, render its fat, and dip the candles with wicks made of slave-grown cotton that Ballard or her neighbors had carded and spun themselves. The raw cotton had journeyed hundreds, perhaps thousands of miles: grown and harvested by kidnapped African men and women or their descendants, whom enslavers held captive in either West Indian or U.S. slave labor camps, cotton fibers were then shipped up New England rivers by merchants and finally purchased by white women who processed the fibers for local and national markets. The tallow fuel of the candle, on the other hand, had traveled no more than a few miles, while the light of that flame helped Ballard and her family do the work necessary to secure and reproduce those same locally knit relations on the margins of a capitalist world economy.[9] In the making and use of such candles, rural communities made and remade their lives with and against intensifying capitalist production, sharing and burning local animal fats around Atlantic cottons to keep the capitalist world at arm's length, but always within reach.

In Peru, meanwhile, the tallow candle flame lit the underground labor of an Indian man hacking or hauling silver ore — South American metal that was monetizing social relations from Europe to China, field to ship, factory to Martha Ballard's farmstead—in one of the massive mines of Potosí. Conscripted into the mine through the forced labor draft system called the *mita*, the man may have come from one of the many surrounding villages raising and sending men to work underground, and raising and sending cattle to be slaughtered for food and candles. The Spanish had transformed the Inca tribute system into a forced labor regime reorienting the whole of Peru into the service of silver, from land to cattle to people. The regime allowed officials and those who rented the drafted laborers to exploit not just the individuals but the communities from which they came. Forced to work well past the "legal" hours of shifts, and with contractors providing only enough candles to light legal work times, the under- or unpaid miner had to illuminate the remaining hours of work with candles he purchased himself. The thousands of Peruvians forced into working in the silver and mercury mines of Potosí spent more than 25,000 pesos a year on candles, money neither they nor their communities could afford to part with. Contractors regularly cheated miners even further by providing them with short candles, and all while demanding that drafted Indians bring to the mines, as additional tribute, cotton or wool that their villages had produced, fibers that candlemakers twisted into wicks. But Potosí's hunger for candles (and people) could not be sated locally or even regionally. The "pull of silver" drew Indian captives from across the Americas into the mines of Potosí and Zacatecas. It also drew thousands of tons of tallow from hundreds of thousands of cattle from New Spain, the Andes, the Pampas, and California into those silver miners' candles. And those cattle, which fed and reproduced themselves on land contested by other animals and people, were raised, slaughtered, processed, and transported by mission Indians, enslaved African and Indian cattlemen, and migrant workers from Hawaii and Europe. This second candle may have been chemically indistinguishable from Martha Ballard's, but behind and before its flame lay vastly different geographies of life, labor, and power.[10]

One of the central transformations that this book explores concerns the radical changes in geographies bringing lights to life. The shift from a world in which most lights were reproduced through relations similar to those yielding Ballard's tallow dips—a social biology that was essentially an extension of a single cow congealed around threads of Atlantic cotton—to the kind of global capitalist geographies governing the reproduction of Potosí candles was an important, if underappreciated, process in the making of Atlantic, Pacific, and American capitalism. The new, brighter, cheaper lights that allowed homes, cities, and factories to push back the night consumed energy accumulated by hundreds of thousands of laborers mobilized to work and die in mines, forest camps, farms, factories, ships, and oil wells.

Melville had it right. "For God's sake," he implored the readers of *Moby-Dick,* "be economical with your lamps and candles! not a gallon you burn, but at least one drop of man's blood was spilled for it."[11] Behind the famous and celebrated inventions and science, the surprisingly violent work and struggles to produce, control, and consume the changing means of illumination over the course of the eighteenth and nineteenth centuries transformed slavery, industrial capitalism, and urban life in profound, often hidden ways. Only by taking whalemen, woodsmen, and needlewomen as seriously as inventors and businessmen can the full significance of these social transformations be understood. This book tells a story of the relationships among the living (and dying) makers and consumers of lights — those American lucifers — and the worlds they made in the process during the first century of the industrialization of light, a century before electricity.

When Senator Platt proclaimed electric light to be the visible manifestation of the expressed thought of man, he was not merely engaging in scientific boosterism, exaggeration, or blasphemy. He was making a political statement. Some of the most predominant tensions in the eighteenth and nineteenth centuries arose from the fierce debates over the relationships between work, progress, technology, and capitalism. Were streetlamps to be for the state, the middle-class public, or the working poor? Should women work by the light of private lamps in private homes for public commerce? Were whaleships factories and were whalemen wage laborers in the new national economy, or were they remnants of a colonial past? What about the enslaved men working in turpentine camps, coal mines, and gasworks? Could they possibly be part of modernity? Could hogs, swineherds, and slaughterhouse floors really be the foundations for advancement in the illuminating arts? Taken together, these were really two questions. First, how would light be achieved and who would do the work? Second, who should be seen, literally, in the making and illuminating of the modern world?

Dodging such political questions and instead focusing on and celebrating *systems* of light seemed to relieve a deeply held anxiety shared by many middle- and upper-class Americans about their reliance on workers, the poor, nonwhites, servants, and immigrants they neither trusted nor thought belonged in a "free" republic. Senator Platt, Thomas Edison, and the other electric boosters who celebrated the ascendance of a world in which thought became reality were espousing not only an understanding of progress and the future by which society would transcend work in the abstract, but a vision in which technology would eliminate the presence or need for the actual workers. One early advertising campaign for the General Electric Company even encouraged well-off women "to solve your servant problem" with the Edison Mazda light, "The Lamp that Lights the Way to Lighter Housework." After all, as one historian noted, "electrical appliances could not talk back."[12]

Senator Platt's declaration of victory over God and Nature was, at best, premature. Electric lighting was around for decades before anything resembling a grid or a universal public lighting utility emerged. In trying to carve out little kingdoms of their own in cities dominated by gaslight and kerosene, electric boosters worked tirelessly to sell their lights as the inevitable and glorious future. That many of these late-nineteenth-century electric visions eventually seemed to come true should not, however, be taken as proof that the boosters were correct.[13] To hold electricity as the inexorable future or, worse, the Promethean beginning of artificial light is to fundamentally misrepresent and misunderstand the nineteenth century. Until the outbreak of the Civil War, the future of light in the United States that most people anticipated promised more light, cheaper and accessible to more people, but not safety. The new lights from the 1830s through the 1870s may have been cheap and plentiful, but they also tended to explode catastrophically. Nor was there any indication that the rising tide of "progress" had placed technology and free labor in the same boat. If anything, when it came to illumination, free labor's boat seemed to have sprung a serious leak. The future of American light from the viewpoint of 1860 appeared inextricably tied to an expansion of industrial enslavement, sweated outwork, and child labor.

When electric light boosters like Edison and Platt came along at the end of the nineteenth century, they claimed to be the inheritors and prophets of a long and accelerating arc of technological and human progress that had witnessed a momentous increase of human freedom, especially freedom from toil. It was and remains a compelling story. It was also a grossly self-serving fiction. Electrification not only multiplied the underground dangers and labors of coal and copper miners; the whole electric edifice was built atop social structures, dreams, and material foundations deposited over a century of steadily industrializing slave labor and the increasingly rigid expectations that unenslaved women and children do certain kinds of work for their parents, husbands, and employers for free or for starvation wages. Any honest rendering of the history of light must show how the good and the bad produced each other. Only through engaging with such contradictions can we begin to challenge the enduring hold heroic stories of light and progress have had on popular discourse.

The best histories of artificial light and night have focused on culture and experience, on the profound effects that the "industrialization of light" in the nineteenth century had on practices like reading and nightlife, architecture, crime, and commercial culture. But they leave out at least half the story. The histories of how making and powering those lights transformed landscapes of energy and reorganized labor and life in mines, factories, and cities have been left untold.[14] As Melville knew, the history of light was a tale of violence and labor, blood and sweat. It was, in other words, a history of work. *American Lucifers* tells that story.

One of the enduring obstacles to constructing a labor history of light has been rooted in the way people have defined "country" and "city." Usually, both Martha Ballard's farm and a North Carolina turpentine camp would fall under the category of "country." But farms working out a rough independence through negotiated and periodic relationships with regional and global markets were not the same as forest turpentine camps, which consumed and transformed and moved nature almost entirely with urban spaces in mind. The eighteenth- and nineteenth-century conquest of circumscribed regions of the country by the city, of the uneven gravitational capture of rural spaces into urban orbits, is easy to miss when the tools of analysis are familiar city/country or metropole/periphery binaries. The majority of Americans in the years of this story did not come to live in cities, nor did spaces like Martha Ballard's community disappear. But in important ways, increasing numbers came to and were forced to live *for* cities. This was as true for those laboring in forest camps, pigpens, mines, and whaleships as it was for those in urban workshops. And changes in how people organized the production and the consumption of the means of light were critical parts of the process. If the work of light forms this book's bones, understanding how struggles over this work brought such city-serving geographies of light into being is the question articulating its skeleton.

Labor histories have gone out of fashion. Even rarer is one that eschews the factory in favor of tenement, ship, and backcountry, let alone one that includes animals, children, men, and women as meaningful actors struggling together in the same story. I think this is a shame. We need more stories about work and workers. We need to recognize that people and nonhuman beings have always made their lives with and against both nonhuman beings and other people. To understand how "capitalism took command" in the nineteenth-century United States means looking, as capital most certainly did, at more than free white workingmen for sources of living labor. Unionized workingmen and white male artisans, merchants, and capitalists were, of course, important, and their experiences and struggles need to be studied, critiqued, and historicized. But by privileging the stories of the already privileged—and I should be clear that organized white workingmen, especially wage laborers, constituted a vanishingly small, privileged class among the producers of light—historians risk rendering a thoroughly distorted picture of the actual power relations of work that have underpinned the last several centuries of capitalism.[15] From 1750 to 1870, capitalists and husbands, herders and slaughterers, mine owners and working mothers directed the monetized living labor of free and enslaved men, marriage-bound women, family-bound children, and property-bound hogs and cattle into transforming grasses, textiles, whales, trees, bones, and minerals into capital, power, and modern illumination. Other similar stories surely exist, if only we have the will and imagination to tell them.

Sometimes finding those stories means looking for workers and relations of exploitation in places we don't normally look and maybe don't want to see. We should resist conflating relations of oppression: enslavement was not the same as marriage was not the same as keeping animals captive was not the same as child labor. But we should also resist enshrining a hierarchy of what-was-worse-and-better without thinking about how all these struggles related to and constituted one another. So long as we allow ourselves to think that being free was "better" than being enslaved, that being human was "better" than being an animal, that leisure was "better" than labor, all we're really saying is that power was better than weakness, which is perilously close to saying that the powerful were better than the weak. It may be counterintuitive, but if we want to learn how to be truly free, we'll find far wiser teachers among those who lived as livestock than among those who lived as farmers, as children rather than adults, as colonized rather than colonizers, as unfree rather than free.

The late Ursula K. Le Guin perhaps said it best in her short story of the aging anarchist Laia Odo. Odo knew waste, cruelty, disorder, and mud; she had grown up with them and known them as the badges of her oppression. But she also knew that freedom wouldn't be clean and orderly; indeed, it couldn't be. And she remembered the disgust: "But will you drag civilization down into the mud? cried the shocked decent people, later on, and she had tried for years to explain to them that if all you had was mud, then if you were God you made it into human beings, and if you were human you tried to make it into houses where human beings could live. But nobody who thought he was better than mud would understand."[16] A full accounting of the history of light demands that, like Odo, we jump down into the darkness and the dirt, way, way below the illuminated halls of power. Odo tried to explain that civilization was made from mud, and that if you didn't understand that, you wouldn't understand anything. We should follow her example and question the shocked decent people who thought their visibly clean hands meant their hands were actually clean, that their bright lights made their homes and streets better than the dark worlds of work that fueled the flames.

T HIS book tracks the hot spots in workers' struggles over the means of light through the tumultuous century that began with the rise of the American whale fishery in the 1750s and ended with both the destruction of slavery in the United States and the emergence, around the Civil War, of the petroleum industry and its primary product, kerosene. There are many excellent books about the history of electricity. This is not one of them.[17] This project began with research into the early history of electrification, and I address some of that briefly in the epilogue; but the more I learned and wrote, the more I realized that to tell the story I wanted to tell—to show how important the Civil War was in the history of light, to truly challenge the misleading teleology that everything that came before was just laying the groundwork for electricity, was prehistory—I needed to end with kerosene. Electricity wasn't the beginning of the story, or even the end of another. But it was the instrument that Americans used to help blind themselves to a labor history of light, to the subject of this book.

In one sense, *American Lucifers* is a story about the changes wrought between oil and oil (whale and rock). To be sure, the story of the change from whale oil to kerosene has been told before, even if whalemen and oilmen have figured as peripheral actors in the telling. In *American Lucifers*, I seek to do more than show the forgotten workers behind a familiar story. I hope to tell a different kind of story altogether. The industries and lights that came between whale oil and kerosene were not "transitional" technologies but contested processes with significant stakes for how slavery, households, agriculture, and industrial capitalism would be organized across the world. It is a story of sometimes astonishing oppression and exploitation, but it is also, critically, a story of those who fought back, and of how their struggles, both big and small, reshaped lives and landscapes.

Beginning with the rise of the American whale fishery in the 1750s, *American Lucifers* explores the accumulation and circulation of whales' embodied energy in the form of oil and candles, the first illuminants produced, graded, and sold at industrial scales. American deep-sea whaling voyages, encircling the globe in pursuit of blubbery means of light, triggered an Atlantic street-lighting revolution radiating from London, while a New England–run candle trade helped illuminate and circulate the people, products, and work processes caught up in colonial transatlantic sugar slavery. Later, American whale oils lubricated the spindles and heavy machinery turning at the heart of an industrial revolution in cotton manufacturing and steam-powered transportation. As these entwined revolutions in night and industry intensified in the antebellum period, they overwhelmed the capacity of the American fishery to meet the demand for both light and lubrication, even as ship masters drove whalemen on harder and longer voyages for less and less pay. But as lubrication-thirsty railroad engines and cotton spindles spun whale oil away from lamps, new antebellum geographies of light and risk emerged.

The next section of the book turns inland to turpentine, coal gas, lard oil, and phosphorus. In the urban cores, monopoly gasworks threaded coal-gas lights protectively in and around bourgeois space, while servants and out-working seamstresses labored late into the night with cheap, explosive turpentine lamps. At the peripheries, mixed armies of enslaved and free laborers worked ever-more-dangerous coal mines, while planters coerced enslaved men into tapping remote southern pines, and all struggled to assert some control over this antebellum empire. In the Ohio River Valley, meanwhile, a pork and candle industry emerged in the geographic interstices of slavery and free white labor to propel millions of hogs from farms and cornfields into a constellation of seasonal industrial death camps centered in Cincinnati. This geography of life and death unmade hogs so successfully that, in combination with the new industrial chemistry of sulfuric acid, wage-worked by-product industries in candles, lard oil, and soap became not only possible but enormously profitable. Sulfuric acid also made possible the other revolutionary lighting technology of the nineteenth century: lucifer friction matches. Using thousands of tons of coal and sulfuric acid, European chemical manufacturers extracted hundreds of tons of elemental phosphorus from the mountains of bones left in the wake of the slaughter of Pampas cattle in the South American port cities of the Río de La Plata (a geography comparably deadly to that of the Ohio Valley) and, later, from phosphates mined on West Indian guano islands. In American cities, thousands of child workers used that European phosphorus to mass-produce incredibly cheap lucifer matches, a process that starkly illustrated the hidden politics and slow violence of producing the means of light. Struggling to work and, even more pressingly, to live in the inescapable ecology of toxic phosphorus, lucifer-making children attempted to change work environments and win powerful allies, all while trying to survive and mitigate an agonizing, degenerative, disfiguring illness from phosphorus poisoning they called "the compo" and officials called "the jaw disease."

The final section examines how the combined onslaught of Pennsylvania petroleum and the Civil War radically reoriented the possibilities and geographies of light in North America on what may have been the eve of slavery's industrial revolution. As military clashes interrupted and destroyed turpentine camps, whaleships, and southern coal mining, the reservoirs of American light shifted their center of gravity markedly northward and westward. The coincidences of a civil war, an oil rush in Pennsylvania that succeeded and was celebrated, an oil rush in western Virginia that was disrupted by war and forgotten, and the emergence of kerosene boosters invested in selling a future of light as a future of freedom all combined to mystify the history of light. These coincidences allowed motivated storytellers to disguise what was really a political and geographic revolution destroying American racial slavery and, instead, make it seem to be a story of nature's provenance, American exceptionalism, and the inevitable destiny of technological progress freeing mankind from danger and labor.

The futures of light that came to be after 1861, and the stories that were told about them, would have been wildly unimaginable to any clear-eyed observer in 1860. They were deeply unlikely, and indeed, many remained more myth than reality. To understand both past and present, we'll need to recover that shock.

Chapter One

DRAGGED UP HITHER FROM
THE BOTTOM OF THE SEA

*It is a land of oil, true enough: . . . nowhere in all America will you find
more patrician-like houses; parks and gardens more opulent, than in
New Bedford. Whence came they? how planted upon this once scraggy
scoria of a country? . . . Yes; all these brave houses and flowery gardens
came from the Atlantic, Pacific, and Indian oceans. One and all, they
were harpooned and dragged up hither from the bottom of the sea.*

HERMAN MELVILLE, *MOBY-DICK* (1851)

Awaiting the last sea-scattered rays of the setting sun, Jonathan Bruce set to
work on his lamps in his high, lonely perch. It was a ritual he had repeated
nearly every evening for the last twenty years. Turning from the greasy lighthouse window, he struck a small flame and carefully lit the oil-soaked wicks.
Sputtering to life, they soon bathed him in a glow that transformed the room
into one of the brightest spaces in all the night. Of course, the light was not
meant for him. The lamps and lenses cast the radiance for miles around
the stone tower, dim by the time it reached other eyes but nonetheless a
beacon in the night for any sailor passing near Boston Harbor. Settling in
amongst his brilliant companions, Jonathan Bruce prepared, that September
evening in 1832, to stand guard over the lamps and the tower in which they
shone. It wasn't the most exciting work, but it was important. Bruce was the
keeper of Boston's lighthouse, and the steady trade of the maritime city depended on him.[1]

Across the harbor in Boston, as the first light from the lighthouse reached the docks, Richard Hixson may have glanced out at the sudden gleam as he shouldered his bag and marched up the gangway of a sloop by the name of *Nantuck*. Or perhaps it would be the gridded glow that drew Hixson's gaze, as smaller versions of Bruce's lamps were brought to life by lamplighters all across the city. More likely, busy preparing to ship out, Hixson gave little thought or attention to either sets of lamps. But Hixson's actions were as much shaped by these lights as Bruce's were. Even as streetlamps and light-houses produced the illuminated terrain through which trade flowed and order was proclaimed, they did so only by consuming work and energy accumulated across spaces far removed from those flames and their keepers. Hixson was beginning a journey to just such a frontier of luminous accumulation. He'd left behind his farm in Sharon, Massachusetts, traveled overland to Boston, and, now, he was to become a whaleman.[2]

When, later that night, the Nantucket-bound *Nantuck* sailed through the light of Bruce's sperm-oil lamps, the two men may not have known how much their work connected them. But connected they were. Hixson and Bruce were counterparts bound together in webs of whale light, the one to accumulate its cetaceous means, the other to channel its consumption into a beacon in the dark. That the illuminated and lubricated spaces of commerce, industry, and public order were emerging through webs of work and energy spinning outward from the deep-sea application of harpoons hardly sounded enlightened. But it was true. Theirs were worlds of sweat, sail, and boiled blubber.

MARCH 2, 1833, off the coast of Chile, the "watches employed as yesterday." It had been six months since Richard Hixson sailed past Boston's lighthouse to go hunting after oil, and he still hadn't encountered a single sperm whale. This was not uncommon. Just getting to the whaling grounds took months, and once there, whalemen's time was usually dominated by sailing, repairing, and procuring food, not to mention finding whales to hunt at all. But that March morning, the men of the *Maria* would finally have a chance to do what they had sailed 10,000 miles to accomplish. For the first time in the voyage, "11 oclock A.M. saw a sperm whale on our weather bow distant 1½ mile, lower'd boats, and rowed for him." In about an hour they pulled even with the whale, and the "waste [waist] boat made fast." Very soon Hixson's boat, the "Stabboard" one, "got 2 irons into the monster of the deep, and after another hour of struggle the whale lay a motionless lump on the top of the water." Hixson later reflected in his journal that it "was to me an interesting scene, to be engaged for the first time in fighting with and killing a large whale, he truly made the deep boil like a pot."

Unfortunately for the whalemen, killing the whale was only the first step in transforming it into oil for lamps and spermaceti for candles. Thrill done, now came the long haul of a fifty-ton carcass miles back to the ship. Muscles and oars straining against the water, Hixson and his boat crew steadily drew their prize closer, and by "5 P.M. the boats all come in alongside ship," having failed to catch any other whales, and "all hands getting ready to cut him in." It had taken six months and six hours of hard labor to catch this whale, and their work was only just begun.

The next day, with the whale secured against the side of the ship and the sails furled to keep the boat from drifting or listing, "all hands engaged in cutting in whale."[3] An innocuous enough phrase, "cutting in" meant, first, the crew secured the whale to the ship using heavy chains. Next, they lowered a wooden scaffolding over the carcass, where the captain and officers cut into the whale with long spades. They separated the head, which whalemen called the junk, to deal with later, and then they peeled the blubbery skin off the whale like ants spiralizing a grapefruit. To begin peeling, the men swung an enormous cluster of cutting tackles and blocks over the whale and lowered it. The second mate then made an incision near the pectoral fin, and one of the boatsteerers, tethered to a man on board by a rope tied around his waist, had the perilous job of being slowly lowered onto the whale. Balanced carefully on the carcass, he tried to avoid being crushed against the hull of the ship on one side, sliding into roiling shark-frenzied waters on the other, and all while inserting a 100-pound iron hook, attached by chains to the blocks and tackles, into the whale's flesh. "This done," Herman Melville described how the shouting crew "now commence heaving in one dense crowd at the windlass. When instantly, the entire ship careens over on her side; every bolt in her starts like the nail heads of an old house in frosty weather; she trembles, quivers, and nods her frighted mast-heads to the sky." The strain of ripping the skin off a fifty-ton whale was tremendous, and "more and more" the ship "leans over to the whale, while every gasping heave of the windlass is answered by a helping heave from the billows; till at last, a swift, startling snap is heard; with a great swash the ship rolls upwards and backwards from the whale, and the triumphant tackle rises into sight dragging after it the disengaged semicircular end of the first strip of blubber." It was a scene that Hixson tried to capture in an illustration at the back of his journal.[4]

As the peeling of the whale continued, down the main hatch went the first strip of blubber "into an unfurnished parlour called the blubber room.

"The Maria . . . with a whale alongside cuting in."
Watercolor by Richard Hixson (1833). Courtesy Houghton Library.

Into this twilight apartment sundry nimble hands keep coiling away the long blanket-piece as if it were a great live mass of plaited serpents." The whale-men then "minced" the blubber, further slicing the chunks of fat cut from the larger "blanket" into strips called "bible leaves." This maximized the surface area of the blubber and thereby squeezed the greatest quantity of oil from the skin of the whale. After all the blubber was cut in, the whale carcass was let go, the meat fetching no price. Following mincing, the men returned to the head and used buckets to ladle out a waxy substance called spermaceti. About one-third of the oil taken from a sperm whale came from the head. Almost pure spermaceti, head oil sold for a higher price than body oil. Offi-cers thus instructed the crew to keep head oils separate on board the ship, even though oilworks in Nantucket would later recombine head with body oil. Indeed, the disassembly of whales was always a process shaped as much by market forces as by the specific chemical needs of manufacturers.[5]

The tryworks were the metabolic centers of whaleships, where whale-men boiled the bible leaves of blubber in specially constructed cauldrons called try pots. Shipwrights had built the very skeleton of the ship in service of the tryworks, situated in the center of the ship, in "the most roomy part of the deck," where the "timbers beneath are of a peculiar strength, fitted to sustain the weight of an almost solid mass of brick and mortar, some ten feet by eight square, and five in height." This massive structure "does not pene-trate the deck, but the masonry is firmly secured to the surface by ponderous

Dragged up Hither from the Bottom of the Sea

knees of iron bracing it on all sides, and screwing it down to the timbers." Covered by a large hatchway, the top of this brick structure contained within it the *Maria's* two try pots, each capable of holding several barrels of oil.[6] Firing up the try pots, the men of the *Maria* boiled all the blubber stripped from the whale until it had been rendered into oil, which they then placed in casks to cool. Trying out was the alchemy at the heart of the whale fishery, the process translating deep-sea blubber into marketable whale oils.

It took the crew of the *Maria* over half a day to skin their first whale and another thirty-six hours to finish rendering all its fat, and it was only of middling size, producing fifty-five barrels of oil. "People would naturaly think," Hixson wrote, "that it took a great quantity of wood to try out so much oil, and that it would be inconvenient for a ship to furnish it but this is not the case, we use no wood, but burn the scraps, and they make an excellent fire, far better than wood." Melville, too, described this efficient practice, although perhaps less admiringly, whereby "the crisp, shrivelled blubber, now called scraps or fritters . . . feed[s] the flames. Like a plethoric burning martyr, or a self-consuming misanthrope, once ignited, the whale supplies his own fuel and burns by his own body. Would that he consumed his own smoke!" Melville lamented, "For his smoke is horrible to inhale, and inhale it you must, and not only that, but you must live in it for the time." The smoke also made the ship incredibly difficult to hide. The smoke curling up into the sky would have been revealing enough for pirates, privateers, or whaling competitors, but so powerful was the stench of smoke and grease that when the tryworks were going, "a whale ship could be smelled over the horizon before it could be seen."[7] Even a midsized whale required several days of continuous, gory, smoky labor to turn it into oil, and usually several more to finish coopering and stowing the oil below. Deep at sea, and quite unceremoniously, the glittering lights of the modern world were being forged from grease fires and butchered blubber.

This chapter tells the story of this relationship, of making lights and wielding lights, of bringing whales and tryworks together, and then joining oil with lamps. It's a story of whole worlds hunted, harpooned, and dragged up from the bottom of the sea. This translation of sunlight to lamplight — of solar energy congealed in the fat of these marine giants into urban, coastal, and industrial illumination — consumed extraordinary amounts of life, labor, wood, and food. Behind whale lights lay weeks stretching into years of careful planning and continuous provisioning, of tedious drifting punctuated by bouts of intense danger and activity, of whales learning and fighting and usually escaping, and following kill after difficult kill, of whalemen boiling ungodly amounts of blubber. Before the lights lay sugar and cotton, slaves and wage workers, thieves and police, fire and war.

The Quaker owners of the New England whale fishery turned the dangerous, prolonged labor of American whalemen and the blubber of whales into fortunes for themselves and a new world order for others. They did so by injecting whale oils into critical junctures in four processes at the heart of a global political economy: the transatlantic slave trade, the manufacture and trade of sugar, the making of the urban poor into the working classes, and the spinning of American cotton into factory textiles. By the time Richard Hixson next passed Boston's lighthouse on his way home to his farm in Sharon in 1836, Jonathan Bruce would be gone and Hixson no longer a whaleman. Briefly, however, both had been bringers of light.

THE worlds making Richard Hixson and Jonathan Bruce into American lucifers were conceived in a five-tongued fire ignited in the middle of the eighteenth century. Each of these five fires burned in a different kind of space, each crucial to the webs of commerce, work, and power forming the Atlantic world. First were the lamp-sprung flames that spread wildly through the old wooden structure of the original Boston lighthouse in 1751, leaving little behind besides the stone foundation and an island-studded harbor shrouded in darkness. The keeper Robert Ball and the man he enslaved may have feared a loss of livelihood, but the ship crews, watching anxiously while the crackling, bursting inferno consumed the lighthouse, knew that a dark ocean passage could mean loss of life. Second were the fires carefully lit that same year deep at sea, which greasy, blood-soaked men used to transform freshly cut whale blubber into oil to be casked and coopered: these were the first onboard tryworks. Third were the smallest but most numerous of the fires, also born in 1751, burning steadily atop the first spermaceti candles in the world; by 1768, Caribbean buyers, the largest market for the candles, were burning over 225,000 pounds of New England spermaceti a year. Fourth were the 5,000 new lamps burning all night, every night on the streets of London, consuming over 8,000 gallons of sperm oil annually. Fifth were the plantation boiling houses burning whale oil to illuminate the twenty-four-hour, six-month-long continuous production of sugar. All five fires stemmed from the shared body of an emerging Atlantic capitalism, while stoking them, constituting these constellations of light, were dark circuits of ships, barrels, oil presses, coopers, and a surprising array of animals, all scattered across oceans and continents.[8]

Dragged up Hither from the Bottom of the Sea

I am not suggesting that the mid-eighteenth century was the beginning of whaling, lighthouses, or street lighting. Whaling had been pursued extensively by Europeans and Indians for centuries. Some have argued that whaling even helped to draw the first European sailors to North American shores as they followed right and pilot whales. And the Indian whalemen conscripted into the colonial fishery with their knowledge and labor were probably the reason the New England and New York shore whaling fisheries succeeded in the first place. Boston Light, Boston Harbor's lighthouse, was first lit in 1716, a key beacon reducing the very real risk of wreckage for the agents and objects of an Atlantic empire circulating in ships. More recently, but still well before 1751, year-round street lighting in London really began in 1736, when, at least initially, the new streetlamps used seal rather than whale oil. And that is partly the point. Prior to the 1750s, whale oil had been one illuminant among many, certainly less important than the beef and mutton tallow used to manufacture most candles, whether made by women on farms like the one where Martha Ballard grew up, or by the artisans producing candles for use in the mines of Potosí, or by the urban chandlers, like the family Benjamin Franklin was born into, who made candles to sell locally in cities as small as Boston and as large as London.[9]

But London's assault on its dark streets, as well as against thieves, pirates, and privateers across the Atlantic, opened new opportunities for American fisheries. In the first half of the eighteenth century, the British navy commenced a devastating campaign of violence and terror against pirates and privateers. In an absolute sense, this campaign did little to increase or decrease the violence of the Atlantic. Pirates may have been masterless, even pioneers of revolutionary and democratic politics, but they were far from pacifists. What changed most for sailors (and whalemen) after the antipiracy campaigns was that violence on the sea now became increasingly organized and hierarchical, and alternatives to the harsh discipline of the navy and merchant fleet narrowed. But for at least a few decades, the campaign produced an oceanic space safer for commerce and longer fishing voyages. Meanwhile, imperial and merchant forces also combined to make maritime movements into and out of Atlantic ports easier and more closely monitored through the erection of dozens of lighthouses burning whale oil. The last decade before the American Revolution would witness especially remarkable changes. Boston Light was rebuilt, this time of stone and metal, and was fitted with a lightning rod. By the end of the Revolution, there were twelve other lighthouses on the east coast of the United States, eight of them in New England.[10] Lighthouses made possible the maritime capitalism comprised of the circulation of slaves, commodities, navies, and merchants in ships navigated mostly by sight. If a ship was caught in a storm or in the dark of a moonless night, a lighthouse might be the only thing standing between survival and wreckage. Expanding into this newly secured marine frontier in the 1750s, Quaker whalers from the sandy, unimportant island of Nantucket began to risk open seas and shipboard fires to gain access to far greater whale stocks than had previously been possible.

Producing oil from whales involved two very distinct processes that were increasingly combined after the 1750s. First, whales had to be caught and cut up. Second, that cut blubber had to be boiled into oil. Prior to deepsea whaling, tryworks and ships were invariably kept separate, and the distance a ship could travel from the onshore tryworks in search of whales was limited by the rapid rate at which blubber spoiled between cutting in and trying out. Tied to the coast, such a relationship between tryworks and ship meant that the vast populations of pelagic cetaceans such as the sperm whale pods following deep sea currents across the globe remained entirely out of reach. It also meant that coastal communities and families remained rooted and relatively unstrained by the shore fishery supporting them.

Although onboard tryworks risked the very real danger of shipboard fires, they revolutionized human relationships with the energy flowing first through the Atlantic, then through all the world's oceans. Using onboard tryworks also stretched maritime communities and families thin across time and space as the hunt for profitable whale oil scattered men on increasingly longer voyages. Suddenly within reach of shipborne humans were vast wells of embodied solar energy formed over decades by whales. It was luminous energy for which growing metropolises like London would pay huge sums in the accelerating campaign to bring law and order not only to the seas, but to the streets and class struggles of cities.[11]

As the first Nantucket ships were experimenting with cutting in and trying out whale blubber in try pots arrayed on specially built brick-floored sections of the deck, the oil streetlamps of London were being met with such widespread praise that urban officials across Europe and the Atlantic would soon be scrambling to replicate the lightshow in their own cities. It is a commonplace among historians of street lighting that the lamps were primarily erected to combat crime and make the streets safe.[12] I think the story needs to be reexamined.

Dragged up Hither from the Bottom of the Sea

I T was about nine at night in Hockley in the Hole when James Daniel, an Irish grocer, stepped out of the Two Brewers. Daniel had stopped at the tavern for a pint on his way home from Islington, just north of the city, and now he badly needed to pee. "I was all alone," he would tell the court on September 11, 1751, nine days after the incident. All alone, "except my shoes tied up in a handkerchief," when he "saw three men standing by a lamp[;] two of them had hats, and one a cap." Not minding an audience, Daniel began to undo his trousers, even as the men "crossed over to me: then I turned up to make water, in a yard." This was a huge mistake. "One of them got hold on my collar, (for they did not give me leave to button up my breeches) the other on my shoulder on the other side." He claimed the man threatened him ("One swore he would knock my brains out if I stir'd") and then robbed him of his hat, one shilling and sixpence, and his handkerchief containing his shoes. "After this they run from me," he said, "two one way, and one another; and thinking to catch one of them, I called out, stop! stop! stop! but I saw no more of them that night. I know the two prisoners were two of the men, *for I saw their faces by the lamp*."[13]

A simple search for "lamp" in the online database of Old Bailey records reveals hundreds of convictions in the middle decades of the eighteenth century based on identifications made by lamplight. "Was it a light night?" "No, but there was a lamp," became almost a scripted mantra in court proceedings. In court, Tim the Taylor, one of the three men whom James Daniel identified, turned on the others. William Newman and James March claimed that they were innocent, that Tim had later given them items of which James Daniel claimed he had been robbed. Their defense fell flat. The lamp identification was enough. Having stolen one and a half shillings, some shoes, and a hat by the light of a streetlamp, Newman and March were both sentenced to hang. On October 23, 1751, both would be "cheated on the Tree."[14]

What paths had led these men to the gallows? Seven days before the state killed William Newman, he was visited by John Taylor, the Ordinary of Newgate Prison. Newman was not alone. With him were nine other men and two women, each of whom London officials were herding toward the same noose, looming one week distant. James March was one of these fellow prisoners. David Brown, the first approached by their new visitor, had also been convicted of assaulting and robbing a man right under a lamp, hardly the association between illumination and crime claimed by the proponents of better street lighting. And these were almost entirely crimes against property. Of the other ten people sentenced to hang with Newman and March, nine had been convicted of theft or smuggling. Only one had been convicted of murder, and that had occurred in broad daylight.[15]

But these accounts only tell us the convicts' paths to death. How were the lives of the London hanged shaped by whale light? John Taylor visited each of the prisoners in turn, recording the stories of their lives and their "final" words. James March, aged seventeen and the younger of the two men convicted of the Daniel robbery, had been apprenticed to a waterman in his youth, when he would have learned the details of the waterways of London as he ferried passengers along and across the Thames. Watermen worked in the freshwater interstices of a capitalist empire quickly centering in London and its river docks—spaces produced in part by the growing number of lighthouses securing British shipping channels throughout the Atlantic world. And as was common among the London hanged, March had broken his apprenticeship, leaving an abusive master, and joined a street crew that made its living through petty street theft and housebreaking.[16] From his time as a waterman, March would have known London and its routes well, a useful skill when working at the edges of the legal economy.

While March had worked at facilitating travel at the geographic heart of the empire as a waterman, his co-conspirator, William Newman, had served in the Royal Navy policing the waves to make space for British commerce and whaling, and fighting against other European navies in the War of Austrian Succession. Like many of those hanged at Tyburn, including his fellow prisoner and night thief David Brown, Newman had been a sailor. After several years of fighting and toiling at sea for the crown, Newman became one of the 40,000 seamen demobilized following the peace of 1748. After losing, or perhaps he thought of it more as escaping, his position in the navy, Newman, like so many of the young men discharged from service, moved to London, where he lived with his sister. There, facing a labor market suddenly glutted with unemployed able-bodied men like him, Newman chose "to rely upon the Industry of his Fingers to procure him a common Subsistence; and he was indefatigable in the Practice of picking Pockets." His preferred haunt was the Royal Exchange, where he relieved men, growing rich on the empire he had—until its rulers abandoned him to his fate—helped secure, of handkerchiefs and other valuables. Newman had only worked with March and his crew for a few weeks before the thief-taker apprehended them.[17]

"As an immediate result," the historian Edouard Stackpole wrote of the mid-eighteenth-century decision to light all of London all night all year, "the demand for whale oil increased one hundred fold. The addition of more street lights resulted in the decrease of crime. It has always been an axiom that crime does not thrive in the light, whether in illumination from lamps or from an enlightened society. Great cities like London and Paris recognized these important facts early, and made provisions for better lighting."[18] Axiom or not, as James Daniel, William Newman, James March, and the hundreds of others robbed, assaulted, and sentenced to death by lamplight well knew, streetlamps were far from obviously making the streets safer. Lamps, which were intended to banish the danger of theft and murder from the night streets, were in fact doing little to prevent either and may have even facilitated those very practices. Instead of protecting night travelers, these lights became instruments and symbols of property law, declaring that the only people who had the right to take things from someone against their will were the tax collectors of the state. Lamps helped the state to violently secure a monopoly on legitimate taking. By providing sufficient visibility for witnesses to make identifications, streetlamps enabled state officials to murder hundreds of people for interfering with the authorized circulation of things, for hats and shillings, for pairs of shoes. None of this meant "crime" decreased. Quite the opposite. Not only were officials calling more things crimes, but, now better able to enforce their new laws through the expansion of lamps and police, they actually *increased* crime, criminalizing ever more practices in their effort to colonize the night. And it was by supplying these lamps with oil that Nantucket Quakers began building their fortunes. Newman's and March's deaths reveal a different police role for streetlamps: making street theft as a survival strategy at the margins, or as an alternative to living and dying by working for wages, too dangerous to risk. As with the British campaign against pirates, the lamp-aided campaign against thieves was less about reducing violence or preventing unwanted taking than it was a statement that confiscating things was the exclusive privilege of the agents of states, and they would brook no competitors.

Twenty years earlier, in 1730, London had no more than 700 streetlights, which were lit for only 750 hours per year. In the five years before 1736, an average of 481 thefts were tried each year at London's Old Bailey Court. In the five years after 1736, when London expanded its streetlamps to 4,679, each burning for over 5,000 hours a year, an average of 452 thefts were tried, a more than 4,500 percent increase in lamp hours and a mere 6 percent decrease in thefts. By the time William Newman and James March were being drawn by cart down Oxford Row, that well-worn path from Newgate Prison to the gallows at Tyburn, there were over 5,000 lights in London. They had not saved the city. War reduced thefts, but after each peace treaty they surged again. In 1749, following the end of the War of Austrian Succession, 561 thefts were tried. In 1764, the year after the Seven Years' War, there were 546 thefts. By 1780, the number of lamps had tripled to 15,000 — which consumed annually 25,000 barrels of sperm oil, the product of about 60 ships, 1,200 whalemen, and around 500 sperm whales — and it "was London's boast that there were more streetlamps along Oxford Row than in the entire city of Paris!" Yet in 1784, when sailors and soldiers returned to the city from the American Revolutionary War, although authorities burned 100 times more lamp oil than they did in 1735, they tried 1,000 thefts, and that number didn't fall below 600 until 1790. The city had grown from about 700,000 in 1730 to around 1 million in the 1780s, and the amount of street light had multiplied a hundredfold, but thefts per capita were actually higher. The only measurable correlation between streetlamps and crime was a positive one. What did bring crime down was war. There were only five years in the eighteenth century when London's authorities made fewer than 300 indictments for theft: 1745 and the four years between 1758 and 1762, all during the height of wartime.[19]

But whether or not lamps prevented crime or made streets safer, the anxious ruling elite rightly understood that they could use these lamps to project their power over the ruled. Over the eighteenth century, the conviction rate for all crimes remained around two-thirds. The involvement of lamps did not alter this. What lamplight did was increase the likelihood a guilty conviction would lead to death. London officials hanged one in ten of the people they convicted of theft in the eighteenth century. But when the term "lamp" was mentioned in court, the chance that the accused would end with their head in a noose more than tripled. Of the 587 persons they tried for committing a theft around a lamp between 1715 and 1800, officials executed 212 of them, over one in three. And as both petty theft and the number of streetlamps continued to rise over the eighteenth century, where crime, light, and law intersected, officials worked to narrow the range of punishments to the most severe. When people stole near lamps, authorities became ever less likely to sentence those they convicted to "transportation" (deportation to the New World and, later, Australia), but more likely to sentence them to hang. Using identifications of the accused made possible by whale light, authorities were legally channeling increasing numbers of the ungovernable working poor toward death. It was no great mystery, then, why smashing streetlamps became one of the most popular acts of defiance among working people, or why the practice was particularly common during moments of uprising and revolution.[20]

Both the campaign by London's ruling classes to light the streets and

Nantucketers' campaign to provide the oil for the lamps were projects to profit from and manage an imperial war machine in the interstices of peace. Whalers, as easily taken prizes, usually did most of their whaling during cessations of open hostilities. And the empire's rulers usually only had the will and the fear necessary to organize and pay for expansions of street lighting during periods of demobilization, when they urgently sought to contain the rising thefts and unrest led by discharged sailors and soldiers until the next war once again removed them from the city.[21]

In Europe, not only did ports build more sperm-oil-burning lighthouses, but cities like Paris and Amsterdam began staging their own streetlamp power plays after the example in London. Between 1768 and 1772, the number of American whaleships tripled to over 300, and in 1765, Thomas Hutchinson of Massachusetts proudly wrote that the "increase of the consumption of oyl by lamps as well as divers manufactures in Europe has been no small encouragement to our whale fishery. The flourishing state of the island of Nantucket must be attributed to it." Supplying states with the means to wage campaigns to colonize night streets and night seas was lucrative business — so lucrative that on the eve of the Revolution, the whale fishery was the most economically important industry in New England, accounting for over half the British sterling entering the northern American colonies. And Nantucket was at the heart.[22]

THE island did not, however, have a monopoly on producing the means of whale light. Oil lamps were not the only path by which New Englanders transformed whales into light, money, and power. The processes of rendering and pressing that turned sperm whale fat into oil also yielded a hard, moldable substance called spermaceti. Nantucket dominated the crude-oil trade, but until the eve of the Revolution, the production of spermaceti candles and graded, processed, marketable whale oil was centered in Newport, Rhode Island.

In 1751, Jacob Rodriguez Rivera, a recent arrival in Newport, set up the first spermaceti candleworks in the world. Newport would soon emerge as the center of sperm-oil processing and distribution, the firms there dominated by Sephardic Jews like Rivera. Not only did candles and candle making provide a profitable outlet for the products of sperm whaling, but candleworks became central stations in the grading and trading of the new stocks of whale fat laid open by the revolutionary union of ship and tryworks. From Newport, spermaceti candles wrapped in blue paper and packed in elaborately labeled boxes circulated through the empire as luxury items and as one of the principal mediums of exchange within the Atlantic trade in slaves and slave-produced goods. Clean burning, perfectly white, and less liable to deform or rot on the shelf than tallow dips, spermaceti candles were practical and symbolic tools for tropical elites on both sides of the Atlantic. Sold to affluent consumers in Atlantic metropoles, sugar islands, and African slave ports, the biggest market for these candles was in the Caribbean. From 1768 to 1772, Newport shipped over 200,000 pounds of candles each year to West Indian planters in exchange for sugar and molasses (for rum) wrenched from the life and labor of enslaved men and women whose transatlantic journeys may very well have begun with a West African exchange of just such New England candles or rum.[23]

But it was in the heart of sugar making that whale light was most inextricably bound up with Atlantic slavery and racial capitalism. For sugarcane to be made into sugar, it had to be processed as soon as possible after being cut, and the only way to do this profitably on large plantations was to keep the boiling houses running continuously. Just as tryworks were the primary sites of oil production, the boiling house sat at the center of cane sugar production. Planters knew well that only under conditions of slavery could they compel laborers to work twenty-hour days, and they took full advantage to make sugar production one of the most profitable industries in the history of the world, and one synonymous with forced labor. During crop time, in the race to keep the cane from fermenting, sugar planters drove the people they enslaved so hard through space and time — through sunlit cane fields and lamp-lit sugar works — that, in the haunting words of the historian Vincent Brown, they "would exhaust black lives as productive capacity, grinding them into sugar," murders amounting to "the digestion of the enslaved to enhance the vitality of the proletariat." Then those enslavers coldly claimed that Africans were inherently lazy because they wanted nothing but rest when not under the lash of the field or mill. The apparent "failure" of free labor in places like Haiti to keep sugar works running at night was held up as further justification for the necessity of racial slavery.[24]

Enslavers drove men, women, and children through the continuous times and spaces of sugar boiling in "spells" of twenty to thirty people. Depending on the number of spells, enslaved workers would move between the dawn-to-dusk workplaces of the cane fields and the noon-to-midnight, midnight-to-noon (or, if three spells, eight-hour) shifts of the boiling houses such that during the whole of "crop time" — which usually lasted from five to six months, but in particularly good crop years could stretch for as many as nine — enslaved cane workers could expect to get no more than three hours of rest a day, if they were lucky. Since boiling only stopped briefly for the sun-lit hours of Sunday, fatigue was overwhelming, and hands torn off or caught in the mill were so commonplace that a hatchet was kept ever ready to sever a ruined limb and prevent the machinery from gumming up. Through the expanded time and concentrated labor of the sugar works, slaveholders were literally consuming the lives, times, sleep, and hands of those they enslaved in the production of sugar.[25]

Illuminating this nocturnal engine of necromancy constituted another terrain of struggle. Several West Indian writers listed copper lamps as among the instruments a planter needed to set up a boiling house, while the Jamaican Edward Long complained that his fellow planters were too reliant on imports of whale oil from the North American colonies. A later recollection described eighteenth- and early-nineteenth-century boiling-house lamps as metal vessels, each "3 or 4 inches in diameter, and about 6 inches deep, with two tubes or spouts, ½ inch in diameter on opposite sides, and a brail with a hook to hang it by. Underneath was another vessel to catch the constant drippings of whale oil, with which the lamp was supplied. A long, large, twisted wick was floated in it, with the two ends projecting from the spouts." As the enslaved workers tended the cauldrons, ladled the boiling sugar from copper to copper, and inspected the liquid for color, clarity, and consistency, they hung the whale-oil lamps "up where light was needed," attaching the hooks to a wall or an overhead bar or rope running above the copper pots, or carried the lamps about by hand. In the seventeenth century, when the colonial Atlantic sugar industry first arose, planters would have likely used a mix of available and cheaper illuminants like seal, fish, grease, and palm oils. But sugar slavery was one of the most dynamic institutions in the Atlantic world. Continually pursuing new technologies and new ways to extract the work and knowledge of enslaved producers, sugar planters readily embraced whale oil when the American fishery made it affordable and available during the eighteenth century.[26]

Tunneling through the steady rise and fall of day and darkness, whale-oil lamps made possible incredibly profitable spaces of continuous sugar production. If the lamps stopped transforming oil into flame light, the sudden inrush of darkness would immediately grind activity in a nocturnal boiling house to a halt. To keep these lamps burning, then, and thereby tunneling through diurnal time, planters secured wicks by paying "some of the poorer whites" on the islands to "spin cotton for the lamps in the boiling houses." Making sure that oil was on hand to fill the lamps demanded no less attention, a lesson that enslavers like Thomas Thistlewood were loathe to learn:

"Mr Hartnole, I hear, was quite drunk, insomuch that the boilers could get no lamp oil for the boiling house use, &c."[27]

But even when all the supplies were in order, enslaved workers struggled to strategically sabotage lamps while slaveholders strove to defend them. "Constant snuffing, which the negroes did with their bare fingers, was required to get any light at all," one Cuban planter recalled, and the lamps "were constantly being upset, suspending the work, until they were relighted, often by blowing the wick against a brand of fire." The planter blamed the technology for these discontinuities, but it seems just as likely that the upset lamps and work stoppages were no accidents. Enslaved sugar makers, driven all day in the cane fields and all night in the boiling house, where watchmen forced them to tend the burning wicks of the lamps with their bare fingers, would have understood perfectly well that the source of their pain was also the means of their night labor. The Jamaican planter Thomas Roughley sought to gain advantage in this struggle by taking lamps entirely out of the hands of the enslaved. "Instead of two hanging copper lamps, which are made use of in the boiling-house at night, close to the lower coppers, and the heads of the people there, to furnish them with light," Roughley wrote in criticism of these more accessible—and probably more useful—lamps, "I prefer a globe lamp . . . hung in the centre of the boiling-house, at a height to prevent its being broken, and sufficiently low to diffuse general good light." This arrangement not only consumed oil more efficiently but would "prevent the thieving of the negroes, who watch every opportunity, not only to steal the oil, but the wick soaked in it. One pint of oil will be enough for the globe burners every night, whereas it takes near a quart every night, when the boiling-house is at work, to supply the lamps for the low coppers and syphons."[28]

Here was the center of the second web of light that the makers and merchants of whale oil spun to ensnare the workers of the Atlantic world. Like the streetlamps of London, the boiling-house lamps of the West Indies were ruling-class instruments of power and death. As in London, workers on plantations fought to subvert and repurpose whale lights. But unlike streetlamps, boiling-house lamps were also deadly engines of extraordinary wealth. During the eighteenth century, New England merchants would arrive in the Caribbean with candles and whale oil, or with enslaved people purchased with candles, and depart with holds filled with sugar produced through whale light. In the process, colonial merchants, whaleship owners, and planters were arranging tryworks, lamps, candles, and boiling houses into webs of power—webs operating across slave pens, ships, plantations,

　　　　　　　　　　　　Dragged up Hither from the Bottom of the Sea

and waves to bind and transform the living labor of whalemen and slaves, whales and cane, sunlight and lamplight into the political-economic geography of the colonial world.[29]

Taking advantage of their privileged position in a mercantilist economy in producing raw materials, the Quaker men and women of Nantucket exploited this arrangement of thieves, sailors, slaves, and oil to transform their remote island into one of the centers of colonial British America. The French traveler de Crèvecoeur described Nantucket as "a barren sandbank, fertilized with whale oil only." Astonished, he wondered "that a sandy spot of about twenty-three thousand acres, affording neither stones nor timber, meadows nor arable, yet can boast of an handsome town consisting of more than 500 houses, should possess above 200 sail of vessels, constantly employ upwards of 2,000 seamen; feed more than 15,000 sheep, 500 cows, 200 horses; and has several citizens worth £20,000 sterling!"[30] It was certainly a land drenched in oil, but no matter the pacifist pretensions of Nantucket Quakers, this fuel was burned at the expense of more than just whale blood.

As the colonial order that whale light helped weave began unraveling in the age of revolutions, so too did the makers of that light find themselves in dire crisis. Nantucket was devastated by the American Revolution. Naval war always dealt blows to whale fisheries, but war with a naval superpower like Great Britain proved even worse. Nantucket was dangerously dependent on the regular processing of whales, and the nature of whaling made that next to impossible during the war. For a voyage to be profitable, the ship had to be nearly filled with oil. This meant whale ships traveled light, and any cannons or firearms on board were potential explosives occupying space needed for, and in close proximity with, highly combustible oil. Trying out oil on board sent up plumes of smoke visible and smellable for miles, such that the only whaleships that evaded detection were those that failed to catch any whales. By the war's end, the island's former fleet of 150 whaleships had been reduced to 30, and damages were estimated at over $1 million. Whalemen, moreover, were experienced sailors and were frequently targeted for impressment. According to one historian, more "than a thousand Nantucket seamen, the majority whalemen, were either killed or imprisoned, creating 202 widows and 342 orphaned children out of eight hundred families."[31]

Slowly, haltingly, American whaling towns began to rebuild after the war. Cities in the new United States such as Boston, Philadelphia, New York, and Baltimore began to copy London's street-lighting plan, and Nantucket built new geographic relations from the wreckage of its transatlantic circuits. As Nantucket men went after whales, grown numerous in the reprieve granted by the war, Nantucket gained back some of its former glory, emerging as the clear center of the American whaling industry in the early republic. Such was the growing demand for whale oil for both light and lubrication that by the time Hixson set sail for Nantucket in 1832, the New England fleet consisted of nearly 400 deep-sea vessels manned by over 10,000 sailors, it had a combined mass of around 150,000 tons, and its book value was in excess of $10 million. By the middle of the nineteenth century, whaling and its products constituted the third most valuable industry in Massachusetts, after cotton and shoes, and the fifth largest industry in the United States.[32] Although much of that capital had been relocated to the rival port of New Bedford, Nantucket remained at the core of the industry, especially when it came to sperm whales. And this was why the thirty-one-year-old Hixson was on his way to the island to try his hand as a whaleman.

N ANTUCKET *Harbor, 1832*. This was not what he had signed up for. He knew he was in for at least a two-year voyage that would take him across two hemispheres, perhaps to India or Japan. He knew it would be hard, dangerous work, and he might not be able to communicate with his family until he returned home. Richard Hixson knew all this, but as he reached Nantucket, he ran up against something entirely unexpected. It had been three days since he had sailed for Nantucket, but he still remained tantalizingly out of reach of beginning his life as a whaleman. Indeed, he and the rest of the passengers and crew aboard the sloop found themselves prisoners within sight of their destination. "Lying in Quarantine in a small sloop," Hixson penned miserably in his journal, "16 passengers we sleep in a small cabin, enough to bread the Cholera of itself." Stuck on a ship in the middle of Nantucket Harbor, the passengers aboard the *Nantuck*—many of whom, like Hixson, were on their way to becoming whalemen—experienced firsthand the at once enabling and disabling effects of such deeply interconnected relationships of light and cotton, plantation and ship, factory and empire.[33]

In a sense, Hixson had been partially incarcerated by the candles he intended to make. Spermaceti candles were more than just lights; they were technologies uniting the terrestrial cotton empire embodied in cotton wicks with the maritime spaces embodied in spermaceti wax. It was not just poor sanitation spreading cholera; the movement of candles, textiles, ships, and oil provided vectors for infection, too. As cholera, and fears of cholera, traveled rapidly through commercial and social routes, it triggered attempts to disentangle deeply interconnected spaces. Port masters struggled to keep cholera at bay so that whale oil could flow uninterrupted into Nantucket,

Dragged up Hither from the Bottom of the Sea

New Bedford, and other New England ports while sailors, manufacturers, and merchants worked tirelessly to turn whales into oil into light.

Cotton plantations and gin houses, the primary sites of cotton production, were also, like sugar works, spaces made directly through slavery, wicks, oil, and candles. In July 1828, the Supreme Court of Alabama heard a case in which a slaveholder and his son had set fire to their gin house when an open glass lamp fell onto and ignited the dry, fibrous cotton filling the room. Using lamps inside a gin house was clearly a bad idea, but was it common? The court asked the jury to determine how "customary" the practice was.[34] Carrying lamps into a gin house was certainly dangerous, and anyone who spent time around ginned and baled cotton would have known the risk. That these enslavers took the risk anyway suggested how powerful the pressure to inspect their cotton and meet the demands of buyers could be.

Light and flame enveloped and threatened cotton all along its industrial life cycle. Greeting the bales as they arrived in the proliferating cotton mills of the Northeast were specially designed picker rooms. In these rooms, low-paid workers unpacked the bales shipped from plantations, removing any remaining seeds or debris and thereby saturating the air with tiny, combustible cotton fibers. Even more than a gin house, a picker room was, as one judge put it, "almost as perilous as a powdermagazine, to use lamps in." Yet use them many manufacturers did, despite the fact that such "is the extraordinary fineness of the cotton fibres and dust which fills the air in that room, in factories in great quantities, that any lamp which has air holes, or an open top and loose cover, (such as are necessary to continue or preserve the light,) is liable to be filled with them and to ignite them, and, unless the building is detached or secured by iron doors, to cause the almost inevitable loss of the whole establishment." But the potential reward for using lamplight to squeeze a few extra hours of labor each day from factory workers was simply too tempting. Many industrialists deliberately risked life and property in order to extend production into the evening.

Night work in cotton factories required a delicate dance between flame and fiber. One factory in upstate New York, built in 1832, navigated this dance for over a decade before burning down in 1846. As recounted in a suit brought by the insurance company, "The fire originated in the picking-room, which was situated in the center of the building, and in which a glass lamp was permanently suspended from the ceiling, and into which room a glass lantern was carried that evening, and placed by the workman on the window-sill which the picker was in operation." The workman continued to pick through the raw cotton by the light of his lamp until he saw a sudden flash above the glass chimney, "as if the cotton-dust had become ignited through the air-holes, and the fire was communicated with such rapidity to the whole cotton he was unable to extinguish it." It was not long before the whole factory was burned to the ground.[35] The very work of seeing at night, visual labor that was packaged into cotton bales and textiles, was itself incredibly dangerous. In nineteenth-century nights, seeing could kill.

Yet seeing, during both night and day, was absolutely necessary for the creation of an empire of cotton. It was also an empire surprisingly dependent on whales. As Hixson set sail for Nantucket, capitalists were hiring builders to construct a spate of new cotton factories across the Northeast, including one down the road from his farm in Sharon. These multiplying industrial spaces not only created new demand for light but consumed tremendous quantities of sperm oil to lubricate the cotton spindles. Sperm oil was necessary for producing the lights of factories, streets, and harbors; it was also the most highly sought after lubricant for the spindles spinning at the center of an Atlantic industrial revolution.[36]

Whale oils seeped into the fabric of industrial cotton slavery in other, less visible ways as well. When the British declared the slave trade illegal, the candles-for-slaves trade that had emerged during the first sugar boom did not disappear; it merely moved underground. According to one study, "In the 58 years during which the illegal slave trade was carried out" from 1807 to 1865, the spermaceti "exported easily exceeded 150,000,000 candles, worth over $9,000,000. Most of these candles," moreover, "were destined for the slave trade, and permitted the purchase of approximately 100,000 slaves on the West African coast." The relation between spermaceti candle exports and the slave trade was so firmly established that British palm-oil manufacturers and merchants publicly argued that developing palm-oil plantations in West Africa and a palm-oil and candle industry in England would both break the American monopoly on high-quality candles and "help force an end to the spermaceti driven West African slave trade."[37] But despite boosters' hopes, such competition failed to dislodge whale oil from global circuits. The New England nexus converting whales into light, lubrication, and cotton continued to draw thousands of West Africans illegally into Brazilian, Cuban, and U.S. slavery; thousands of young women into cotton mills; and thousands of young men like Hixson into the whale fishery.

Hixson, drawn toward the fishery by the demands of this geography of cotton and oil, had been ensnared in Nantucket Harbor by fears of a microscopic stowaway. He was vulnerable, however, to more than the risk of infection. "Riding out Quarentine everything on board goes rong," Hixson

Dragged up Hither from the Bottom of the Sea

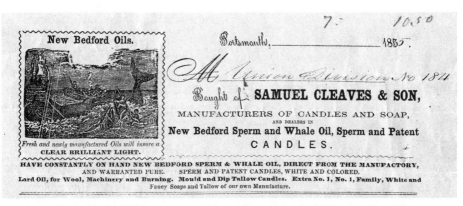

The packaging. Scenes depicting the work and dangers of whaling, usually of men harpooning and battling whales, were commonly printed on wrappings and invoices for commercial whale products. Whale-oil merchants were the only sellers of light in the nineteenth century to regularly represent the origins of illuminants in terms of work and nature, although, perhaps revealingly, such illustrations almost always foregrounded the hunt rather than cutting in or trying out. Whale oil invoices, Warshaw Oils, Archives Center, National Museum of American History, Smithsonian Institution.

grouched the following day from a crowded, probably foul-smelling cabin. "The Capt. drunk, and nothing to eat but bread and salt beef." Three days later the quarantine was up. "Happy in leaving a small vessel, and a very disagreeable master," Richard Hixson ended one voyage shaped by predatory, non-human actors and was ready to begin another in which he and his fellow crew members would be the hunters.[38] Riding currents long plied in the pursuit of a reliable means of light, men like Hixson were after whales, renown, and the freedom of the high seas. At least that was what it said on the packaging.

Whaling in the nineteenth century was more than an industry made in the deep between whales and whalers; it was a practice translated and reinvented in candleworks, dry goods stores, newspapers, and literature.[39] Representations of the manly labor of the fishery could be found on nearly every package of sperm candles or advertisement for oil. Such images helped to sell commodities; but they also contributed to the creation of an imagined fishery, and such imaginings served to recruit men like Hixson, who had little connection to Nantucket or New Bedford, to ship out on three- or four-year whaling voyages.

Other kinds of recruitment stories circulated as well. For black men, both enslaved and free, the true stories of slaves such as Prince Boston winning freedom through the Quaker fishery and Absalom Boston becoming captain of an all-black crew helped to reorganize the geography of freedom in the United States. Whaleships became refuges for many black men fleeing beyond the geographic reach of slaveholders' power, fugitive routes into saltwater worlds where enslavers' claims, backed as they were by the terrestrial resources and authority of the U.S. federal government, weakened considerably. The most famous enslaved American to escape through the whale fishery was John Thompson, who wrote and published a narrative of his experience in 1856. Pretending to be an experienced steward, Thompson shipped out on a whaleship, and his ruse was not discovered until he was deep at sea. There, outside the reach of U.S. law, the captain demanded to know why a man who had never been at sea before would try to ship out as a steward. Thompson explained his deception in the clearest possible terms: "I answered, 'I am a fugitive slave from Maryland, and have a family in Philadelphia; but fearing to remain there any longer, I thought I would go a whaling voyage, as being the place where I stood least chance of being arrested by slave hunters.'" The fishery, with its Quaker masters, was one of the most antislavery industries to employ black men, and Thompson gambled on this political affinity in so boldly and honestly stating his case. The captain kept his secret and trained him as a whaleman, and by the time John Thompson returned to his family in Philadelphia with the money he earned on his two-year voyage, the slave hunters had either given up or lost his trail.[40]

At least one man who would ship out of Nantucket with Hixson had followed this path to freedom. Levi Smith, whom Hixson later taught to read, "was a slave in North Carolina" who "was sold and transported to New Orleans from whence he made his escape and came to Boston." On Nantucket, and aboard nearly any whaleship, Levi Smith would have encountered a thick maritime community of New England Indians and free and fugitive black whalemen. The *Maria* was no exception. In the back of his journal,

Dragged up Hither from the Bottom of the Sea

Hixson recorded the names and occupations of the crew; as was common at the time, of the sixteen seamen onboard (not including officers, coopers, carpenters, cook, or steward), he listed seven as "Coloured," a category that encompassed both Native and black people in New England censuses.[41]

By continually asserting the right to thread their lives and labor into and out of the processes of producing whale light on their own terms, whalemen helped make the fishery into a fugitive geography. Fugitive slaves like John Thompson and Levi Smith smuggled themselves on board and into the work rhythms of whaleships, while others fled the law from different directions. Having escaped the New York police, and finding himself at odds with the captain and officers after a near-death experience in a stoved boat, a different Smith—whose flight from the law had led him to ship out with the fugitive John Thompson—deserted in Madagascar with three others from the ship, only to be betrayed by the locals, recaptured, flogged by the captain, and then escape again, leading multiple ship captains and crews in pursuit, triggering more desertions, and precipitating collective struggles over movement, labor, and authority.[42]

Some of the young men on board whaleships had actually been placed there *by* the law. From 1827 to 1850, the New York House of Refuge, a reformist "juvenile delinquent" prison and workhouse for poor boys and girls who had run afoul of the law or been taken away from "dissipated" parents by city officials, indentured out at least 240 of its older boys on ships leaving from New England and Long Island ports for two- to four-year whaling voyages. While some seemed genuinely excited to ship out, many others, like J.B.C., who had "made his escape four times, and made two unsuccessful attempts to escape" from the House of Refuge, were sent whaling in a last-ditch effort to geographically discipline boys who had consistently rejected the authority of adults, masters, and agents of the state. For the young men of the New York House of Refuge, the fishery became more of a carceral than a fugitive geography, but they, too, developed strategies within this strange community of young seamen to control their own movements. House of Refuge reports were littered with stories like those of C.D., T.S., and J.L., "one of our hardest boys." Each had shipped out on whaling voyages only to be "left for sick" on, respectively, Maui, Tahiti, and the Marquesas as soon as their ships reached the Pacific. Remarkably, after escaping from or being abandoned by the whalemen, they had then succeeded in making their ways back thousands of miles to the House of Refuge to share their tales with the other inmates. As antislavery fugitives, men on the lam, "juvenile delinquents," and Melville's "meanest mariners, and renegades and castaways" all struggled into and out of the watery and wooden worlds of the fishery and told one another their stories, they nurtured a politics counter to the pro-slavery, pro-industrial political economy being made through the circulation of the very products of their labor. And the telling of stories like these, full of all the tragedy, irony, and possibility that has always inhabited the real political lives of the weak, helped to guide workers into Nantucket.[43]

They did not come for the sights. Although desperate to finally get off the sloop, Hixson was hardly ecstatic to reach shore. Unimpressed, within a few hours he was ready to leave: "6 oclock" and "have seen all that I want to of Nantucket." Not wasting any time, Hixson spent the next day looking for a way out. Touring the docks and anchored ships, he was soon "engaged to go on board of the ship Maria Captain Alexander Macy, bound on a voyage round Cape Horn in pursuit of whale oil & bone." At last, he was going to be a whaleman.

Or not. A week later he was still stuck in a town he had had his fill of in half a day. They were supposed to have been at Martha's Vineyard to outfit the ship by then, but "on account of a North East storm I am afraid the vessel will not leave till Sunday." Hixson's unwanted stay onshore demonstrated the extent to which a whaling voyage was a collective enterprise dependent on assembling a band of light bringers, some more human than others. The ship itself was an old and experienced vector of whale oil. Built in 1822 in the shipyards of Haddam, Connecticut, the *Maria* had, over the previous decade, made three voyages around Cape Horn to the Pacific, channeling 6,592 barrels of sperm oil into Nantucket. Just returned to port in June, the *Maria* now found itself with a new captain and an entirely new crew, but the labors and lives of the previous three crews were embedded in every repaired timber, in every cared-for corner of the ship.[44]

As the crew readied to leave Nantucket on the *Maria*, twenty-five men crammed onto a deck 100 feet long and 27 feet wide, with small compartments below for sleeping, knowing they might live this way for up to four years. Living conditions on whaleships were worse than just cramped, however. "The forecastle," where most of the crew slept, "was black and slimy with filth, very small and hot as an oven," wrote one journalist who shipped aboard a New Bedford whaler in 1842. The room, which had twelve small sleeping cubbies, "was filled with a compound of foul air, smoke, sea-chests, soap-kegs, greasy pans, tainted meat . . . in a hole about sixteen feet wide, and as many perhaps, from the bulkheads to the fore-peak; so low that a full-grown person could not stand upright in it, and so wedged with rubbish as to leave scarcely room for a foothold."[45] Such were the vessels of whale light.

Dragged up Hither from the Bottom of the Sea

And these vessels were sites of truly extraordinary consumption. If a voyage produced 4,000 barrels in fourteen months but sank a mile from Nantucket Harbor, all that labor and energy would be lost. This was why outfitting a voyage could take so long. Ports like Nantucket and New Bedford were clearinghouses for the hundreds of thousands of tons of American-made food, wood, metal, and rope mustered annually on board hundreds of whaling ships being readied to leave for the Pacific whaling grounds. Each year the American fleet consumed millions of barrel staves and barrels upon bushels of flour, beef, pork, molasses, rice, and dried apples. Whaleships were loaded with tens of thousands of boat boards and oars and hundreds of whale boats so crews could replace, repair, and refit these floating factories over their four-year journeys. To even get to the whales, therefore, took tremendous labor, matter, and energy. Yet it was still not enough. In the Pacific, whalers had to continually seek out ports, plantations, and islands to replenish their rapidly depleted food and water supplies.[46]

The *Maria* finally left Nantucket on September 24 and "got into Edgartown [Martha's Vineyard] at dusk." There, Hixson and the crew labored in the rain for two weeks to load and ready the ship for its voyage. On October 8, having "Hauled ship into stream" in preparation to depart to sea, Hixson made his final arrangements with the land. He said goodbye to family and friends: "10 oclock P.M. just wrote home the last letter before I go to sea." And he purchased products of the soil: "have been on shore this evening and bought myself some peppersauce, mustard, apples . . . to take to sea." They departed at sunrise the next morning, "Edgartown fast receding from view."

After weeks of preparation and waiting for the *Maria* to finally ship out, the impatient Hixson soon realized that getting to the Pacific whaling grounds would be no speedier. Yet neither was it going to be lonely. The *Maria* sailed through channels dense with other whaleships. On the journey to Cape Horn, which they reached around January, at least one sail could be seen on the horizon nearly the entire time. The *Maria* spent a good while sailing, going after blackfish (a very small whale whose oil they would use to trade for provisions with Pacific islanders), and socializing, or "gamming," with the *Charles Carroll*, a new whaleship that had sailed with the *Maria* from Nantucket. Not only did the crews of the two ships frequently mix and share labor and company during their regular gams, but many had been fellow travelers before reaching Nantucket. Aboard the *Charles Carroll* one evening in October, Hixson "had a very pleasant time with George Knapp, J. C. Edmond & Charles C. Lincoln three passengers with me in the sloop

Nantuck from Boston. Staid on board of the C.C. till dark." Eventually, the crews of the two ships would be divided by a leak. The *Charles Carroll*, as was not uncommon for ships that had never sailed before, sprung a leak that forced it to turn back and try to repair. Hixson believed, falsely it turned out, that the *Charles Carroll* had sailed back to Nantucket. Unbeknownst to Hixson, the crew managed to patch the leak and reach Talcahuano in Chile — where many of them promptly deserted — but the *Maria* was already long gone.[47]

Sailing, of course, was more than just avoiding leaks. And whaling was mostly just sailing. Although the exciting and terrifying moments when crews in tiny whaleboats pulled hard after the beasts of the deep were what the people like Hixson who dreamed of going whaling saw on candle packaging and read about in novels, the vast majority of the time, energy, and work of a whaling voyage was spent moving and maintaining the ship. It, too, involved a good deal of danger and risk. Men climbed up rigging, dangling over the deck one minute and the choppy waves the next as the ship heeled over in the wind. The same day the *Charles Carroll* sprung a leak, Hixson, while "reeling topsail," lost his hat overboard. As he watched it fall, he no doubt realized how lucky he was that it was just his hat. William Johnson, another member of the crew, was not so fortunate, when "in attempting to up the leach of the fore sail fell and was considerable hurt." When storms struck, the *Maria*'s sails ripped, masts snapped, and mountainous waves threatened to sink the whole ship. First the crew had to work to survive. Then they had to repair the damage. Perhaps nothing illustrated better how much importance was placed on the integrity of the ship and its cargo than that there were two master carpenters and two master coopers on board, but only one cook and no trained physician.[48]

By late November, while the crew readied the *Maria* for passage around Cape Horn, the men, too, were being prepared and readied, made from Atlantic sailors into Pacific whalemen. "Dead calm through all this day," Hixson wrote, staring at the thermometer, which "ranges at 90 degrees." At three in the afternoon, before an audience of five other ships, the men "lower'd boats and there respective crews man'd them, to exercise. We went through all the manauvers of rowing for throughing irons, lancing, and finely taking the whale." For most of the crew, this was probably their first time attempting such practices, which "made quite a display before the ship keepers."[49] They were hardly experts by the time they reached the Pacific, but these exercises provided important training without wasting whales.

Dragged up Hither from the Bottom of the Sea

S EVEN months after setting out from Boston, Richard Hixson was cornering an enormous beast in the middle of the Pacific. It was not a whale. In fact, in all the months at sea, he had gone after only a single sperm whale, the one with which this chapter began. No, this was a tortoise, and Hixson was on a mountain, not a ship. Since setting out in the *Maria* as a deckhand, Hixson had been frequently sick, lost five hats overboard, nearly collided with an iceberg around Cape Horn, repaired a mast destroyed by a storm, and been robbed in Peru, and now he and the crew had been tasked with carrying hundred-pound turtles three miles back to the ship.[50]

New England whalers gathered on the Galapagos off the coast of Ecuador to exchange information and mail, to resupply, and most importantly, to collect tens of thousands of the turtles living on the islands for food. The production of lamplight was as reliant on the caloric metabolism of proteins, fats, and carbohydrates in human muscle as it was on the wind in the sails, the squid in the sperm whales' stomachs, and the fires in the tryworks. Galapagos tortoises were a favorite means of meeting those caloric needs, prized for their ease of care: the turtles could go months without food or water without losing weight or tastiness. But getting them on the ship was no easy task. "All hands called at 2 oclock this morning to go after Turpine we pulled about 6 miles and landed," Hixson wrote at the end of "the hardest days work I ever did." No sooner had they stepped onshore than they "immediately commenced ascending a mountain. About 2½ or 3 miles up the mountain we came to the place where the Turpin or Turtle live, they weigh from 50 to 100 pounds. We have to sling them on the back like a knapsack and then," he grumbled, "travel down the mountain over briars, thorns, rocks, & everything else that is bad." The official logbook of the voyage kept by the impious and wry Charles Murphey, the first mate of the ship, described the turtle hauling rather differently: "Went after tortoise. Got 40 found it remarkable easy work and consequently am in great haste for the Sabbath to pass away that we might go after them for be it known our principles are such we cannot conscientiously labor on this Holy Day."[51]

Nor were turtles the only nonhuman (or nonwhale) organisms to support and be entangled in whaling. "I have taken this day an account of the live stock on board," Hixson wrote one afternoon, and "it is as follows, viz. 2 Sows with 7 pigs each. 1 Boar. 2 Pigeons male and female. 1 Dog. one spanish rabbit, a female, she is in daily expectation of becoming a mother. The male was accidently killed a few days since. 2 Cats male & female." In Tumbes, Peru, they picked up three hens, a goat and its kid, and an unruly monkey, which they carried to Hawaii. In Hawaii they added eighteen more goats, and everywhere they stopped, they stocked the ship with potatoes, "vegi-tables," and fruits including "mellons," bananas, plantains, pumpkins, and coconuts grown by "human beings at work on there beautifull plantations." But in summing up, Hixson pointed to another set of living actors inhabit-ing the spaces of the fishery: "This is all our live stock, with the exception of mice and cockroaches."[52]

The combination of whale blood with these menageries made whale-ships into exceptional havens for unwanted pests like rats and roaches. Rats "were more numerous on whaleships than on any other vessels, probably because of the profusion of blood and oil that soaked the decks, despite the regular scrubbings. They were more than any ship's cat could cope with, and then as now, there was nothing that could cope with cockroaches." Cock-roaches were so endemic to whaling vessels that "for many seamen, the roaches were a more predominant aspect of a whaling voyage than whales." Smelly, loud, sharp-toothed, and impossible to exterminate, the roaches, some nevertheless claimed, served a useful if unpleasant purpose: they ap-parently ate the fleas right off whalemen's sleeping bodies.[53]

Roaches also congregated deep inside the *Maria*, inducing battles with the whalemen over food and water. Traveling from off the coast of Japan to California, the crew stopped in Santa Barbara to refill the freshwater casks they needed to live. There, in the holds, they confronted their competitors. The insects had drunk the remaining fresh water in the casks and then made the containers into new homes. The crew took their revenge, drowning the roaches, but their victory came at a cost. When the casks "were filled in St. Barbara there was hundreds of cockroaches in them, and no pains was taken to get them out previous to filling them with the water that we now drink, the consiquence is that the tea and coffee smells almost as bad as the in-sects themselves." And the war wasn't over. "Since we commenced wetting lower hold," Hixson wrote, "the cockroaches have come from thence to be-tween decks in great numbers and all hands from the cabin to Forecastle have bottles (with molasses in them, for a decoy) in requisition taking these troublesome fellows and by this many vast numbers were destroyed."[54]

The webs of turtles and goats, rats and roaches, ports and plantation workers were not exclusive to whaling; they sustained and were sustained by other sailing voyages as well. But without the hundreds of multiyear whaling voyages in the Pacific, these webs would not have been nearly as extensive. And if provisions ran out — whether consumed by humans, rats, roaches, or rot (especially the fruits and molasses-soaked potatoes to prevent scurvy) — there could be no whaling, no oil, no light.

Dragged up Hither from the Bottom of the Sea

W HALESHIPS were massive tools designed to transform whales into oil that could be sold as a commodity. They carried sailors and whaleboats; housed tryworks and tackles; stored barrels, food, and livestock; and thereby made it possible to translate and transport living whale fat from the Pacific into barrels of whale oil on New England docks. They were the social mouths and stomachs that provided nourishment for the flames of oil lamps thousands of miles away. Following the first kill, over the next three years, the crew of the *Maria* would repeat the process of catching, cutting in, trying out, and storing below thirty-six more times. Some days they caught as many as four whales before they began cutting in. By voyage's end, the *Maria*'s assemblage of men, boats, harpoons, spades, tackles, hooks, tryworks, barrels, and holds would consume into storage a total of fifty-two whales.

Organizing this all-important work required something not far from military tactics and precision. It wasn't hard to see why. In a very real sense, the whalemen were at war with the whales. While the *Maria* cruised for whales, officers generally stationed two men aloft in the crow's nest to scout for whale spouts. In the normal course of the voyage, Captain Macy divided the crew into watches so that "part of the crew are always on the watch, while the others are asleep below." Yet when a whale was spotted, all hands would be called on deck and divided into boat crews. When the boats were lowered after whales, the captain tended to be at the head of the starboard boat, the first mate commanded the larboard (or port) boat, and the second mate had charge of the waist (or as Hixson spelled it, the "waste") boat. The third or fourth mate remained on the ship using signal flags on different masts to communicate with the boats and coordinate the battle using information gathered by the men scanning the waves from on high. The choreographed marking of work, space, and status set in motion on the ship continued in the tight confines of the whaleboats. Once the boats were lowered, each hunt began with "the officer steering, while the harpooner, who is termed the boatswain, rows the bow oar, until the whale is fastened with the harpoon, which operation is performed by this person. This being done, the boatsteerer goes aft and takes the officer's place, while that person goes forward to kill the whale."[55] And these whales were hardly passive sacks of oil waiting to be plucked up by whaleships.

Far more often, the whales escaped the *Maria*'s boat crews. But even when the men did succeed in killing a whale, he or she rarely went down without a fight. Hixson's boat was repeatedly slammed by the massive tails of sperm whales, causing leaks and tossing men overboard. It could be a har-

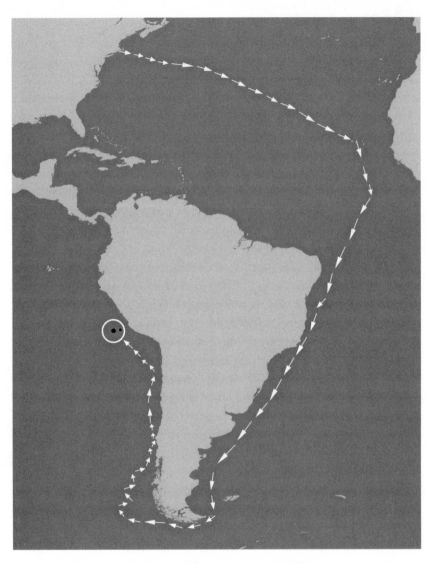

The voyage of the *Maria* to the first kill.
Each arrow represents approximately three days.

rowing experience. During one hunt, after the whale was made fast, "the whale came up directly under" Hixson's boat "and grounded her on his back so that for a few moments we could not get the boat off, at last he gave a spring, [threw] the boat off, and struck her with his flukes or tail and stove in the boats bottom so that it kep two men bailing continuely." Miraculously, the men managed to remain in the boat as they "had another fine ride

Dragged up Hither from the Bottom of the Sea

"**A view of the Maria's boats, battling a large whale February 11th 1834: when the Starboard's boat's crew were thrown of the boat by the whale and the boat badly stove.** The Ship is seen at a distance with her signals set at the main Top Gallant head to inform the whale boats another whale is up. The Starboard crew remained in the water a considerable time after which the waste boat picked them up." Watercolor by Richard Hixson (1834). Courtesy Houghton Library.

after the whale," dragged along by the harpoon line; "he took us through the water at a rapid rate, for two hours or more, he went in every direction, but keep within two or three miles the ship."[56]

At other times the men were not so lucky. About a year after they caught their first whale, the *Maria*'s crew "raised two sperm whale on our leebow." The hunt began normally enough when, after six hours of rowing for the whale, at "4 P.M. we through into him." This proved to be an even more dangerous action than usual. The angered bull "struck the boat" with his massive flukes "and threw her entirely out of the water and everyman out of the boat with the exception of Captain Macy and stove the boat." Remarkably, heroically, the "captain held on to the whale, altho the boat was more than half full of water," until Hixson and the rest of the starboard boat crew swam through the churning waves for "the waste boat, which picked them up" and returned them to the captain. "The other boats coming up the whale attacked them and for a while seam'd determined to destroy them," a still-terrified Hixson wrote, "but after a while, say half an hour, we succeeded in getting an opportunity to lance him which tamed him very much, and in a short time he died." This whale's violence was so awesome that "three of the waste boat crew," a completely different boat's crew, "got frightened and jumped out."[57] So powerful was the experience of being tossed into the churning sea by an enraged whale that the scene inspired one of only four paintings in Hixson's journal.

The furious flailing whale appeared commonly in the writings of whalemen, but it was not the only kind of whale behavior that they suspected to be conscious resistance. The overwhelming majority of pursuits ended with the boats returning to ship and whales swimming free. This was usually attributed to either human shortcomings or the speed of the whales. But sometimes whalers were thwarted by what seemed to be organized tactics. For four days in August 1834, midway between Hawaii and Japan, "no whale was taken altho' the ocean was cover'd with them." Day after day, the crews of the *Maria* and three other ships chased a pod of young bulls from sunup to sundown, but though "there were whale in all direction," they were "so wild we could not make fast altho' we got almost on." The wind was low, and the whales were too aware of the boats to be surprised. And the "wild" movements of the whales thoroughly confused the human crews. After four days of continuous pursuit, only one whale was caught, and not by the *Maria*. The rest of the whales managed to escape in the night.[58]

Herman Melville went so far as to claim that the apparent decline in whale populations was actually an illusion produced by the shifting paramilitary strategies of whales. Denying that sperm whales would suffer the same fate as American bison, Melville conceded that "in former years (the latter part of the last century, say) these Leviathans, in small pods, were encountered much oftener than at present, and, in consequence, the voyages were not so prolonged, and were also much more remunerative." But the longer, farther voyages of the nineteenth century were, he claimed, less the consequence of overfishing than of whales learning new ways to counter the practices of whalers, for "those whales, influenced by some views to safety, now swim the seas in immense caravans, so that to a large degree the scattered solitaries, yokes, and pods, and schools of other days are now aggregated into vast but widely separated armies. That is all." The impact of American whalers on global whale populations remains a question of serious debate. Some scholars claim that it was not until the twentieth-century Japanese and Norwegian fisheries that whale stocks truly plummeted, the Americans being limited by the technologies of sail, muscle, and harpoon. Others argue that the trophic cascades triggered by American whaling were likely even more devastating than ecologists' highest estimates.[59] But whatever the reason, Atlantic whalers were forced to become Pacific and then Arctic whalers, with demand rising, voyages lengthening, profits declining, and perhaps sperm whales—the animals with the largest known brains, living or extinct—teaching one another new strategies to hide and fight back.

Ship, whaleboat, and tryworks have received a great deal of scholarly and cultural attention, associated as they were with the romance of the hunt. Barrels have been largely ignored, despite their tremendous importance. Barrels were more than just containers; they, too, were social organs working together to sustain these webs of light over vast geographic and social distances. The entire forehold of the ship was devoted to storing unassembled barrel staves and hoops. Coopers continually led the rest of the men in assembling these staves into barrels in anticipation of a catch. This way, when the oil was tried out, it could be "quickly" stored in air- and oil-tight barrels, each holding around thirty-one and a half gallons. It typically took several days after trying out to cooper and stow all the oil from a whale in the holds below deck. If times were particularly busy, the crew left the barrels on deck so that more men could go after whales. There were two master coopers aboard the *Maria*, but as Hixson described in his journal, all hands were involved in coopering and stowing oil in the belly of the ship.[60]

And storage was only the beginning of a much longer process. The crew of the *Maria* periodically hauled up the oil-laden barrels to check for leaks and to reseal them. They were solidly and expertly built, but these barrels were still prone to the same kind of warp and wear that affected all wooden structures on the ship. They needed to be watered (more often, in fact, than the turtles) so that the staves could swell slightly and form a better seal. Yet too much water could ruin them, as Hixson recorded when preparing for a weeklong process of seeing to all seventy-five tons of oil: "had everything ready to commence hoisting up and cooper the oil. 7 A.M. commenced raining. . . . Watch went below. It will not answer to cooper oil in a rain," Hixson explained, as "the cask swell and when stowed down they shrink again, which will cause the oil to leak." Later that week, Hixson witnessed how dangerous this work could be: "while hoisting empty cask out of fore hole, for the purpose of filling with oil, William Magee one of the coopers fell down the whole distance and was badly bruised, but no bones brocken."[61]

ALL these transformations, all this labor, and all these contests between whales and boats, waves and ship, roaches and men emerged through a complex and contested government of the fishery. Whaleships were circumscribed spaces in which all on board had to work together for a profitable voyage. But this cooperation was never just going to happen on its own. The bringers of whale light hadn't come together as equals with identical interests. Like everywhere else in the capitalist world, those who actually did the work of turning whales into oil had only the loosest ownership over the products of their labor. Nor were the men obliged to obey the captain in order to survive. Paid only at the completion of a voyage, which took years — and the size of the payment dependent on the success of the catch — whalemen could desert in nearly every port, and they did so in large numbers. The hard-earned movement of men in and out of the fishery across its entire global span helped to create a spatial politics of labor that was both relatively egalitarian and a constant challenge for officers to manage. Whalemen could sign up on different voyages to renegotiate their "lay" (the percentage of the voyage's proceeds) and navigate the uncertainty of unlucky or incompetent ships. Mutiny was not unheard of, and neither was arson (sometimes deep at sea), from which the crew would escape in whaleboats. Captains and owners had to find ways to make crews do what they wanted.[62]

While still in the North Atlantic, an increasingly wage-based geography of labor onshore and on merchant and naval ships eroded the leverage whalemen previously had in negotiating a lay, when many could just go back to their farms or trades. But in Pacific ports like Talcahuano and Honolulu, where whaleships and whalemen gathered in large numbers, whalemen could desert one ship and negotiate better lays with another. In the Pacific their leverage returned, and friction with officers grew as well. Still, whaling went on, and as captains tried desperately to make voyages successful, crews learned new ways to resist.

Eliza Brock, the wife of the captain of the *Lexington*, watched from the decks at Talcahuano as yet another fire erupted into this story. The port of Talcahuano, Chile, where whaleships frequently stopped to resupply, was a site of unusual resistance. Based on the tables published at the end of the nineteenth century by the whaling historian Alexander Starbuck, it would appear that no port saw more whalemen desert than Talcahuano. What Eliza Brock witnessed was less common but far more frightening to the owners and capitalists of the fishery. "Yesterday," Brock wrote on March 17, 1856, the *George Washington* was "set on fire by four of the sailors. They towed her on shore and scuttled her. Today is still burning. They are in safe keeping tied up in the rigging." Later that night "the fire from the George Washington burst through; she is all burning up; her masts fell this morning at 6 o'clock. An awful sight to see that noble ship perishing in the flames; all by the recklessness of depraved sailors."[63]

Shore leave was another point of friction. On the *Maria*, only half the crew was allowed onshore at a time, and Captain Macy strictly forbade returning with alcohol or prostitutes. As these prohibitions were not uncommon, whalemen seeking relief from strict shipboard discipline congregated in the brothels, bars, and markets of Pacific ports. That these were also the same spaces targeted by missionaries led to conflicts over interaction between Native Hawaiians and American whalemen. Hawaii was a particularly contested space. As Hixson perhaps somewhat exaggerated, in Maui "a female is not allowed to go on board a ship when at anchor. This is not the case at [Oahu], we went on board the W.L.P. and her decks were cover'd with abandoned females, who swim off to every ship that comes to anchor at this island, and become bedfellows with the officers, and sailors, if the Captain will permit it." The Christian missionaries living on the islands made considerable effort to stop such sex and drinking. After a sailor was jailed for drunkenness in 1852 in missionary-controlled Honolulu, the frustrated whalemen onshore rioted, and only a lucky wind kept the 150 whaleships in the harbor from catching fire.[64]

Gamming between ships was another practice that captains and officers used to coordinate voyages, and it, too, could generate friction. Months after their first pleasant gams with the *Charles Carroll*, Hixson wrote that "I for one (and I am not alone) think it wrong for the captain to be visiting all the ships he sees. We came here after oil, not for the purpose of going, day after day, and night after night on board other ships to have a chat, making the men pull the boat at 10 or 12 oclock at night after there Captain, when he ought to have staid on board his own ship." It was not simply that Hixson and the crew found these meetings tedious; "it is very unpleasant for men (very likely after a hard pull after whale) to lower boat at 10 or 11 oclock at night for the purpose of fetching the captain on board." Gamming, he crescendoed, "is a great damage to the voige, we are not so likely to get oil, it is bothersome to the men, and it ought not so to be. I think if it had not been for this practice we should have had more oil than we have got. We are all sufferers by this foolish practice, and no one is benefitted by it. It is wrong! wrong!! wrong!!!" Through the circulation of news, letters, and stories of home and fishery, these meetings played central roles in the articulation of informed authority among captains and served to facilitate long-distance community building for crews; but many whalemen would have preferred to just skip the ritual and get home faster.[65]

Getting everyone to work together in the economic interests of the voyage took considerable skill that not all captains or officers possessed. The crew of the *Franklin*, one of the ships the *Maria* encountered, told Hixson of how their "first mate proved to be a tyrent and they knocked him down a few times, which helped him very much. The Captain gets frequently intoxicated and knows not what he is about. The Franklin has lost 4 men by accident, 2 fell from aloft one was killed instantly the other had both legs brocken and was sent home. 2 were taken out of the boat by the line and were not seen afterwards." A badly governed ship could lead to desertion as well as death. On shore near Tumbes, Hixson saw "the grave of *Collins* one of the ship Lopers crew of Nantucket. He was shot by the mate of that ship in this place in the year 1831 when in a passion." The officer, understanding the uneven geography of law, "made his escape from Nantucket after the ship arrived there to avoid coming to tryal." More poignantly, Hixson noted, by "the side of Collins grave, is another one, where lies the remains of one of the Kingston's crew, of Nantucket Capt Sherman of that ship abused him so while crossing the *barr* that he jumped out of the boat and was drowned. This last accident, if accident it can be called, took place a short time previous to our arriving here."[66]

Discipline generated the most intense struggles in the fishery. From the crew of the *William Thompson*, Hixson learned that "at St. Francisco Capt. Potter was one day about flogging the steward when the crew interfered, the captain called a number of spaniard from the shore, they came on board and by the instigation of Potter, one of the spaniards ran a man through with a cutlass and he died immediately." "The crew represent Potter as a great villain," Hixson wrote, and an "investigation of this affair will take place when the ship arrives in the United States as the crew are determined to inform against him." And the crew knew Potter might not even try to prevent them from testifying. On a previous voyage, they told Hixson, Captain Potter "stuck the cook of his ship with a brand of fire and put out both his eyes which cost him on his arrival in America 5000 dollars." Five thousand dollars was certainly a considerable amount of money; but it was not jail, and it was apparently not career-ending. In contrast, Hixson usually described Captain Macy as an effective and fair administrator. "A foolish affair took place," Hixson noted, "in which six of the people living in the steerage were concerned." He was vague on the details, "but they all acknowledge that they had don wrong and were sorry for it with the exception of one." Using the rhythms of shipboard work as a disciplining tool, "as a punishment Captain Macy has stoped for the present his watch below in the daytime and given the officers orders to keep him at work." Denying the offending whaleman his customary time off, his "watch below," he would be kept at hard labor for two weeks before being allowed to resume his usual daytime breaks from work.[67] Cap-

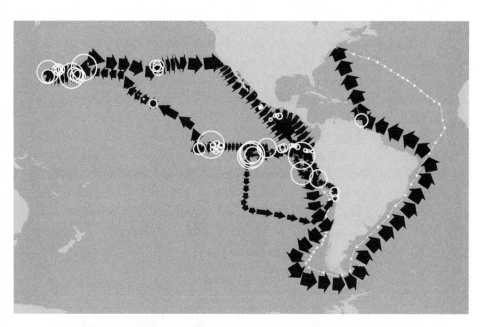

The complete voyage of the *Maria*. Each arrow represents approximately three days. White arrows show outgoing voyage. Black arrows track the voyage after the first catch. Thickness of arrows reflects total amount of oil accumulated. White circles indicate whales caught, and diameter reflects amount of oil tried out.

tain Macy was flexing his authority by declaring this maritime custom to be a privilege, not a right, under his command.

As the voyage wore on, crew, ship, and sea continued to negotiate complicated social and ecological relationships in the process of accumulating oil. By the end of Hixson's journal on December 31, 1834, the *Maria* had turned close to fifty whales into around 1,500 barrels of sperm oil, weighing more than 175 tons; it was not an impressive voyage. It was an average catch but a longer than normal trip. On each of the *Maria*'s previous three voyages, the ship had returned with around 2,000 barrels and had done so in less time. The voyage continued another fifteen months after Hixson's last surviving journal entry, yet the official logbook indicated that they captured only a few more whales in all that time. With whalers paid by the barrel, and each whaleman given his lay, it should have been no surprise that after months of hardly a drop of oil added, several men deserted the ship in Talcahuano, Chile.[68]

On March, 12, 1836, Christopher Mitchell, the agent for the *Maria*, began writing from his Nantucket office to "Mr. Thos. Folger." "You will have seen by the papers that the Chas. Carroll has arrived with a good voyage and that she reports the Maria with 1,500 bbls. This is what Capt. Macy writes us he has got; he also writes that he shall use up two months in looking for whales on his passage home, & that we need not look for him until after the middle of April & we were in hopes that he would add something to his oil in that time." Yet before Mitchell could finish writing the letter, at "half-past four P.M. Capt. Macy has just arrived & informed us that he was obliged to make the best of his way home on account of the Scurvy, being entirely out of vegetables. He has one man down with the scurvy now. He took one 40-bbl. whale off Trinidad."[69] Limping into port, the Maria had added only 150 barrels in fifteen months. Hixson never recorded his lay or his reason for going whaling, but as he returned to New England, he likely left the fishery with his proceeds (and possibly debts) and returned to Sharon via the newly constructed Boston-Providence Railroad. Thus ended the journey of one lucifer, but the products of his labor were still not yet lights.

G ETTING barrels onto Nantucket docks marked the end of a whaleman's direct relation to the production of the means of light. It took another series of relays before that oil could end up in a streetlamp or domestic candle. When Hixson had first arrived in Nantucket four years prior, 5,000 barrels of sperm oil had just been deposited on Nantucket docks, 2,170 by the ship *Loper*. Most of the oil from the *Loper* had been immediately bought and transported to the candleworks owned by the wealthy Starbuck family. While Hixson had waited to ship out in pursuit of barrels of his own, he'd likely seen men and horses wheeling these containers about town, moving the distilled products of human labor and cetaceous life that would become the oil burning in streetlamps and lighthouses down the American coast.

As the Starbucks recounted in their memoirs, dockhands would have hoisted the "black, greasy, bulging casks" from the *Loper*'s hold and rolled the barrels "onto the 'dray,' two long planks balanced on and fastened to the axle of two wheels. A horse was harnessed between the upper ends of these planks, while the lower ends could be tipped down to the ground; three and sometimes four casks were rolled up the planks and balanced so that one horse could move a very heavy load." Plodding along cobblestoned streets past an idle and impatient Hixson, these horses had pulled oil-laden barrels to one of the few dozen oil and candle works in Nantucket. According to one study, by 1832, the year Hixson shipped out, "there were forty-three oil and candle works in Nantucket. Annually, 250 workers produced 1,400,000 gallons of sperm oil and 1,200,000 pounds of candles. Turning the raw materials of whaling into finished products, Nantucketers came to dominate the manufacture of whale oils as they dominated whaling."[70]

By the time the *Maria* returned to Nantucket in 1836, and Christopher Mitchell, the ship's agent, was busy contracting with men to carry those 1,500 barrels over cobblestones and into candleworks, the owners of these long sheds watched "the try-house come to life. The wood was wheeled in, the fires started, and lines of men drawn from the cooper-shop and the cordage-shop passed in and out carrying heavy buckets of crude oil." These laborers arrived with every new shipment of oil and disappeared back into making barrels and rope once operations were under way. Before they left, however, the "kettles were filled; the heat became intense as the oil bubbled and seethed in the kettles, throwing off steam and sending particles of blubber and other impurities to the surface to be skimmed and fed to the fire." These activities took place year-round, but they were only one step in the process.

Once the oil was purified, men carried it in buckets to a cellar or storehouse, where they placed it in casks until a cold day in winter. The seasonality of this storage was necessary because the oil had to be brought down to a sufficiently cold temperature to be further separated and graded. Because the *Maria* returned to Nantucket at the beginning of spring, after this initial processing, the oil from the voyage would have remained in storage until the winter. Winter workers began by stuffing the semisolid oil cakes into bags and loading them into the massive press that dominated each candleworks. In the Starbucks' try-house, at each end "stood a spermaceti press. Huge beams, sixteen inches square and over thirty feet long, hung in the air along the sides of the shed. These were the timber-levers. One end was held between two massive upright posts." In front of these gigantic levers were the press boxes, into which workmen "placed the bags of 'black-cake' separated into layers by heavy wooden 'leaves.'" Once they had filled the boxes, the men lowered the post end of the beam "until it rested on the topmost of the 'leaves.' The very weight of the beam-end made the oil start from the bags; but when" the force of the beam was increased by application of weights to the other end, "there gushed from the bags clear and limpid oil, the first and finest product of the whole process."[71]

The product of this first pressing, which exerted "two thousand pounds of pressure per square inch," was called "Winter Strained Sperm Oil," and because it was produced at such cold temperatures, the liquid pressed from the cakes would resist freezing or congealing even in New England winters. Oil makers thus used seasonal time as a chemical sieve. By forcing a complex mixture of organic molecules through fields of pressure and cold, they separated the amalgam into a purified liquid unrivaled for streetlamps, lighthouses, and lightweight lubrication. About two-thirds of the crude was converted into winter oil. The owner of the works used his sensory knowledge to ensure its quality; he "dipped his finger into this finished product; he rubbed it in the palm of his hand; he sniffed its odor and touched it with his tongue for its taste." The remaining solid was then stored again until temperatures reached about sixty degrees. Once more, the cakes were pressed; the lower-quality oil squeezed out was called "spring oil." This accounted for about 10 percent of the final product and was sold as a lower-quality illuminant for about 5 cents less than winter oil. Finally, in summer, the cakes were "pressed a third time, at one hundred thousand pounds per square inch." From that extraordinary pressure trickled the cheapest, "summer oil," 5 percent of the total product and useful only in climates where the temperature didn't fall below seventy degrees, when summer oil would congeal.[72]

By late summer or autumn, when many ships returned from their voyages, "only one man was at work" in the Starbuck factory, but before "the year ended, eight would be needed." For the final stage in the production process, Joseph Starbuck "watched the man as he lifted the flattened bags when they came from the press. As the spermaceti left the bag and fell into the kettle, it was yellowish, dry and brittle." To transform this into the white, uniform candle wax of such fame, under "the heat of the fire it would turn to oil," and to the mix "water and potash would be added, then hardening substances. Vapors would rise carrying off the water and potash and the mixture would be ready for the molds."

But a cylinder of spermaceti was not by itself a candle. A wick was required, too. "In the 'yarn room' a woman was twisting cotton yarn for the wicks," one visitor recalled. Starbuck "picked up one of the wicks and untwisted it; it had six separate strands." Meticulously, he "compared it with the pattern; he frayed out an end; rubbed the soft bloom he had made against his cheek" and patronizingly lectured the woman, "'Never forget that you are doing the most important job; the candles can be spoiled if you fail to see that these wicks have the proper texture and size.'" Molding and wicking took two days, and then "the candles would be inspected, counted, boxed and shipped."[73]

Schooners and sloops continually transported these finished products from Nantucket and New Bedford oilworks to agents and wholesalers in Boston, New York, and Philadelphia. The *Nantuck*, which had first brought Hixson to Nantucket, was likely one of these transporters. And when he returned to Boston, he likely traveled on a vessel laden with the products of the fishery. For decades these oils had flowed from sea to ship to works to lamp, prompting Melville's declaration that "though the world scouts at us whale hunters, yet does it unwittingly pay us the profoundest homage; yea, an all-abounding adoration! for almost all the tapers, lamps, and candles that burn around the globe, burn, as before so many shrines, to our glory!"[74] But even as Melville wrote these words and Hixson sailed past Boston Light for a second time, this relationship was beginning to unravel.

T HE imperial and industrial revolutions of the mid-nineteenth century accomplished what neither piracy nor war had managed to do: mortally cripple the New England whale fishery. The combination of colonial Atlantic transformations in sugar, slavery, and urban nights had accelerated a thirst for light that American whalers desperately attempted to quench, plundering the far corners of all four oceans for over a century. They sailed in hunt until that thirst for oil grew so great that the worlds the whalers made began to fray along their seams. By the 1840s, the flows of whale oil were no longer able to slake a thirst that had found new offerings in turpentine and coal gas. It was not that whale oil was replaced in some evolution of technology. Rather, the hundreds of thousands of whales sacrificed in Atlantic lamps and candles had inspired visions of light that expanded and intensified faster than whalers could replenish the altars.

As cotton plantations and gins spread southwest into newly conquered lands and mills proliferated in New England in the 1830s and 1840s, their demand for oil began to further overwhelm the fishery. Cotton mills thirstily devoured sperm oil, and not for light. According to the census of 1860, cotton factories "consume large quantities of sperm oil, each spindle using about half a gallon." Sperm oil's unrivaled properties as a lightweight, long-lasting lubricant meant that lamps were forced to compete with spindles for a limited resource. The proliferation of cotton mills in New England not only eclipsed the formerly privileged place of the whale fishery but meant that publicly illuminated cities were struggling to meet the rising costs of sperm oil. By the 1840s, spindles at home and abroad had grown so thirsty that "very little sperm oil was available domestically for purposes of illumination."[75] Under the assault of these spindles, webs of whale light were fast being respun into dark webs of lubrication.

To make matters worse, Nantucket was burning. Fire had always been a threat in a town literally flowing with oil, and extraordinary measures were sometimes taken to keep these flows from igniting. In 1838, the *Maria* having again returned to the Pacific, the ship's agent, Christopher Mitchell, wrote reassuringly to his insurers, "We sent you this morning's paper in which you will find some account of the late fire in this town. The loss is not so great, as was first anticipated." More importantly, Mitchell wanted the insurers to know that everything had been done to keep his oilworks from burning, such that "several buildings were blown up by the fire department, . . . all in the immediate vicinity of our oil-works." Fire departments using gunpowder to blow up nonburning buildings might have sounded rather extreme, but it "seemed to be the opinion of everyone that if the fire could be prevented from Communicating with our premises, a valuable portion of the town would be saved." In July 1846, the town would not be so "lucky." "The flames spread with such rapidity, as to baffle every exertion to repress them," the *Boston Daily Atlas* reported. Desperately, firemen tried to halt the fire as they had done before, and "blowing up houses was resorted to, as the only means of arresting the conflagration; but it would seem, with little success." Still, more "than TWENTY buildings are thus destroyed; and, indeed, many more would have been demolished, but that all the powder in the place was consumed." According to a recent study, "Over a million dollars in property was destroyed, one third of the island's buildings were gone, and 800 islanders were homeless. All along the waterfront the fire, violently fueled by burning whale oil and tar, leveled wharves, counting houses, ropewalks, sail lofts, warehouses, and cooper shops."[76]

After the fire of 1846, the fishery became clearly centered in New Bedford. This shift followed whale oil's change from illuminant to lubricant, as cities, again following London's lead, began to replace sperm-oil lamps with far cheaper and far more reliable coal-gas burners (at least cheaper once all the gas mains had been laid). Between 1840 and 1860, the number of U.S. cities with gas lighting increased from seven to forty-one. Cotton spindles, meanwhile, continued to consume the flickering remains of the once-luminous sperm whale fishery. According to a recent study, by the 1850s, the lubrication requirements for New England's cotton spindles exceeded 100 percent of the sperm oil available domestically.[77]

All across the Atlantic world, spaces that had been born, sustained, and greased by whale lights seemed to be systematically prying loose their tethers to the fishery. In the "juggernaut of the Cuban sugar economy," planters embraced a fully industrialized slavery of rail, steam, and chemistry far ahead of most of the Atlantic capitalist world. And part of this modernization involved planters abandoning whale oil in favor of gaslight. With spiking sugar prices, and "considerable outlay of capital," Cuban planters, according to

Dragged up Hither from the Bottom of the Sea

an 1841 report to Parliament, were expanding production amidst the post-emancipation ruins of the British West Indian sugar industry "with an organization and completeness far exceeding anything heretofore attempted and perfected." But perhaps most terrifying of all was the Cuban embrace of "the modern invention of coal-gas to obtain that artificial light by which the labours of the Cuban pandemonium, the crushing and boiling-house, might be carried on so long as human physical endurance, forced to its extreme extent by the lash of the driver, could carry it; the roads or paths from the cane-fields to the crushing and boiling-houses were even lighted with gas to enable the overwrought African slave to see his way to this human hell."[78] Gaslit plantation factories were not so much revolutionary breaks as modernizing extensions of processes begun generations earlier with the oppressive nocturnal union of sugar, slavery, and whale oil—a union held together and enthusiastically exploited by British planters, slave traders, and factory owners for decades until antislavery movements tore it apart.

Rocked and strained by the rippling effects of war, industrial revolution, and the exhaustion of whales, webs of whale light were also unraveling from within. Building on the generations of fugitive practices that had made the fishery into such an unusual geography of freedom, during the 1850s, in ports all across the Pacific world, whalemen deserted in rising numbers. In 1859, the *Whalemen's Shipping List* complained bitterly, that "our Courts should make an example of the desperadoes visiting New Bedford and securing berths on board our ships, is evident from the following statement of the burning of whale-ships, and mutinous conduct. The loss of property at their hands has been immense." The article went on to list mutinies, desertions, and thirteen deliberate burnings of whaleships by their crews over the previous decade, concluding that "whaling masters in these days must go well-armed, and, expecting no favors at home, must exercise their own judgment for the maintenance of order, the preservation of peace, and protection of life."[79] By the antebellum period, the American whale fishery, already struggling to meet the insatiable thirst of cotton spindles, began to unravel as labor relations on its deep sea vessels grew increasingly antagonistic. Although the incredible demand from cotton manufacturing was driving up the price of oil, lays were growing increasingly unequal, desertion became more common, and voyages grew longer as whales became harder to find. Finally, fewer Americans were willing to invest their lives, labor, and capital in the fishery, given the lower risks and better returns of working and investing in textile and machine industries.

The fishery persisted into the 1870s, in much-reduced form, but it had become a dark industry. No longer were whales, tryworks, and lamps spun together by the movements and labor of whaleships and whalemen; a new set of relationships had emerged. The dreams and spaces of light, first triggered by the revolutionary union of ship and tryworks, survived and continued to expand. Woven into the niches first occupied by whale-oil lamps, new luminous webs, with new geographies of labor, energy, and power, began to transform the antebellum United States again.

Chapter Two

PINEY LIGHTS

It was night in New York City, and so like tens of thousands of other women, Mary Clark and Ellen Cooley were at home sewing for their lives. Gathered around the shared light of a single lamp, the two women carefully plied needle and thread. On the table before them lay the assembly pieces of men's shirts, precut to industry standards by men in the "cutting departments" of New York's clothing houses. Clark and Cooley, along with countless other New York women that June evening in 1858, were the sweated outworkers who sewed the shirts together. They did not so much make clothes as assemble them, and in this gendered division of labor across space, their job was simply to stitch. And stitch. And stitch. To put together just one of these shirts took around 2,000 stitches and at least six hours of work. For their trouble, they could hope to earn (and sometimes hope was all they had) from 4 to 12 cents. These were starvation wages, but their families desperately needed the money.[1] With laws, customs, and class politics so tightly circumscribing the respectable options for women's paid work, Clark and Cooley had little choice but to sew, and so they were making, one stitch at a time, shirts that might let their households pay rent, eat, and survive one more week.

Theirs were the needles stitching a revolution in men's ready-made clothing. Contracting with poor women like Mary Clark and Ellen Cooley to sew garments at home for men to sell to other men in enormous clothing houses, clothiers disguised a revolution in production by keeping working women in their gendered place and out of public view. Mass-produced clothing was, until much later in the century, an exclusively male market, women's dress remaining rooted in practices of homespun and expensive tailored fashions. It was a process whereby thousands of northern women, working in households for wages far below those of their husbands, fathers, brothers, and sons, made inexpensive clothes for hundreds of thousands of American men—strangers who otherwise could claim no right to exploit their housework—to wear on the street and on the job.[2]

By eleven o'clock that night, if Clark and Cooley had worked steadily without mistake for eighteen hours, each woman might have been on her third shirt of the day. More likely, given the incessant domestic demands of husbands, children, cooking, and cleaning, they were still struggling through their first or second. Their straining eyes, aching necks, and sore fingers may have demanded that they rest. But rest was not an option. To reach even the starvation-level wages they aimed for, Clark and Cooley had to stretch their labor deep into the night.

That meant relying on cheap, (relatively) bright light. "Not many years ago," *Scientific American* observed in 1858, "the only fluids employed in our country for household light were animal oils obtained by perilous adventure on the stormy sea with monsters of the deep." Yet with cotton spindles and heavy machinery spinning these fluids into industrial lubrication, "whale oils are in comparatively limited use for illumination, and are becoming more limited every year." Sperm oil may not have had any rivals in quality, "but it has become so dear that cheaper substitutes have been sought and obtained." Chief among these was a new synthetic illuminant called camphene. A liquid mixture of spirits of turpentine and highly distilled alcohol, by the 1840s, camphene had become the dominant lamp fuel in the United States. The camphene Clark and Cooley were burning that night had cost them about 50 cents a gallon. Whale oil would have cost them nearly twice as much; sperm oil, more than three times. Camphene was not only cheaper than whale oil but brighter and cleaner, leaving little of the sooty residue that oil lamps deposited over a room. It was a light of bourgeois sensibilities and working-class thrift. And unlike some other new illuminants, such as pure turpentine, lard oil, or coal oil, the addition of the volatile alcohol allowed camphene to be burned in any ordinary oil lamp. Only tallow and stearine candles could compete in terms of cost, and few women chose to sew for hours only by candlelight. For Clark and Cooley, camphene really was the only option. Burning it meant they could meet the relentless cash demands of landlords, merchants, and families a little more reliably. It also meant risking their lives.[3]

Having worked frugally for hours with only one wick burning, Ellen Cooley decided she needed more light to see her needlework. Reaching out "with the intention of lighting one of the wicks not then burning," she picked

Piney Lights

up the camphene lamp and tilted it down for the unlit wick to catch the flame of the other, "when a quantity of fluid ran out and ignited." Terrified and singed, Mary Clark leaped up and threw the lamp into Cooley's lap, "where it exploded." As liquid fire shot over the room, Cooley somehow managed to smother the flames consuming her dress and hair by wrapping herself up in some bedclothes. Clark was not so lucky. Her clothes immediately ignited, and she ran shrieking out of the house and "through the street enveloped in flames." Rushing to her aid as her "piercing screams alarmed the neighborhood," a small group seized Clark and extinguished the flames, but not before she was severely injured. After being rushed to the city hospital, she lingered in agony for over a month before finally succumbing to her burns. She was twenty years old. It was an altogether too common tragedy.[4]

Six years earlier and hundreds of miles to the south, in the backwoods of North Carolina turpentine country, Jack and Willis Parmerly prepared to turn pines into the means of light that killed Mary Clark. In the early morning autumn air, the two enslaved distillers began the work of transforming pine resin into spirits of turpentine at the two-story still, work they knew could lead to enormous explosions. They, too, were trapped in an ecology of violence. But explosions weren't the danger they faced today. A team of enslaved men, perhaps including the Parmerlys themselves, had built the still several years before by a creek that emptied into the Cape Fear River, a waterway that carried precious provisions to the camps and barrels of distilled turpentine to Wilmington. But it could also flood. That Monday morning, the sun dawned on a river threatening to steal months' worth of light and labor. "Jack found the river had rose so high that it was taking off some of our spirits casks that was beneath our Platform on the River Bank," wrote James R. Grist, enslaver and owner of the camp, to his father. As the only two "hands that was at the still early monday morning," Jack and Willis Parmerly sprang into action to save the casks. They bravely hopped in a shallow-draft pole boat, but according to James, the overseer "orderd Willis not to go but headstrong like would do so." The current quickly overwhelmed the boat and flipped it over, and "Jack come near drowning & Willis not being able to swim was drowned."

Piney light killed Willis Parmerly every bit as much as Mary Clark, but his death was translated through very different structures of blame. James R. Grist was mostly interested in making sure that he was not liable for Willis, an enslaved man he had rented from another enslaver, and so he contended that, although he was sorry for the loss, "you see at once we have not worked or employed Willis by water & on the contrary he was orderd not to go by

Mr Skiles our agent, therefore Mr Parmerly cannot expect us to pay for him." Besides, what Grist really cared about was transforming life and labor into turpentine he could ship to Wilmington and then to New York, and production was going just fine. As he "rode over a portion of the boxes yesterday," he had found the trees "well faced & well chiped. I see nothing going on rong in the business." Men might drown, and some spirits might be lost, but for Grist it was no more than collateral damage so long as the turpentine kept flowing. "I really think there will be at least 8000 bls of Turpentine to get off & dip," he concluded, and "the team looks very well indeed."[5]

Mary Clark and Willis Parmerly were casualties of class, gender, race, and light, but it was a species of light that history has largely forgotten. Forgetting camphene has also meant writing the white women who used it and the black men who made it out of the stories of who and what really mattered in the making of antebellum history. White women working in the household, but for a system of mass production, and black men working as slaves, but in an industry and place bearing little resemblance or proximity to a plantation, have appeared neither similar enough to nor different enough from the stories contemporaries and historians have told about industry, gender, and slavery to attract much attention. But theirs were not sideshows. They were entangled in processes whose very invisibility allowed them to quietly underwrite the northern and southern worlds of white men in the 1840s and 1850s, quietly except for the moments when the materiality of piney light erupted violently into public view.[6]

The explosive violence imperfectly contained in the era's cheap new camphene lamps made possible a political economy in which men's safer days expropriated the work of women's perilous nights. Without a cheap, portable illuminant like camphene, a revolution in urban domestic night work would have been inconceivable. But dependency didn't mean desire, or even consent. The violence of camphene wasn't a "cost" women "chose" to pay. It was danger they had no choice but to risk. E. Meriam, a reformer from Brooklyn, tried to quantify this violence by searching through New York newspapers. Excluding the eleven "children" and forty-five "persons" listed in his widely circulated report on the dangers of camphene, from October 1855 to October 1856, he recorded seventy-seven women and twenty-two girls burned or killed by camphene lamp explosions, while only twenty-six men and five boys suffered the same. Women, according to these figures, absorbed more than twice as much of the violent risk of camphene as men, while girls were more than four times as likely as boys to be injured by burning fluid.[7]

Camphene's explosive materiality was something almost no one had any familiarity with before 1840. Wood might snap or crackle as it burned, and when oils and tallows were spilled, they could certainly cause intense fires. But they didn't explode. Oil lamps wouldn't instantly envelop someone in liquid flame who was refilling their basins or lighting a wick. Camphene would. Because of its extreme volatility, handling a camphene lamp became, by most measures, one of the deadliest activities of the antebellum period. In the words of one outraged editorial, the "progress of the age, and the ingenuity of man, have introduced no engines of destruction so potent as camphene" and all so-called burning fluids, "and could the yearly victims of these latter-day monsters be gathered in one pile, it would present a mammoth hecatomb, compared with which the heaps slain by steam explosions and railroad accidents, would be as ant hills to the Egyptian pyramids." Editorial hyperbole aside, the deadliness of camphene was real enough. Stories of gruesome deaths by camphene circulated through the nation's newspapers, usually accompanied by an outraged and incredulous demand to know, "When will people cease to use this infernal stuff?"[8]

But the outrage captured only part of the story. To accumulate the astonishing quantities of turpentine used to make camphene, planters brought new kinds of lands under the dominion of slavery in North Carolina's piney backcountry.[9] Camphene connected the enslaved North Carolina men who tapped longleaf pines in remote forest turpentine camps, to produce nearly every drop of turpentine in the country, with the seamstresses burning that turpentine in nocturnal domestic workshops. In the decades before the Civil War, enslavers, merchants, manufacturers, and clothiers came to dominate crucial circuits of work and energy between forest and city, South and North, to wrench profit and power from the labor of enslaved woodsmen, factory hands, and outworking women. Through this national process of producing and consuming camphene, southerners reinforced the institution of racial slavery, men reinvented their power over women, and northerners and westerners clothed a new popular politics for all white men. The making of piney light enabled, at one end, white workingwomen to chase a modicum of security and respectability as they sewed the public clothes of white men, while at the other end planters dislocated black men from their tidewater communities and forced them to transform North Carolina's piney woods into living engines of turpentine and enslavement. This chapter tells the story of the sudden emergence, and eventual sundering, of the surprising antebellum relationship between New York seamstresses and North Carolina turpentine slaves, the American lucifers who dipped, distilled, and stitched themselves together in the hidden making of piney light.

B Y the 1850s, more than 400 clothing businesses had been established in
New York City. Prices were plummeting while demand for respectable
white male clothing expanded from Broadway to Bowery to the South
and West. At the same time, as the city's landlords used their monopoly over
property to charge exorbitant rents, working-class New Yorkers found their
homes and workshops increasingly difficult to afford. In the 1840s, clothiers
took advantage of working-class desperation to replace male in-house tailors with outworking seamstresses paid below-subsistence piece rates. By
the 1850s, the enormously profitable clothiers of New York were employing
hundreds of skilled male cutters and tailors but tens of thousands of women
like Mary Clark and Ellen Cooley.[10]

A deliberate, managed market politics was producing a gendered
poverty. "There is many a song of the shirt sung in the garrets of our metropolis," one *New York Daily Times* writer lamented, asserting it was "a fact
proved, that the majority of our sewing women are working for starvation
prices, with a fair chance of being defrauded at that." Indeed, there was "a system about it; these fellows do everything by rule." Whenever a merchant
manufacturer needed some clothes to be made, "he tells his sharp clerk to
get them done in the cheapest possible manner. They advertise, and women
needing work, lured by the advertisement, apply." These women, desperate
for cash and forced to compete for work and wages with thousands of other
women, first had to scrounge up a dollar or two, as they "are required to leave
money as a pledge for their honesty, — the thing seems fair enough, for they
are strangers to the advertisers, and besides, they are glad to get work on
any terms, — so they leave their dollar, make the coarse garments and return
them to the store."[11]

Tied to the store by their deposits, these women had to make the clothes
in their homes, outside the supervision of their employers, and provide the
rooms, needles, thread, fuel, and light, all at their own expense. The domestic
workspaces of seamstresses were cramped, noxious, and expensive. According to contemporary accounts, they generally lived and worked in "a single
room, or perhaps two small rooms, in the upper story of some poor, ill-
constructed, unventilated house in a filthy street, constantly kept so by the
absence of back yards and the neglect of the street inspector," or in "an attic
room, seven feet by five, . . . in which we found, seated on low boxes around
a candle placed on a keg, a woman and her oldest daughter . . . sewing shirts,
for the making of which they were paid *four* cents apiece," less than half that
earned by the lowest-paid man in the industry.[12]

To secure the rights even to use rooms such as these—which had to function as both garment workshops and living spaces — "the tenants never pay less than three to four and a half dollars per month — and *pay* they must and do. Some of the very worst garrets, destitute of closet or convenience of any kind, and perhaps lighted only by a hole cut in the roof, rent as low as two dollars a month." With piece rates ranging from about 4 to 12 cents per shirt, and even the swiftest hands only able to complete three shirts "by working from sunrise to midnight," the most a seamstress could hope to earn in a week was never more than a third of a month's rent. Then, of course, "there were fuel and lights to buy" and, especially in the case of camphene, "with it all the terrible chances of sickness and accident."[13]

When seamstresses returned to the central clothing house depots with the finished shirts, pants, coats, cravats, and hats, they had to negotiate and carefully navigate their way through a blatantly uneven tangle of power relations to get the money they had been promised in exchange for their labor. In the large stores, the giving and receiving of work, sometimes for as many as 4,000 women, took place on only one day each week. On the designated day, each clothing store transformed into a strange market manufactory. "No seats are provided; a long file of women are waiting their turn to be served," one observer described. Meanwhile, a "single clerk is detached from the large corps of assistants in the store to attend" to the line of women seeking work and pay. Approaching the clerk, the first woman in line "tremblingly offers her work for his inspection; he picks up the neatly made garments, handles them roughly, grumbles at the stupidity of these women in not doing something which he never has told them to do, perhaps swears a little by way of variety, hands the frightened woman another bundle of work and her money, and passes on to the next." Merchants and clerks managed this theatrically performed domination and resistance of women, repeated over and over, until the "clock strikes four; no more work can be received, and those who are unfortunate enough to come after that time are forced to trudge home again to wait another week." By thus controlling and dividing the spaces of exchange (clothing houses) from the spaces of production (tenements), a typical clothing house employed and disciplined between 500 and 600 outworking seamstresses, who took home, assembled, and returned over 3,000 shirts each week; many larger houses employed and produced triple that number.[14]

Pressed into domestic outwork by a spatial revolution in the clothing industry, women like Mary Clark and Ellen Cooley were consigned to *night* work by an equally powerful temporal politics, one fought through the uneven times and spaces produced around two new, and unequally distributed, species of light: camphene and gaslight. During the antebellum decades, city governments and industrialists collaborated to build gas lighting systems that assertively illuminated the commercial and bourgeois cores of American cities. Embedded in the cityscape with retorts, factories, gasometers, and pipes, gas lighting systems were certainly impressive. But their direct reach remained relatively limited for most of the century. Gaslight, while ultimately cheaper for those who could afford it, was a privilege principally restricted to property-owning classes, as gas companies would only pay to dig up streets and lay mains to neighborhoods where consumers were wealthy enough to afford the upfront costs of connecting buildings, fitting pipes, and installing fixtures. The tenements, boardinghouses, and shanties into which urban regimes of property crowded laborers also became illuminated longer and more brightly than previously, but with an important exception: the domestic night spaces that workers appended to the diurnal day were produced with camphene rather than gas.[15]

Nowhere was this process clearer than in the clothing industry. The clothing houses themselves, usually narrow, deep rooms extending back from the street, would have been dark or dim but for the scores of gas lamps brilliantly illuminating the shopping floors and in-house cutting departments staffed by male clerks and cutters. The owners of Lewis & Hanford, a major New York "clothing palace" that employed 4,000 people in a typical week—mostly as outworkers—prided themselves on having 112 gas burners to light the premises. But the vast majority of the work of manufacturing clothes took place far removed in time and space, and many a New York needlewoman was "compelled to toil from dewy morn—not to dusky eve, for of that she might not complain—but to the tired hour of midnight; and out of it all gain a scant supply of the mere absolute wants of life, at the sacrifice of all company, relaxation, and with the fearful penalty of broken health."[16]

The extension of the working day into the evening pushed more than leisure into darkness. Many women worked as servants during the day and were forced to put off the vital domestic work for their own households until well past sundown. Many others, like Mary Clark, faced an inversion of the working day, scrambling to keep house during the day while trying to accomplish their waged piecework at night. Cramped and noxious, a seamstress's domestic workshop also had to function as a nocturnal space, and as one writer described, the "light by which she had been working still burned on the table, a little camphene lamp, so faint that I could hardly have read a small type with it." Notwithstanding its dimness, this camphene lamp was the light by which she "worked from *five* in the morning 'till *eleven* at night. ... Her eyes had failed her, she said, during the last winter, from working so long by lamp-light." Yet considering that the alternative was an even dimmer tallow or stearine candle, a camphene lamp—even a dim camphene lamp—was something many seamstresses worked hard to obtain. Teams of related needlewomen sometimes coordinated to relay the work of sewing over night and day by forming night shifts. As the women's advocate Virginia Penny observed, in places "where there are two or three or more women or girls engaged in this enterprise of making shirts to enable gentlemen to appear respectable in society, they absolutely divide the night season into watches, so that the claims of sleep may not snatch from the grasp of the shirt manufacturers an iota of their rights. In this way, by working about twenty hours a day, the amazing sum of $2.50, and sometimes $3, is earned per week," a wage still no more than a fourth what a male cutter could command, and less than half the $6 considered a living wage. In these domestic night spaces, outside the glow of the new capital-intensive gaslight systems, cheap and portable light was a necessity. They knew the risk, but households and businesses that couldn't afford access to the still-rare gas lighting systems overwhelmingly used camphene anyway.[17]

The contest over the meaning and control of women's domestic work in an industrializing city certainly contributed to Mary Clark's death. But she was killed specifically by camphene, and the story of camphene was about slavery as much as about outwork, and about industrialization in southern frontiers as much as in northern cities. Seamstresses navigating dangerous domestic night work in New York and enslaved woodsmen trying to survive labor camps in North Carolina forests were not, no matter how contemporaries and historians have treated them, separate struggles. Both were parts of the story of piney light.

C AMPHENE breathed new life into eastern slavery. To keep urban lamps filled with cheap turpentine, planters brought a new sort of sandy, swampy, piney terrain under slavery's dominion, opening up a new frontier of accumulation that came nowhere near reaching its limit by the eve of the Civil War. Practically all turpentine in the United States originated south of the Mason-Dixon line and, with the exception of some small subsistence producers in North Carolina, depended almost entirely on the labor of enslaved young black men raised in the plantations of the coastal plain of the Carolinas. For piney light to exist and be an engine of power and profit, slaveholders would have to compel groups of thirty or more enslaved men, separated from their plantation communities for months, to live and labor alone in the woods extracting the resin of millions of trees.

To produce turpentine, teams of laborers first had to "box" pines in the winter (hack a collecting cavity into the base of the trunk), then "chip" them in the spring (scrape off the bark above the box and cut into the sapwood to make the resin run down into the box), "dip" the resin from the boxes into barrels, get the barrels to a distillery, and finally distill out the volatile spirits of turpentine.[18] For a turpentine camp to function as enslavers envisioned, they had to coerce the men they enslaved into keeping the solar energy captured and congealed in resin moving at a coordinated pace between the four energy reservoirs of boxes, resin barrels, stills, and spirit barrels. Successfully building and maintaining these connections through widely variable and uneven terrain was never assured and was always a deeply political process. The particular spatial configurations of these webs meant that some links were more (or at least differently) vulnerable than others, especially those joining boxes to barrels. It was in the making and unmaking of such tenuous links where the struggles between enslavers and enslaved grew the most intense and the most violent.

In the 1840s, North Carolina planters transformed a marginal backwoods industry worked by small, poor, mostly white producers into a booming, slave-based engine of light. Not particularly profitable in colonial America, the "naval stores" industry, which consisted mainly of tar, pitch, turpentine, and other products made from the wood of resinous pines—products mostly sold to the British navy—had centered early on in the piney woods of North Carolina, where the sandy and swampy soil supported little agriculture. The demand for tar and pitch by navies and maritime industries was considerable, but it was the discovery in the 1830s that spirits of turpentine could be mixed with alcohol to produce a bright, cheap illuminant that catapulted naval stores to prominence. By the 1850s, naval stores constituted the third most valuable export from the South (after cotton and tobacco), with North Carolina producing 96 percent of that total. *De Bow's Review* estimated that in 1847 around 5,000 "hands" extracted more than 800,000 barrels of resin, valued at $2 million, from North Carolina's piney swamps. By 1860, camp operators coerced nearly 8,000 enslaved woodsmen into producing $7,409,745 worth of resin across the whole South, $5,311,420 in North Carolina alone.[19]

By bringing the once-marginal piney woods fully under the dominion of slavery, North Carolina planters believed they were finally regaining control over a geography of labor dominated by the massive gravitational pull of the

Cotton Kingdom. In 1849, a North Carolina paper claimed, "The 'crop' of naval stores, in proportion to the capital and labor employed, is a far greater and more certain one than either sugar or cotton, and is gathered without the use of any thing else than an axe to tap the trees, and a tub to collect the turpentine." By the 1850s, turpentine producers were leasing enslaved men from other enslavers at rates from $150 to $200, and sometimes as much as $300 per year, whereas "hands for the tobacco factories are only offered $75 to $100; on public works $140 to $150. On farms $120 to $130." Instead of selling the people they owned west to Mississippi, now coastal planters could hire their chattel out to the piney woods, where the "discovery of the value of our pines, aided by our plank roads, has worked a wonderful change within the last few years. Formerly many moved off to the South and West, and none came from abroad. Now, many come and none go." The North Carolinian planters who so enthusiastically pursued turpentine imagined themselves engaged in a counter-cotton project to save slavery from itself, and for themselves. They believed that by making turpentine into a burgeoning commodity, they had made their human property more valuable, thereby securing their power in the coastal plain while reversing a population loss to the cotton belt they argued was weakening North Carolina politically and economically.[20]

James R. Grist was one of a handful of large producers who came to dominate the turpentine industry as it tied itself to urban lamps in the 1840s and 1850s. Owning and leasing well over 100 enslaved woodsmen, the Grist family was matched by only a few others. Most woodsmen would have worked in smaller camps, but, whether directly or indirectly, large producers began to take increasing control over the thousands working in North Carolina's 1,600 turpentine operations. The Grists and a handful of others built their turpentine empires by renting lands and enslaved men, and by monopolizing ownership of the much smaller number of distilleries scattered through the woods and concentrated in Fayetteville and Wilmington. Spanning half a million acres of pines, 50 million boxes, 150 stills, and producing nearly 1 million barrels of resin annually, these antebellum turpentine camps formed the frontiers of piney light. To properly understand the strange geography of turpentine requires close attention to people and place and work, and so I tell the story through the Grist family business, which not only was one of the frontier industry's wealthiest but left one of the richest archival records. Only by carefully following the struggles among enslavers, enslaved, and landscapes in time and space can the full contours of resistance, power, and freedom in the camps be seen.[21]

As the spaces joining trees, barrels, stills, and ports were drawn, erased, and redrawn in the piney woods, resin and workers were coerced into motion and reined to a stop. And nothing did more to determine the rhythm and shape of this churning geography than the rising and falling of the rivers linking and separating frontier turpentine camps and Atlantic coastal centers. Controlling how barrels and provisions circulated in the unpredictable rivers could bring enormous profits, and this desire for control led to fierce competition among shipping lines and river towns. Wilmington had supplies and wanted turpentine. The turpentine camps had turpentine and wanted supplies.[22] Promoters and shippers who saw riches in resin organized competing and complementary projects of canals, steam, railways, and plank roads to bring the forest to the market and the market to the forest. These visionary projects of "regional improvement" pulled hundreds of enslaved men into the backbreaking work of cutting pathways through thickets of scrub, roots, trees, and vines—and all in order to begin the equally unendurable tasks of dredging canals, clearing rivers, and building roads in the hot, swampy, mosquito-ridden terrain through which turpentine moved. In the 1850s, Fayetteville, located about 120 miles upriver, and Wilmington, situated at the mouth of the Cape Fear, emerged from this brutal wave of frontier-making as the two major poles of turpentine country. Fayetteville was as far up the river as a steamboat could travel before it hit the falls, and in order to amplify the town's geographic significance, Fayetteville's promoters conscripted still more enslaved people to build a network of plank roads radiating out into the piney woods.[23]

In the 1850s, Wilmington manipulated steamers, rail, and Fayetteville's plank road network to gain control of the spaces and energy of the region. The Tar and Neuse Rivers that fed into the original North Carolina naval stores centers in Washington and New Bern were too shallow for steamships. Wilmington exploited its advantage on the deeper Cape Fear River by pushing production farther upriver and inland. After a letter of request, James R. Grist convinced his suppliers to use the steamboats of operator J. Banks. So grateful were the Banks brothers to have secured a portion of this circulation of energy that a week after they wrote the letter from Fayetteville, the newly rebuilt *Douglass* arrived in Wilmington renamed as the *James R. Grist*. At least as important was the development and manufacture of copper stills. Like the tryworks on board whaling ships, copper stills made possible the accumulation of turpentine farther from processing and trading centers on the coast, allowing producers to push camps deeper into the forest where enslaved distillers could distill on-site. Steamers like the *James R.*

Grist, then, were able to gather spirits from an extensive region onto massive docks and into dozens of stills and warehouses, making Wilmington into one of the greatest reservoirs of artificial light on earth.[24]

Steamers also helped make Wilmington into the flammable epicenter of an already extraordinarily flammable geography. In and around Wilmington, distilleries, warehouses, and wharves regularly caught fire, destroying tens of thousands of dollars' worth of turpentine. Camps, too, could erupt in flames, devastating entire regions as accumulated resin and wood chips ignited like camphene. Even the steamers and trains moving turpentine from camp to market faced the continual risk of fire.[25] And as seamstresses in New York and the enslaved makers and movers of turpentine could all attest, the flammable materiality of making piney light fell hardest onto those who could least resist, while the potential rewards were reserved for those who had pressed them into peril. But first they had to be pressed. Before the merchants, manufacturers, and enslavers of Wilmington and Fayetteville could dominate the volatile circulation of light, slavery had to be reproduced in the piney woods, and that resinous energy had to be pulled out of the trees.

D URING the winter of 1850, thirty enslaved men began to transform a section of forest into the newest colony of the Grist turpentine empire. Axes in hand, they fanned out alone over dozens of acres to cut boxes into the trunks of tens of thousands of pines. They were the advance guard of a new frontier of slavery, and James R. Grist had sent his cousin Benjamin Grist 100 miles upriver from Wilmington to oversee the development and discipline of the nascent turpentine camp. They would call it Gristville.[26]

Re-creating the power relations of slavery was an immediate and continual concern for camp overseers. They began by dislocating their workforce. New turpentine recruits faced an unfamiliar terrain, arriving in the remote, uncultivated camps with little knowledge of how to find food, water, and shelter in such landscapes. Believing that enslaved woodsmen with wives and children would be harder to discipline and keep in the camps, producers preferred young men, torn from their plantation communities and more isolated than ever.

Benjamin Grist's first task was to transform these young men into turpentine slaves, and pines into boxes. "If we enter, in the winter," Frederick Law Olmsted wrote of his journey through the southern states, "a part of a forest that is about to be converted into a 'turpentine orchard,' we come upon negroes engaged in making boxes, in which the sap is to be collected the following spring." Boxing surgically modified living pines, forcing the trees to collect their own vital energies in easily accessible gouges. To effectively reroute and lay claim to the sunlight flowing through the pines, the enslaved men had to thread their labor between the trees' vital cycles. This meant boxing was winter work, beginning sometime in November and ending in March. Winter was the time to make men into boxers and pines into solar tools, forced to give up the products of their biological work for the rest of their spring and summer lives.[27]

Boxing required strength, precision, and time. But most of all, it required enslaved men and training. "I am cuting boxes with all of the beast hands," Benjamin Grist wrote to James R. Grist's father and business partner, Allen Grist. He was also having some trouble with "a good manny Green hands," as they were "hard to learn to cut boxes." Yet he was confident that by the time Allen or James visited, "I will have all of the Green hands larnt how to box."[28] When Olmsted toured the region, he observed "the green hands doing 'prentice work upon any stray oaks, or other *non*-turpentine trees they can find in the low grounds." The new woodsmen were kept at the low grounds for supervision and education, to make use of their labor in the still, and, as Olmsted emphasized, to keep them away from the valuable pines. If the lucifers-in-training were going to botch a box, it was better that they ruined an oak (which was more valuable dead, anyhow) than cut short the life of a longleaf pine. While alive and properly modified, such a pine could act as a spigot of liquid light for up to ten years. As Olmsted witnessed with the green hands learning on sacrificial oaks near the still, boxing took practice. The boxes, each of which would hold about a quart, were cut into the trunk a half-foot or so above the roots and were "shaped like a distended waistcoat-pocket." The point was to steal life and not, as in the timber industry, to kill, and so the "less the ax approaches towards the centre of the tree, to obtain the proper capacity in the box, the better, as the vitality of the tree is less endangered."[29]

Boxing, like most turpentine work, was solitary task labor, and each man labored to create a 100-acre, sunlight-channeling engine of resin and enslavement powered by himself and several thousand trees. "Green hands to commence cutting boxes, say the 1st of November, would cut by the middle of February," one writer estimated, "from five to six thousand boxes, which are about as many as they could tend well the first year," and while there "are many hands in North Carolina who tend 7,500 to 9,000 boxes for their tasks, making 300 barrels" or more of resin, "they are the brag hands of the country." Experience meant precision as well as speed, and an "expert hand will make a box in less than ten minutes; and seventy-five to a hundred — according to the size and proximity of the trees—is considered a day's work."[30] But this was never just about most efficiently coercing trees to give up their resin. It was always also about preserving and reproducing slaveholders' power over their human property, keeping the enslaved from congregating by carving an individualized task system into the landscape itself.

In the camps, as elsewhere in the geography of slavery, there was no distinguishing between work and politics. Enslavers' need to maintain the camps' system of isolated exploitation continually generated and shaped struggles between enslavers and enslaved, struggles that introduced considerable contradictions into the labor process. Little illustrated this tension better than the problem of oversight. Overseers like Benjamin Grist only ever knew a small portion of what was going on in the camps, as the enslaved men responsible for boxing, chipping, and dipping the resin out of crops extending over thousands of acres could never be monitored all at once. Dividing the forest into crops, each to be tended by a single enslaved man, M. Jones, the overseer at Grist Depot, used an account book to record the outputs of each crop. At one operation, enslaved men designated as "drivers" were responsible for organizing and monitoring the daily labor of about ten other men, each of whom was tasked with cutting fifty to sixty boxes a day. Overseers managed and monitored this labor through weekly quotas and with daily measurements. Every day, drivers or "tallymen" rode through the forest making note as each enslaved woodsman sang out a word upon completing a box.[31]

The change of seasons produced other tensions. Boxing took place in winter, when running away had to be weighed against freezing or starving to death. Producers realized that the advantage they held during winter could rapidly evaporate with spring thaws, and so they pushed hard to complete boxing new crops before the weather warmed and a new geography of slavery opened. Given the seasonal terrain of violence, labor, and life, camp overseers believed whipping enslaved boxers for cutting boxes either too slowly or too poorly was an incentive to work harder and better. Writing in September 1850 to the owners of two enslaved men who had been whipped for running away, James R. Grist claimed they were "the only two that has been thrashed since we quit cutting boxes" months earlier.[32] In other words, whipping for poor work, at least during the winter, was something James R. Grist preached and practiced.

Slashing boxes through the woods began the process of collecting piney light, but operators still had to get the resin from the trees through a complicated, shifting geography to the stills. This transportation was the labor

Chipping a turpentine face. The "box" can be seen at the bottom of the turpentine face, which a woodsman would have hacked out with an axe when the tree was first tapped. The hatches on the face marked where chippers had cut away the bark and gouged the wood to make the tree leak resin down the face into the box. Each new chipping had to be above the last one. When faces were new, chippers would have to bend over all day to chip—back-straining work many woodsmen tried to avoid. In this photograph, the man was approaching the maximum height of a face, the limit of how high he could reach with the "round shave." In North Carolina, the time it took to reach that limit was usually around ten years. Black and White Photographic Print 0017 (Sampson County, 1923–1939), in the Commercial Museum (Philadelphia, Pa.) Collection of North Carolina Photographs (P0072), North Carolina Collection Photographic Archives. Courtesy Wilson Library, University of North Carolina at Chapel Hill.

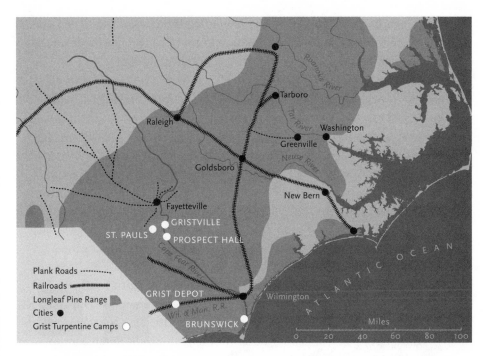

The turpentine region of North Carolina, showing Grist camps in operation from 1850 to 1860. Based on map printed in Percival Perry, "The Naval-Stores Industry in the Old South, 1790–1860," *Journal of Southern History* 34, no. 4 (November 1968): 515.

dominating the warmer months of the year from about March through October. To coordinate the transfer of energy, managers first reordered the forest to make it legible to both overseers and the enslaved. "Before proceeding to dip," *DeBow's Review* recommended, "each task, where there are no natural boundaries, should be marked off by blazing a line of trees. And every task should be further divided by rows of stakes, fifty yards apart, crossing it both ways, from side to side, which will cut it up into squares of about half an acre." This was done for reasons of power as much as efficiency, for without "this the overseer of several hands cannot possibly inspect their work with any accuracy, nor can the hands, however faithful, avoid skipping a great many boxes."[33]

Boxing and staking were like the work of building, labeling, and priming an engine, but that engine still had to be turned on. The pines would only give up their resin if they were injured. The precious resin was a thick, energy-rich substance excreted by the tree via ducts in the wood to protect it from diseases and insects when the bark and outer layers were damaged. To trick the pines into continuously bleeding resin into the boxes, the enslaved woodsmen had to keep flaying and wounding the trees, a process called chipping or hacking. Resin was a response to continually renewed violence, and this fact fundamentally shaped the labor of the industry.[34]

If chipping were to stop, the trees' wounds would clot and resin would cease to flow. Camp operators couldn't let this happen. They expected each enslaved chipper to move and labor carefully and precisely through the gridded pines with a hatchet-sized tool called a hacker or a shave, cutting furrows a few inches deep through bark and sapwood in a cross-hatch pattern angling toward the ground. It was swift, skillful labor, meant to be accomplished in just two mirrored strokes for each turpentine face. Repeating this process, a chipper was expected to chip through his whole crop of around 2,500 trees in a week, "and, as soon as over, he returns to where he began, and goes over them again and again until his boxes are full. The filling is generally done with four to six 'chippings,' or four to six weeks." Because each gouge had to be placed above older wounds in order to yield resin, turpentine faces steadily grew taller over time. Trees were abandoned when the men could no longer reach high enough to chip, a process that took five to ten years.[35]

The energy path connecting boxes to sunlight—made possible by boxing and kept open by chipping—was arranged to facilitate the process of dipping, the crucial moment of expropriation in which enslaved woodsmen pulled resin out of a forest ecology and into an industrial one. As the weather warmed, the hundreds of thousands of chipped trees surrounding the Gristville still began to leak resin into the boxes, and it was soon "necessary to commence dipping, or the removal of the turpentine from the boxes to barrels."[36] For the next several months, Benjamin Grist pressed fifteen to twenty-five enslaved men to dip a crop of several thousand pines each.

Dipping involved taking an iron or steel ladle and transferring the resin from boxes to buckets, then carrying those buckets to specially made resin barrels. Dipping may have sounded easier than chipping, but the stickiness of the resin and the strain of hauling eight-gallon buckets from boxes to barrels in summer heat made dipping dirty, unpleasant work. After the Civil War, *DeBow's Review* lamented that it "is very difficult now to find any hands willing to execute this branch of the business. Their hands and clothing become smeared with the gum, and even two dollars per diem will not now induce a piny woodsman or freedman to dip much turpentine."[37]

From spring to autumn, an alternating pattern of chipping and dipping steadily pumped resin out of the trees and into barrels. One strategy to keep the channel open between sun, tree, box, and barrel was to divide the enslaved into chippers and dippers. But this division of labor was probably rare.

According to Olmsted, the "other way—and this is more common—is to give each hand a task of trees, each of which he is required to both hack and dip statedly. Twenty-five hundred trees give a man five days' employment hacking, and one day dipping, in a week." Managers thereby divided the labor of dipping and chipping in time instead of in workers, and they also tied individual woodsmen more tightly (and less collectively) to individual crops of pines. One enslaved man would box, chip, and dip one section of the forest—a less efficient division of *labor* but a far more effective division and measurement of *laborers*.[38]

In this way, the very work the enslaved did (or did not do) in the forest became part of their surveillance. As overseers rode through the woods, they might not see or even hear the men, but they could read a history of work and movement in the landscape. The division of a crop into rows was like pre-printing lines on a page suggesting direction and location, while the trees— whether or not they were chipped or dipped—silently betrayed a woodsman's presence or absence. The surveillance net, however, was far from perfect. "Dear Sir," M. Jones hesitantly penned to his employer, "i would say to you that Miles clarke is runaway from the fact that he lackes about thirty five hundred boxes of chipping his crop over this weeke and i could not finde him Satredy in them & have not seen him yet." Jones's uncertain message revealed what were normally carefully obscured fault lines in the geography of labor in the camps. First, Jones only believed Miles Clark had run away. He could not be sure, and so he went looking for him in his crop. This meant that it was not unusual for overseers to neither see nor hear from some of the men they enslaved for days at a time, and that some of the men must have slept in their crops. But the clearest evidence was not even the fact that Jones could not find Miles; it was that about half of the trees in his crop had started to heal over their wounds. Betrayed by the trees, but not before slipping through the loosely woven net of boxes, quotas, tasks, and overseers, Miles Clark managed to stay out for at least a month.[39]

The work of transforming pines into piney light always emerged through a seasonal struggle to get resin out of the trees without letting too many men escape enslavement for too long. In September 1850, an enslaved man named Richmond, hired from the Latham plantation, ran away from the camp in Robeson, so James R. Grist "took him to Brunswick." Trying to use geography as discipline, he had Richmond work with men he knew from his home plantation, but again, "he runaway from me & could not be managed, was caught in onslow county, & I sent him up to Roberson again & you see he is determined not to work any wheres." Fearing that some of the other hired

Latham slaves would try to run, Grist admitted, "it is true that I suffered Boson Boston & Daily & Lewis to work in Brunswick," but only "because they all had wifes there & Boson said he could not [chip] low boxes." Boson refused to work in a newly boxed turpentine camp where chipping meant continually bending over low to hack the new faces. Grist kept the rest of the enslaved men up in Robeson, as "the other boys preferd working up" there and "because the woods was decidely more pleasant & better to work in." This was evidence of neither benevolence, mercy, nor accommodation. By trying to meet some of the demands of the men he enslaved, Grist was really trying to prevent further labor losses, while those men were using the threat of flight to shape how and where they worked in the woods.

It came as something of a surprise to James R. Grist, therefore, to hear that his brother William had brutally whipped Joe and Abner, two enslaved men who had always done good work, simply for running away. "We work altogether by task in our business," James explained, and "those same boys always gained Saturday," meaning that both Joe and Abner finished their work quotas a day early every week. Yet, James wrote, "I understand that some of the negroes said Joe received 500 lashes; Will whiped him & Abner with a small leather strap." He saw this as a foolishly counterproductive punishment. Not only was James "satisfied there has not been 500 licks struck in the business this year," but Abner and Joe "were whiped for running away & not there work."[40]

With 500 lashes, William Grist had all but committed murder. Whether this violence was deliberate or a result of William losing control of himself in frustration over losing control over the workscape, as he nearly whipped two men to death, he was also violently marking the limits of white power in the camps. But they were limits he probably refused to see. What William saw was two slaves casually disrespecting his rights as a white man to control absolutely the movements of black bodies, rights he had come to expect from decades living and working on plantations. But the piney woods were not the plantations, and it was the insubordinate men who better understood the spatial and seasonal politics of the camps.

While the boxing season persisted, overseers confidently whipped enslaved boxers to assert their power, inspire fear, and punish poor work. But when running away became easier with the arrival of the chipping and dipping season, these dynamics shifted. This shift did not mean that slaveholders refrained from violence in the spring and summer. Far from it. According to the author of a treatise on turpentine farming, to "keep hands in order" during the chipping season, "my hands required whipping every time after dipping when chipping was commenced." Routine and invariable torture was apparently the only way to keep the men he enslaved chipping up to speed. Running away was a different matter. If James R. Grist had believed violently punishing chippers and dippers for running away would have been effective, he would have done so without apology. But forced by turpentine workers to recognize the limits of slaveholders' power in the camps, Grist wrote, "We have never whiped any negro that *come in to us* when runaway."[41] Alongside the murderous overreaction of frustrated overseers, the reluctance to punish men for returning to the camps, for running *in* with their badly needed labor, underscored the power (and its limits) enslaved woodsmen held in the geography of turpentine country. A more effective labor tool for overseers was probably the task system itself, partly in the way it made the forest legible, and partly by allowing the enslaved to "get over" or "gain Saturday" by exceeding their quota and earning some money.

Keeping between the whip and the reward of Saturday was more, however, than many men could handle. Whether feigned or not, illness was a consistent product of the physical and mental strain of the task system, and an obstacle impeding the circulation of laboring men and resin between boxes and barrels. A month after being whipped, "Selvester was sick last weeke & last two days & he falde to get over," while "John grist is very sick he lackes about two thousand boxes of getting over last week. he give up friday morning."[42] Broken down or simply fed up, enslaved people's bodies were sites of continual struggle in the piney woods; enslavers sought to keep them laboring compliantly while the enslaved fought not only to stay alive but to gain some measure of control over their lives.

One of enslavers' chief weapons in this struggle was controlling where and how food was locked down in the camps. Almost all of the food and fodder in the camps was delivered by steamship and railroad. For the enslaved, the major problem was what happened to the food once they unloaded it. Overseers would often order the men to concentrate all the food in locked storehouses, as at the M. Jones operation, where "four hands [were] helping Jim Ganer house corn" for a couple weeks. On most plantations, at least some food was usually produced on-site, and hungry enslaved people had, compared with those in the camps, more access to that food. As long as the enslaved worked in plantation smokehouses, milked cows, tended gardens, fed chickens, gathered eggs, or picked corn, enslavers could never completely prevent them from raiding or keeping some of the food for themselves. Not so in the turpentine camps. There, enslavers kept the food locked up and kept the enslaved working mostly on material they couldn't eat. Enslaved woodsmen did have chances to hunt, fish, and forage in woods and swamps that supported both an abundance of wildlife — including large predators like bears, wolves, panthers, and alligators — and free-ranging hogs and cattle, but the demands of the work regime kept their stomachs tied to the centers of the camps and dependent on the rations provided by white operators.[43]

Even when this arrangement worked smoothly, powering and sustaining enslaved woodsmen between boxes and barrels was a constant struggle against summer heat. The hot, dry summer drew the resin out of the trees, but it also leached precious water from the exhausted men. As one manager wrote desperately from South Carolina, "Last weak it was sow warm out hear that the negros fainted down in the woods I had sevon down last weak but I have them all out at work this weak." During a remarkably hot season at the newest Grist turpentine colony in Alabama, the rains failed, "the wells is dry," and for months, this extreme heat threatened to overwhelm the industrial circuits between boxes and barrels. While dehydrated men struggled to keep up with the resin flowing from the trees, Benjamin Grist forced a different form of vital circulation through the woods, as "it was necessary 'to keep Dave hauling water with the carte all the time in the woods to the hands.'"[44]

It was no surprise, then, that beaten down, starved, dehydrated, and miles from loved ones, the enslaved tried to escape their piney frontier prisons. They seem overwhelmingly to have abandoned the camps during the dipping season, when both the labor process and summer heat combined to make leaving more attractive and more feasible than at any other time.[45] Owners and overseers knew this and fought hard to capture, terrorize, or otherwise drive in men who had escaped the confines of the camps. Controlling food, violence, and "rewards" constituted white power over the enslaved in the ecology of the camps, but in the forests and swamps surrounding these camps, enslavers had to rely on other weapons.

Miles Clark may have labored alone, and he may have escaped alone, but as he left his coagulating crop of trees behind, he entered a strangely social landscape that was as much a part of the geography of piney light as the rivers, stills, and boxes. His first concern would likely have been to get as far from camp as possible before nightfall. He also needed to find water, both for survival and to hide his trail from the dogs that might soon be following him. In the swampy interstices between languid rivers and pine savannas, North Carolina runaways made a durable constellation of maroon spaces. Turpentine workers continually moved in and out of these alternative geographies, helping to preserve them by exchanging goods, information, and safety as they crossed and recrossed the piney boundaries of slavery and freedom in the camps. In the middle of the Green Swamp, near Grist Depot and south of Goodman Swamp, runaways had carved out both a grazing area and a garden and built eleven cabins, maybe more. Some of the escaped men may also have encountered the radical networks of enslaved watermen piloting vessels between Fayetteville and Wilmington, that maritime "asylum for Runaways."[46]

Even those still trapped between boxes and barrels were aware of the fugitive geographies of the surrounding swamps. Despite the absence of direct evidence, the recorded behavior of both overseers and the enslaved strongly indicated the existence of hidden networks of enslaved communication. In spite of their isolation, enslaved woodsmen could still speak in the journeys to and from the camps. They shared company on Sundays, in the short but regular congregations in the camp centers when weekly quotas were set and provisions, violence, and rewards were meted out. Dippers and chippers must have stolen brief moments of conversation in the pines during occasional encounters with enslaved haulers collecting resin barrels. As the stories of the enslaved trickled through the interstices of work and movement in the camps, Miles Clark likely heard that a year before, a group of runaways who had established a maroon camp a few miles north had been hounded and terrorized by John T. Council (another Grist camp manager) and his dogs.[47] Yet it was because of just such stories that Miles Clark was likely heading in precisely that direction. These stories, however, are lost to historians. There are no known slave narratives of life in the turpentine camps, isolated outposts of slavery that may have been porous but from which slaves escaped North only with tremendous difficulty.

Sometimes the enslaved coordinated these complicated fugitive movements through relationships they had cultivated before ever arriving in the camps. Enslaved turpentine workers led an often solitary existence, but they almost never arrived at the camps alone. Miles Clark was sold to the Grists with at least eight other enslaved men from the Clark plantation. They were put to work at Grist Depot under the management of M. Jones, about forty miles south of Benjamin Grist's Gristville camp. Sometime in April, as chipping and dipping got into full swing, Ben Clark, Jack Clark, and Tom Clark escaped with three other men named Selvester, Griffin, and Anson. Whether they escaped all at once or one at a time, by May they had formed a camp of their own in Goodman Swamp, close to the northern turpentine operations.

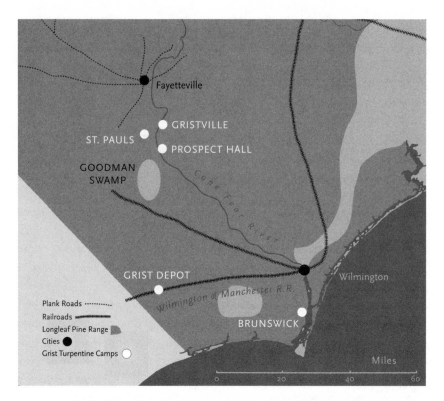

**The Goodman Swamp maroon camp, shown in relation to the Grist
turpentine operations along the Cape Fear River.** Based on map printed
in Percival Perry, "The Naval-Stores Industry in the Old South, 1790–
1860," *Journal of Southern History* 34, no. 4 (November 1968): 515.

This may have been the same camp that John T. Council had tried to dis-
perse the year before with his dogs. Then, too, the camp had drawn fugitives
from several operations.[48]

John T. Council faced even greater troubles the year that the Clark men
escaped. Three months before Miles Clark ran away, Ben and Jack Clark narrowly thwarted Council's attempts to reenslave them. "Mr Stone my over-
seeor found Ben & Jack Clark and caught Ben," Council wrote. After tying
Ben to a tree, the overseer tried to follow Jack but soon lost him. Returning
to where he had tied up Ben, he found that "in making an effort to catch Jack
Ben got away which I very much regret." Council, relying on information circulated through networks of dogs, masters, overseers, and recaptured slaves,
felt there was "little doubt but they are in the neighbourhood. Thos. J. Purdie
sent word to *Stone* that your boys had a camp in Goodmand Swamp below
John Monrain. he had caught Lyon's Boy who had been runaway and he said

he had been with them in their camp." Now that he knew their general location, Council was optimistic that "we will be able to get them altho we maid a bad start to let *Ben* get off, but we will give the matter the strictest attention &c."[49] Though all the managers and overseers in the area undoubtedly tried, it seems that they met with only partial success, as most of the fugitives remained free.

And the few men who did return voluntarily may actually have been disguising attempts to recruit new runaways. "Selvester cum in tusday morning he ses that he has not seen griffin & anson but [once] sence tha went to bladen [Goodman Swamp]," M. Jones wrote on July 4, 1858. A week later, Jones wrote that all of the men except Selvester had "got over" and that "he is no better now then he was before he runaway." One week avoiding dipping, another spent chipping indolently, Selvester was sabotaging his engine of resin by wounding the trees too little and letting them run to waste. Some time after Selvester returned, Lewis Latham and Clem Clark, while out dipping and chipping, escaped up to the Goodman Swamp camp. Miles Clark followed soon after. As some men returned (only to ruin crops by working slowly and sloppily), others ran away, and it is difficult to see this as anything other than evidence of considerable coordination and of well-established and well-protected escape routes and hiding spots kept and accumulated over more than a decade by thousands of men attempting to navigate and escape the slavery of turpentine. These were strategies developed by the enslaved to invert the carceral landscape of trees, distance, swamps, and hunger into a temporary geography of freedom.[50]

The managers were certainly convinced of conspiracy and the power that this hidden geography had to grind turpentine production to a halt. "I cant push as i would like to for the handes will not beare it," M. Jones wrote in utter frustration, "so i have to doo the best i can to ceep them hear. the dam clarke negros has nearley ruend yours." This continued until at least November, with some men moving back and forth between the freedom of the swamp camp and the enslavement of the turpentine camps, while others remained continually on the run. John T. Council, who months earlier seemed so optimistic, now admitted, "I have had one out for three months and cannot hear from him. but my opinion is that they are in the neighbourhood," and pointedly suggested that Grist "send dogs and make some active steps to catch them in which I will give all the assistance I can." A few weeks later and after no apparent action (or at least no apparent success), Council wrote again to strongly advise, "You moove your negros which you intend to moove to *Ala* before you close your years buisness in Columbus, and the quicker you do it the better," for, he emphasized, *"I have had a hint from one of my negroes. That is a noughf for you to know.* be on your lookout."[51]

As Council hinted, the 1858 acts of defiance and flight were more than just products of the internal struggles of the camps. Only a few months before the Clark men deserted Grist Depot, Benjamin Grist had ordered the men he enslaved to dismantle the Gristville camp he had overseen the birth of eight years before. It was, after all, no longer functioning as a productive engine of either turpentine or enslavement; the chipped faces were too high, the long-injured trees were sickened and sluggish, and the enslaved had grown too skilled at escaping and staying out. But as he abandoned his Gristville camp, Benjamin Grist was far from done with turpentine. And the same was true for the enslaved woodsmen under his command. James R. Grist had ordered Benjamin, along with most of those enslaved at Gristville, hundreds of miles south to Alabama. When the Clarks ran away in North Carolina, they made sure they escaped the journey to Alabama, where more productive pines and unfamiliar terrain meant more brutal work and a tighter carceral landscape. Yet their very success at staying in place may have convinced James to dislocate others. By the start of the next season, with Benjamin requesting nearly ninety hands, James decided to ship many of the rebellious men he owned to Alabama, displacing a spatial politics that was slipping out of his control by expanding his turpentine empire into new lands. It seemed winter ended the 1858 confrontation, narrowing the possibilities of finding food and water in the terrain between the storehouses, while Christmas drew the fugitives back to the plantations to be with their families. Whatever the reason, the Clark men no longer appeared able or willing to sustain a geography of freedom against their exploitation in the turpentine camps. Perhaps dogs were finally sent, rewards posted, and captured fugitives forced to reveal their secrets. But as the runaway woodsmen knew all too well, the friction of terrain made both permanent escape and permanent survival in the swamps next to impossible, especially if they hoped to see their families again.[52]

In the antebellum piney woods, the deceptively simple act of pulling resin from a tree to a barrel produced an extraordinarily political space. As overseers and slaveholders spun dogs, ledgers, torture, trees, and markets into entangling webs of surveillance and domination, slaves struggled carefully and skillfully to turn the labor process to their advantage—to redirect a knowledge of the landscape and longleaf biology learned through years of work and pain into a marginally more free existence. And then there was the formidable terrain of struggle to get resin from boxes to the distillery.

W ITH boxes at one end and barrels at the other, the story of dipping was also the story of coopering. As soon as resin left the trees, the precious spirits of turpentine began to evaporate in the heat. Enslaved coopers provided the vessels enveloping and partially arresting the dissipation of that fleeting energy on its passage to the still. For this they made a crude kind of barrel from pine collected at the camps (the best, and far more expensive, barrels were made later from oak staves shipped upriver to store the extremely volatile and leaky distilled spirits of turpentine). Made at the cooper shop near the still, these pine barrels would be hauled out and placed throughout the forest, "thirty-five or forty to the task, at convenient distances, all ready to receive the turpentine." So important were coopers that in "a gang of hands getting turpentine every fifth man may be a cooper, and will be employed the year through in providing his own materials and keeping the others supplied with barrels."

While enslaved dippers carried buckets to and from the boxes and barrels, a smaller crew of workers was supposed to circulate through the forest creating an intersecting circuit of barrels. The operator tasked one laborer (sometimes but not always enslaved) to move through the crops, leading a team of mules yoked to a cart, until he found a full barrel, where he plugged the top and rolled the nearly 300-pound container onto the dray. Repeating this difficult process over the day, each team of mules and man was expected to "haul the turpentine dipped by ten hands an average distance of three miles, with spare time for hauling provisions, empty barrels, &c." Haulers ferried food and empty barrels (and likely messages and stories) out to the laborers and carried resin (and new information) back to the stills.[53]

Keeping barrels circulating was critical and formed a bottleneck that drew the intense attention of managers, the enslaved, and hired workers. Fearing sabotage and the mobility that hauling afforded the enslaved, some managers were even willing to pay free workers to haul. In his first spring at the Gristville camp, Benjamin Grist wrote his cousin that he had to "pay Alf Jackson $50 for waggen & C. B. Tyson $30 for haling done last winter." One manager who tasked enslaved men to haul suspected "sometimes that they hunted for logs, so that they might drive the cart over them, and indeed, take every possible advantage, even to neglecting to water and feed the animals, fix the yoke, etc."[54] While this kind of constipated movement could back up the flow of resin to the stills, hauling too much at the wrong times could be just as problematic.

Absalom Davis, like many smaller producers, discovered for himself how lacking proprietary access to a distillery could render hauling a desperate misadventure. Davis had gotten into the turpentine business at Benjamin Grist's suggestion with the understanding that he would sell his resin to the Grist still near St. Pauls, "but much to my suprise, soon learned that you could not take my Turpentine at the Robeson still." Stills in other towns were "buying freely," but it was "of no avail to us. We cannot reach them." Stranded with hundreds of barrels of resin he could neither sell nor save "for want of a River or a Still," Davis complained that without Grist's assurance "in the outset that you would take my Turpentine this year . . . we would have put up a still at all hazard But now it is too late, and we are doomed to a ruinous loss."[55]

Even with secure access to a still, hauling remained a vulnerable space into which turpentine might yet disappear. At the end of a long career working for various Grist operations, it seemed W. G. Whitfield had decided to take what he considered was due to him. And he knew how to hijack the geography of piney light. John T. Council had just learned to his dismay that Whitfield had sold "a grate quanty" of resin to competing buyers by secretly hauling "Tirpintine in the night with my Teems to Robeson Landing and took my flat and floted it down the river and put it on [a steamboat] and sent it to Wilmington." Stealing resin away from one still and sending it to another could be hugely profitable and apparently not even that difficult, as "it would surprise you to Know the amt of Terpintine he took in that way and it is more surprising that I had not caught him at it but every body appears to know it."[56]

It was rain, however, that posed the most insurmountable obstacle to circulating barrels between crops and stills. Rains made rivers navigable and resin run. They could also turn the ground into a boggy mess. Hauling in rain was not only more difficult but could actually injure the trees, for "hauling heavy loads through the forest in wet weather . . . skins up the roots, and breaks loose their hold." Experienced producers therefore cautioned that hauling "in wet weather should be carefully attended to, and cart-paths made in the thinnest part of the forest." While hauling through mud threatened the long-term productivity of a camp, heavy rains presented even more pressing short-term problems. The year before a heat wave would compel Benjamin Grist to have water hauled out to his enslaved dippers, a deluge of rain threatened to break the new Alabama camp in two. The boxes were linked with the resin barrels and the stills with the spirit casks, but as mules and carts failed in the mud, a rift had opened between these two poles of production. Although the men had dipped 125 barrels ahead of schedule, they could not haul it. Benjamin Grist, determined to reestablish the link be-

Piney Lights

tween crops and stills, set "10 hands all the week [to] mack-ing roads to try & hall turpentin." But thwarting his efforts, the rains continued to fall "like never was known before," rains that "over flowed every thing, even to the bridge here at the house is all gone, & now it is impossible to do any halling out of the woods."[57]

Back in April 1852, while Benjamin Grist prepared the new Gristville camp to be ready when the pines began to run, an enslaved mechanic named Castor was setting up the still. Once Castor had finished building the two-story structure, Jack and Willis Parmerly, the two enslaved distillers who began this story, started expertly weaving furnace fires and cooling creek waters around the copper still, thereby channeling resin taken from the trees into pure spirits of turpentine and rosin. These stills, which ranged in capacity from five to twenty barrels, were "usually placed in a ravine or valley, where water can be brought to them in troughs, so as to flow, at an elevation of fifteen feet from the ground, into the condensing tank." Placed on the high ground of a sloping bank, "the still is set in a brick furnace. A floor or scaffold is erected on a level with the bottom of the still-head, and a roof covers all." When the resin that the pines had made and enslaved men had diverted into boxes, dipped into barrels, and hauled for miles finally reached the low-lying stills, Frederick Law Olmsted watched as the "still-head is taken off, and barrels of turpentine, full of rubbish as it is collected by the negroes, are emptied in. When the still is full, or nearly so, the still-head is put on, and the joint made tight with clay; fire is made, and soon a small, transparent stream of spirits begins to flow from the mouth of the worm, and is caught directly in the barrel in which it finally comes to market."[58]

The process of separating out the spirits would go on for around two hours, enslaved experts like Jack and Willis Parmerly carefully regulating the heat, tasting the liquid from the worm for water content, and trying to prevent either rapid cooling or overheating from causing the distillery to explode. When the stream from the worm began to dry up, and "all the spirits, which can be profitably extracted, are thus drawn off, the fire is raked out of the furnace, a spigot is drawn from a spout at the bottom of the still, and the residuum flows out—a dark, thick fluid, appearing, as it runs, like molasses." This was the rosin. For each barrel of spirits, the source of camphene, distilling would produce about five barrels of rosin, the source of rosin oil and gas.[59]

Given the power concentrated in the hands of shippers by the difficult terrain, most managers attempted to bypass barrel merchants by having enslaved coopers manufacture as many spirit barrels as possible in the camps.

"A North Carolina Turpentine Distillery." Distilleries were some of the only places in turpentine camps where enslaved men gathered and worked together. In this illustration of a two-story still, haulers, mules, coopers, porters, and distillers all coordinated to keep the distillery transforming resin into turpentine. Engraving in *Harper's Weekly*, April 1, 1876, 265.

One writer in *DeBow's Review* observed that the "distiller incurs great expense in the single article of spirit barrels. These must be iron bound, made in the best manner of seasoned white oak, and well coated within with glue, to prevent evaporation. They should contain from forty to forty-five gallons, and when ready for use cost little short of $2 a piece." At this price, expenses added up fast, so the writer urged operators to "let these, by all means, be made at home."[60] As the rivers rose with the spring and summer rains, turpentine camps became accessible to the coasts, and producers and shippers strove to move their traffic before the flowing routes dried up. But the powerful currents that shippers navigated to push barrels, corn, and life to the camps could just as easily wrench them away. These were the currents that claimed Willis Parmerly's life. Perhaps James R. Grist was telling the truth when he claimed his overseer had ordered Jack and Willis Parmerly not to try to rescue the casks, but the importance attached to these barrels was so strenuously articulated in the camps that the warning, if it happened, was likely met with skepticism.

Without enslaved coopers and the barrels they continually produced, the whole geography of labor would unravel. To combat this threat, managers readily deployed violence in an attempt to discipline coopers. Jack and Willis Parmerly had almost certainly seen this done. Upon discovering that a cooper named "Fred had not done any thing & old Jake had just got afire under the glue kettle," one of Benjamin Grist's overseers gave "both a lite whipping." Soon after, Fred ran away. By running away and slowing down work, Fred and Jake showed the power (and suffered its limits) that coopers wielded in turpentine camps. For although the Gristville dippers were accumulating weekly "from 69 to 89 bls[,] the way the cooopers is going on the bisiness must stop." With Fred a runaway, Grist had only "3 cooopers under the shop [and] the stils is behind on account of not having spirit bls in time."[61]

T URPENTINE camps were living ecologies that enslavers forced into being and coerced into survival until each camp grew old and died. But as they were living, they could also be reproduced. An ecological complex that began in North Carolina, turpentine camps spread rapidly across the pine belt in the 1850s, driven overwhelmingly by "Tar Heels" long steeped in the industry. Decades of intense production had so damaged and stressed the ecology of North Carolina's piney woods that by the eve of the Civil War, few stands of pines remained. Boxing did much of the damage, first by seriously injuring the trees and second by spraying flammable wood chips about the forest floor. The fires and insects that tore through stands of wounded, working pines were bad enough, but it was the abandoned turpentine forests — the ones scarred to the limit of production and human reach — that caused the most damage. The tall, slave-wrought faces of hardened resin were like accelerants waiting to be ignited. This confluence of death and fire led to what one historian termed turpentine's "Suicidal Harvest on the Move." Beginning in North Carolina, and by the late 1850s expanding to South Carolina, Georgia, and Alabama, enslavers drove enslaved woodsmen and pines in a Pyrrhic march of turpentine camps across the South.[62]

As old camps died, operators coordinated the disassembly of the camps' social and mechanical structures, uprooting and relocating human and mechanical laborers. The men from the Clark plantation who fought so hard to remain outside the grasp of slaveholders and slave traders during 1858 knew full well that for the Grist family operations, this ecological reproduction was focused most extensively in Alabama, the southwestern frontier of the pine belt, where land and labor had become remarkably cheap. Before the heat and before the rain, and before the Clark men ran away, in the winter of 1857, Benjamin Grist and at least fifteen enslaved men began the work of disassembling the Gristville camp. Benjamin felt he had something to prove to his cousin, writing, "You say I am worne out & ar brocke down & will not attend to any thing but," he asked, "I shod lick to know whoe has beat it on 8 yars ould boxes?" He believed he was being blamed unfairly for the failing trees. "I feare you ar going to desert me now becase the bisness is ould & warne out," but, he argued, "it is the pines is warne out & not me for god knowes I have all wase done my duttey in the bisness."[63] James did not desert him, but as Benjamin oversaw the final resin harvest of the Gristville pines, he would find himself exiled in a sense. Scores of enslaved woodsmen had drained these aging pines for nearly a decade, their chipped faces extending as far as a man could reach with a hacker, and these crops were being abandoned for newer shores, newer rivers, and newer pines encircling Mobile Bay in Alabama.

First, they broke down the still that Willis Parmerly had died defending five years earlier, and they loaded it onto a flatboat resting in the waters that had claimed his body. But they did not ship the still directly to Alabama. First, they shipped themselves. When Benjamin Grist arrived at the mouth of the Fish River in Baldwin County, Alabama, he came with enslaved men, axes, hackers, buckets, and a plan to transform the forest. Perhaps a third of the woodsmen came from North Carolina, their expertise and turpentine experience balanced against the lower cost of slaves in the Cotton Kingdom.[64] Losing no time, Grist drove the men, sometimes deep into the night (presumably lit only by pine torch and moonlight), to box the Alabama pines, beginning the process of their surgical transformation.

Reproducing these camps also reproduced their contradictions, and enslaved boxers kept running away. Benjamin Grist wrote defensively to his cousin, "You say I have bad luck [but] I shod lick to know whoe it is that" had not had any "negroes to runaway?" According to Benjamin, "The negros runaway becase I made them cut boxes at knight"; only one-third of the newly acquired men had any experience cutting boxes, and "8 of them never has got more than 60 a day yeat." The forced reproduction of an extractive landscape violently transformed enslaved people as well. Pushed day and night to cut more and cut faster, field hands were made into turpentine slaves, muscles retrained, skills repurposed, and bodies forced to reveal their labor potentials to exploitative overseers. Some ran away; others were crippled before they had the chance, like "Stave Ellison [who] cut his foot last friday & he will not doe any worck in 4 weeks." By April, when W. G. Whitfield (before his final theft) rolled into the camp "safe with Sol Grange & Jack, 8 mules & 2 waggens," Benjamin Grist wrote proudly of his almost newborn turpentine complex that "I shall quit cuting boxes after this week [and] shall have 192000. Whitfield ses thay ar the beast boxes he ever saw." He had also built a "spirits house & still house & am macking the tubes," and he was "now reddey for the stills."[65]

The stills finally reached the Fish River camp by late May and were immediately put to work. The reproduction was complete, and the spirits and rosin began pouring out. The usual tensions flared, although even with sixteen chippers out, "we have none runaway." There was, however, nonhuman resistance. "That mule is the one thay call Rock," a humbled Whitfield wrote from Alabama; he "went after him last week but cont git him he is hard to catch I ame goinge to try that mule a gaine the last of this week I think I git him." Rock was a wily mule, and a week later Benjamin Grist wrote, "Whitfield he has a [boil] on his ass . . . [and] I have not got the moule yeat." Ass boil or no ass boil, that mule continued to evade capture and continued to deny the camps his labor, and although Whitfield "saw him last Sunday . . . I cont ketch hime [for] he is with thos wild Poney and he is hard to Catch."[66]

As the year wore on, enslaved men ran away, got sick, and were injured, but the frontier labor camp continued to churn out turpentine. By the start of the next season, Benjamin Grist had overseen the production of 1,109 barrels of spirits and 5,508 barrels of rosin, had employed fifty-eight enslaved men in cutting an additional 244,000 boxes, had bought three new mules, and was requesting a total work force of eighty-nine enslaved woodsmen: "44 hands to schip the boxes & 21 to dip & head up the turpentin 9 at the stills 8 to cooper 2 to hall 4 to git timber & cut wood."[67] In the years ahead, first rain and then heat would threaten to unravel this geography of labor, but it survived, and even thrived, along with similar camps expanding across the South.

As turpentine operators compelled woodsmen and trees and land into reproducing these enslaving piney engines across the southern pine belt, streams of ships and railcars ferrying spirits and rosin downriver to depots like Wilmington, Mobile, and Savannah continued to expand and intensify. From these depots of piney energy, enslaved dockworkers loaded schooners bound for New York, Boston, Philadelphia, New Orleans, and Liverpool. In 1850, James R. Grist's New York agents, Hussey & Murray, recorded $1,120.08 in profits, most of which came from selling seventy barrels of spirits received from the schooner "Lamartine from Wilmington." The process of moving and commodifying spirits and rosin accrued considerable costs, totaling $152.53. There were marine insurance and fire insurance to pay for, charges for freight, and charges for carting barrels around docks. Then there were "storage & labor," coopering and gauging, fees for lighterage, and finally, fees for advertising.[68]

But before workingwomen and men could fill their lamps, the spirits of turpentine had to be transformed into camphene. This happened at camphene manufactories, where workingmen further refined spirits of turpentine and then mixed it with alcohol, usually distilled from whiskey on-site. The mixture, marketed variously as "camphene," "burning fluid," or "spirit gas," was then measured into cans for retail. Some camphene manufacturers and dealers also manufactured lamps. Through experimentations with a variety of illuminants, Robert Edwin Dietz, a carpenter by trade, had developed a booming camphene and camphene lamp business in New York City by the 1840s. Primary among Dietz's inventions was the "Doric Lamp," which he advertised as "superior to all other Lamps for burning Camphene. This lamp is simple in construction, easily trimmed, and gives a great deal of light at a small expense." Doric Lamps were cheap, mass-manufactured vessels of camphene, which, according to Dietz's memoirs, "in those days, produced the cheapest artificial light known in the world, and was widely used in New York by reason of its brilliancy and economy, by tailors, shoemakers and thousands of persons who could not afford to burn gas."[69]

In Brooklyn alone, the camphene industry was worth over $2 million by 1859, almost as much as the city's three capital-intensive gas companies, which had a combined value of $3 million. In Philadelphia, the industry produced 1,654,250 gallons of burning fluid worth over $1 million in 1857, while nationally camphene manufactories transformed over 5 million gallons of turpentine (from 1 million barrels of resin) and 25 million gallons of whiskey (from 12 million bushels of corn) into 20 million gallons of camphene worth from $9 million to $16 million, selling between 45 and 65 cents a gallon. By the eve of the Civil War, most of the whiskey pouring into the purified alcohol industry came from Ohio River Valley corn distilled in Cincinnati—where hogs that would become candles feasted on distillers' waste mash—and over 80 percent of that twice-distilled alcohol went into making camphene.[70] The piney lights that antebellum workingwomen reluctantly embraced to survive were thus world-making nexuses, dragging into relation North Carolina turpentine camps, Ohio cornfields, and New York distilleries.

As social and economic pressures to work later into the night increased alongside the price of whale oil, the use of camphene continued to grow, explode, and violently consume some of the life and property its light was helping to circulate and produce. But far and away, camphene most commonly exploded in domestic settings. Wealthier families and clothiers displaced this violence by paying servants and seamstresses like Mary Clark to assume the risks of domestic work in such households and establishments. The gendered domestic work of sewing and mending garments around a light had developed around fire hearths, candles, and perhaps the odd oil lamp. Yet by the 1830s even poor households were forced by landlords and fuel prices to replace open hearths with more efficient but entirely closed and dark iron stoves.[71] Hearth light was no longer an option for evening family labor. Gathered around camphene, this necessary labor became suddenly far more dangerous. And with little choice, thousands of workingwomen and children suffered burns and death in attempts to thread domestic work through increasingly limited urban spaces and times.

Because this violence was translated through newspaper discourses in which there seemed to be little doubt that "almost without exception, females and children are the sufferers," camphene-related deaths became a rallying cry for a kind of consumer politics that sought legal protection for the supposedly weak and ignorant from the hazards of amoral market forces. The Brooklyn activist E. Meriam exemplified this trend in his plea that if "men who deal in this dangerous compound care nothing for the results, nothing for the loss of precious human life, nothing for the agonizing and most painful of deaths, nothing for the pains, the sufferings, the unfortunate victims endure, nothing for the loss of millions of property consumed by fire originating in the use of burning fluid," then the New York state legislature should "prohibit its use by the severest penalties," and fire insurance companies "should refuse to underwrite for buildings or other combustible property, in every case where camphene or burning fluid is used in the building or in any adjoining building." It was not only life that was under assault by camphene, but property and the urban environment. And reformers refused to absolve businessmen of social responsibility, for when "we take into account the small number of persons engaged in the manufacture of burning-fluids, and compare their number with the great number of deaths these manufactures have caused, each one's share in the work of death will be found to have been fearfully great." Yet even reformers as critical of camphene as Meriam were silent on its ties to slavery, and abolitionists never boycotted camphene as they did cotton. Instead, campaigns against camphene were commonly and explicitly tied to temperance movements (no doubt influenced by the strong material connection between camphene and the alcohol industry), emphasizing individual moral responsibility.[72]

These explosive lights were politicized into a discourse focused on using public resources to eliminate the need for personal lamps. *Scientific American* helped to lead the charge to replace camphene, writing hopefully "that the time is not far distant, when every private house, as well as the public ones, in our cities, will be illuminated with good, safe gas, publicly manufactured at a cheaper rate than either, oil, candles, camphene, phosgene, oxygen, or all the phenes (fiends) of spirit gas whatever." The violent toll of camphene was such that *Scientific American* could plausibly claim that every "improvement which tends to cheapen gas light is an incalculable boon to the human family."[73] The virtues of gaslights were far less clear in actual practice, and in any event, they spread slower than the need for working-class lights. Until the Civil War, camphene lamps continued holding women hostage, while the geography of turpentine continued to expand, enslave, and transform frontiers across the American South.

THOSE who made and used piney light had not asked for it. The enslaved woodsmen, wage laborers, seamstresses, and domestic workers who assumed the sometimes terrible risks of working with turpentine did so not because they wanted to, but because they had to. It was others (husbands, employers, slaveholders) who reaped the rewards of all the dangerous unpaid and underpaid work in home, shop, and forest. As the turpentine industry spread across the South, it transformed entire regions, uprooted and destabilized communities and ecologies, and until the eruption of the Civil War, showed no signs of slowing its centrifugal expansion of new frontiers of accumulation. These frontier armies of enslaved light makers were the ignored and eclipsed counterparts to the swelling, ever-brighter antebellum industrial cities. Pushed out of farms and villages from all over the world, the new urban working poor had limited options with which to support and sustain the basic social processes of life. The materials necessary for shelter, food, clothing, water, heat, and light were securely controlled behind bulwarks of property, accessible only through money or theft. Making a living in these cities thus usually meant working for money and building and illuminating a life out of the cheapest and quickest materials available. The way cities grew, and the speed with which they did so in the antebellum United States, would have been close to impossible without cheap light. Working-class families, stretched thin by low wages and long workdays, could never have gathered and survived in sufficient numbers for cities to industrialize had they been unable to meet their own vital and social needs during the dark hours left to them.

The making and consuming of piney light aligned for a few decades the exploitative geography of naval stores with the outsourcing forces of industrialization and urbanization, internally relating turpentine camps with tenements, and the labor of boxing with that of sewing. This alignment was hardly permanent, and indeed its precarious spatial relations were continually threatening to unravel piney light. In the end, it wasn't the rain, the floods, the explosions, or even the escapes that would extinguish the lights of camphene. It was the Civil War and the coincidental advent of a chemically similar and securely northern illuminant in petroleum-derived kerosene that severed these relations. But this was still in the future, a future of light few could have imagined. The lamps of labor had found a new cheap fuel in camphene. The lamps of property had found coal gas. Both were built on slavery, and both made deadly explosions into the stuff of everyday, modern life. As the enslaved geography of turpentine continued to multiply and camphene filled ever more workingwomen's lamps, coal gas continued its march outward and downward: out from enriched industrial urban cores, down into coal dungeons, and deep into the dens of the gas dragons that gaslight hatched under antebellum America.

Chapter Three

DUNGEONS AND DRAGONS
AND GASLIGHTS

It's dark as a dungeon and damp as the dew,
Where danger is double and pleasures are few,
Where the rain never falls and the sun never shines
It's dark as a dungeon way down in the mine.

MERLE TRAVIS, "DARK AS A DUNGEON"

The explosion of hydrogen in a coal-mine, he calls the ferocious
rage of a fiery dragon — the safety-lamp a muzzle to the
dragon, which too often leads the miner to his destruction, as it
induces him to work where the hydrogen has accumulated.

NEW MONTHLY MAGAZINE, OCTOBER 1, 1821

Samuel Gouldin began his last day alive with a premonition. In the darkness before dawn of March 19, 1855, he told his wife he had dreamed his death and gave instructions in the event he did not return from the mine. A white overseer at the Midlothian Coal Pits outside Richmond, Virginia, Gouldin was due in the mines by six in the morning, working underground in the glow of his Davy safety lamp. Raising the odd, wire-mesh-wrapped lamp to the ceiling as he made his rounds, he watched for signs that the shielded flame was flaring or changing color. He was searching for methane gas, or "firedamp" as the miners called the flammable gases that seeped out of all coal, but especially bituminous coal like that of Virginia. If Gouldin found any signs of firedamp, he was supposed to mark off that section of the mine as a warning to anyone who might pass nearby. Even if all the miners carried safety lamps (which they did not) instead of the open-flame oil-wick cap lamps that were standard in antebellum American coal mines, the mesh was still an imperfect protection, as any contact between flame and firedamp could lead to a catastrophic explosion. There's no evidence Gouldin spotted any firedamp that morning, but it wouldn't have mattered. The firedamp that would kill him and forty-one others that afternoon lay hidden where he never could have detected it, not until far too late.[1]

Historians have long noted that the steam engine made possible the industrial revolution by both harnessing and accumulating ever more coal, an energy loop whereby coal-powered steam pumps made it possible to extend coal mines deeper than ever before. Less appreciated in this revolutionary relationship with nature was the role played by another technology: the Davy safety lamp. For no matter how dry the new steam pumps kept the deepening mines, without a new kind of light, the lurking firedamp would make the pits either infernal slaughterhouses or just empty holes in the earth. This was why Samuel Gouldin was carrying a Davy lamp.[2]

The wire mesh was like a selective membrane that enclosed the metabolic processes of oil, wick, and flame, letting in some methane but preventing fire from spreading back across the mesh, and all while allowing enough light to escape so that Gouldin wasn't completely in the dark. Safety lamps divided space and energy such that men, mules, and coal—hacking, blasting, and light—could more easily circulate through gassy mines. Only the deepest mines could accumulate enough gas and ventilate so poorly that firedamp would explode rather than just burn away in small, survivable lamp and candle flares. In the antebellum United States, only the Chesterfield County pits outside Richmond, Virginia, regularly reached such perilous depths.

But for all their fanfare, Davy lamps did not make coal mines any safer. The real purpose of these "safety" lamps was to extend mining space into methane-rich drifts and chambers. Understanding how these safety lamps were employed to endanger miners (whether deliberately or not) is to understand the relationships of space, danger, light, and modernity in new and important ways. As Gouldin and dozens of other miners were about to discover firsthand, life and labor in coal mines—and therefore the steam engines, rails, and gaslights so often celebrated as materially marking the birth of the modern age—hung tenuously on threads of artificial light. This precarious gathering of labor, flame, and gunpowder was earliest and most

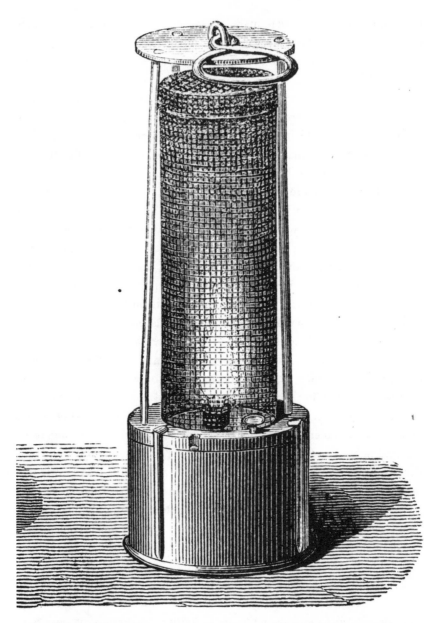

English safety lamp. The wire mesh allowed air into the lamp but prevented the flame from spreading beyond the wire enclosure. This allowed colliers to illuminate (and test for) methane-filled areas in the mines, but safety lamps also meant colliers would now be expected to work in or near those dangerous pockets of firedamp. The wire mesh also blocked some of the light of the flame, and miners, weighing between the risks of darkness, failing to reach quota, and explosion, sometimes removed or cut away the wire. For this reason, later in the century, mine owners began substituting such safety lamps with new locked models that could only be opened with a key. Engraving in Louis Simonin, *Underground Life* (London, 1869), 166.

strikingly obvious in flame-lit coal mines like the Midlothian Pits, but the deadly spatial regime could be found multiplying in turpentine stills, cotton mills, gasworks, and anywhere a camphene lamp or gaslight was burning. This was an industrial revolution built over a powder keg and illuminated by the light of a fuse.

More than cave-ins, smashed fingers, flooding, or darkness, miners feared the coal's deadly breath. The deeper the mine, the more difficult the ventilation and the greater the accumulation of firedamp. And there were no mines in the antebellum United States deeper or (fire)damper than those of Chesterfield County. Yet this same deadly breath was what made Midlothian coal so valuable to gasworks' engineers and was an important part of why Samuel Gouldin was underground that morning. To manufacture gas from coal, stokers shoveled it into clay ovens called retorts; workers tending furnaces superheated the retorts until the coal decomposed into hydrogen gas, other gasses, and the nearly pure carbon remains known as coke; and then condensers and scrubbers purified and collected that hydrogen gas in giant tanks to pipe throughout a city. Gas was the unquestioned light of an industrially enlightened future. First developed in Britain in the early nineteenth century, by the 1840s European and American cities were enthusiastically adopting the technology. All over the United States, from New York, Boston, and Philadelphia to New Orleans, Baltimore, Richmond, and even whale-crazed New Bedford, gasworks were sprouting up, expanding, and thriving in cities still overwhelmingly illuminated with camphene, oil, and candles.

One of the most insistent proponents of gas was the New York publication *Scientific American*. For writers in *Scientific American*, gas was, or should have been, an agent of democracy, equality, and freedom. Complaining of what they perceived to be unfairly high gas rates in 1852, they argued that if "gas was $2 per 1000 feet," as it was in Philadelphia, instead of the $3 it was in New York, "all our working people would use it, and it would prove a blessing to them. There would be no accidents from camphene, and there would be less fires. Community, in *toto*, would be the gainers." Even $2 was still beyond the means of the great majority of Philadelphians, at least given the uneven distribution of gas infrastructure, but the point was a fair one. And so *Scientific American* argued it was the civic "duty of all to exert an influence in bringing about a reform in the gas line." Nor was this an unreasonable goal, the article noted, for every "mechanic in Manchester and Glasgow has his domicil lighted with cheap, convenient, and clean gas light. Why cannot our people, as a whole, have the same advantages?"[3]

The gasworks multiplying in cities across the world, as readily in a slave city like New Orleans as in Philadelphia or New York, became monuments to the modern age. They were material representations of what nineteenth-century middle-class men and women meant when they talked about a liberal, well-regulated city serving some kind of "public good." City governments invested millions chartering corporations to construct and operate gas systems to light streets, bourgeois homes, workshops, theaters, and department stores. This new urban infrastructure made it possible for some people in some sections of these cities to begin to disregard the cycles of moon, clouds, night and day, and solstice and equinox. But this gaslight was not everywhere and for everyone. Through gaslight, nineteenth-century urban elites began to locate modernity in particular places. The liberal, modern city made itself with gaslight; the laboring, nonmodern city burned camphene and candles—or so many people told themselves. But as middle-class Americans tied their dreams ever more tightly to the built environments of gaslight, they also came to depend on the unstable worlds of work and workers that made that light possible, from the Irish immigrants and enslaved mechanics operating the gasworks to the coal pits where enslaved men, free miners, and mine owners battled over labor, wages, safety, and life in deadly underground caverns.

If eastern cities could have manufactured gas from the cheap, abundant anthracite coal being mass-mined in eastern Pennsylvania, they would have, and the dangerous Richmond mines would have fast become a colonial relic. But anthracite coal, with its dry, high-carbon purity, was useless for making gas. Cities and industrialists needed tarry, gassy bituminous coal for gaslight. And so they needed the coal extracted from the Chesterfield underground, where a very specific array of social and mechanical technologies, including Davy lamps, slavery, and life insurance, converged to coerce and smuggle workingmen, animals, and lights to and from such deep, valuable seams.

Cities in the Ohio and Mississippi Valleys could make use of the western Pennsylvania bituminous coals, but with the costs of overland (and overmountain) transport making Pittsburgh coal prohibitively expensive in east coast markets, for those on the eastern seaboard, ship-transported coals from Richmond provided the only affordable and tariff-free bituminous. They never rivaled the Pennsylvania colliers for sheer volume or profit, but the Richmond mines remained some of the country's most important through the end of the Civil War, especially in relation to gaslight. Of the 150,000 tons of coal annually raised from the Richmond basin in the 1850s, at least 15 percent went solely to supplying the Philadelphia and Baltimore gasworks. Add in the major gasworks of Richmond, Charleston, and Savannah, all of which primarily used Chesterfield coals, and New York, Brooklyn, and Newark, where gas consumers demanded corporations substitute the more expensive English coals for Richmond ones, and the forces pressing Samuel Gouldin and the dozens of enslaved men under his command into the Midlothian Pits appear only more thoroughly bound up with gaslight.[4] The result was that so-called free cities ran gasworks fueled by coal primarily mined by enslaved Virginians and British debt peons, while the slave societies of the lower Mississippi Valley employed slave-operated gasworks to light their cities with coal mined by free men in Pittsburgh. By telling the history of American gaslight from way, way, below, I hope to trace the dark chains of coal, gas, and capital that lashed together dungeons and dragons, enslaved and waged laborers, and North and South into powerful engines of antebellum futures. The processes and power relations sustaining and dependent on the continual transformation of coal into gas into flame were not confined to the gasworks themselves, or even to the city, any more than the story of bread began at the bakery. Our search for the political ecology of gaslight must begin in the dark, carbon dungeons that gave coal and life to industrial fires, to gas dragons that were supposed to hatch in cities but were all too often born prematurely into subterranean nurseries.

S TARTING at six in the morning, at the pit head of the Midlothian mines, the mixed crew of around 150 white and enslaved miners climbed, one group after another, into two tubs. Engines slowly lowered the buckets of men 800 feet into the deep darkness. Several minutes later, they hit the shaft bottoms, oil-wick lamps hooked onto their caps shedding the faint light that made the work of mining possible. As was typical in Virginia mines, they were likely burning New Bedford whale oil from right or bowhead whales. A neighboring mine, with less than a quarter the labor force and output of the Midlothian, consumed in a year over 400 gallons of whale lamp oil, thirty pounds of wick, and dozens of lamps at a cost of nearly $250.[5] Indeed, behind every gaslight was not only a miner to hack and haul coal but a whaler to light his way. The dim flames guided men through the uneven terrain and dangerous work of the mines, protecting fingers from being smashed, ankles from twisting, and minds from madness. But in the Midlothian Pits, that precious light could be even more dangerous than darkness.

For the next six hours, men, mules, and lamps trekked miles through the corridors of the mines, drilling, hacking, blasting, shoveling, and hauling coal back to the base of the shaft. Lying on their sides with palm-sized lamps illuminating the glittering black rock, men hacked under the coal face, hoping it would not collapse on them, then drilled holes for a powder charge. Lighting the fuses, the men retreated while the powder blasted the undercut face, collapsing it down into loose coal. Men and boys then advanced back to the blasted face. They shoveled the loose coal into mule-drawn carts, led the mules back to the shaft, and transferred the coal to baskets to be lifted out of the mines. Meanwhile, other machines and workers kept water at bay and air flowing. Steam engines continually pumped water out of the pits so that instead of flooded they were merely damp and puddle-ridden. A furnace lit at the base of a second shaft used convection to circulate air through the mine passages, boys opening and closing carefully placed wooden doors to direct the airflow from surface to coal faces and back. Normally this would go on for hours yet. But by noon that day, the force of around 100 free white colliers — having finished their work in the mine — followed their haul to the surface by retracing their descent.[6]

Not everyone left the pits, however. Forty enslaved black men remained underground accompanied by the overseer Samuel Gouldin and nine other white workers, three of whom were no older than fourteen. Several hours later, after the enslaved colliers had repaired timbers, cleared passages, and fed, watered, and reshod the five mules that were permanently stabled underground, Gouldin directed the men to bore holes into a section of wall in what they called the "North Level" of the mine. John Gray, an enslaved collier who had been working these pits for years, was preparing a nearby section of wall with his partner "when two men came to them and borrowed their implements for blasting, saying that they were about to set off a blast." With the borrowed tools, the second team cut into the soft coal with hand drills, packed gunpowder into the openings, arranged the blasting paper, and then waited for the order to fire. These were old mines, the oldest in the country, and the Virginia coalfields were littered with miles of unmapped, abandoned, and flooded tunnels. Gouldin and the colliers were preparing to blast open a connection to one of these abandoned pits. Gray later told reporters that shortly after he'd loaned the blasting implements, "the men cried out to them to be on the look out, as they were about setting off the blast." Based on the feel of the rock, the sound of the striking picks, and knowledge of the underground environment built up through years of experience, the men thought there might be water behind the wall, and so they had planned for it to come rushing out once they blasted through. Unfortunately, they had guessed wrong.[7]

For it was not water lurking in the old ruins, but a sleeping body of firedamp. The instant "the blast was fired, and the fissure made in the wall be-

Dungeons and Dragons and Gaslights

tween the two chambers," the *Richmond Enquirer* wrote, "the explosion followed, and the awful destruction of life which we record took place." As soon as the partition wall was breached by the powder blast, an enormous volume of gas rushed out, engulfed the nearby miners and their cap lamps, and ignited into a torrent of fire that "swept as a besom of destruction through the various avenues, dealing death with an unsparing hand, on all that came within its course." John Gray, alarmed by the rushing sound of escaping firedamp, turned to see "a solid sheet of flame proceeding from the place where the blast had been made, and covering his face with his hands, threw himself upon the ground, his partner following his example. For a moment or two they were insensible, but upon recovering made for the mouth of the pit—walking a path strewn with corpses." They were the only survivors from the North Level. The teams sent in later to attempt a rescue found that the men had died as they worked, "the flesh charred on their bones"; some "held their shovels in their hands, others were holding to their picks and drills." Alfred and Archer, two men enslaved by the president of the Midlothian Coal Mining Company, were buried under "several tons of stone and dirt" thrown down upon them as the inferno tore through the mine. Those who escaped death in the Pyrrhic birth of this gas dragon were then faced with drowning in its toxic corpse, as "there can be very little doubt that many" of the thirty-four who were later found dead "were suffocated by the 'after damp' [carbon monoxide], rather than killed by the explosion."[8]

As the explosion overflowed the underground space, the wooden shaft heads were blown off "as if they had been paper." At the western shaft, the two cable chains making entrance into or exit from the pits possible "were broken in two as easily as if they had been pipe stems." So tremendous was the force of the firedamp combusting to life that it "caused the earth, for miles around the pits, to wave and rock as a twig in the wind." A mile away, a man crossing the railroad reported feeling "the rails reel under him," while another traveler "passing the road on horse back, declared that his beast staggered and trembled, as if suddenly shocked by a tremendous galvanic battery." The three enslaved carpenters working above the eastern shaft were so startled that one leaped thirty-five feet to the ground, reportedly suffering no injury, while the other two held on for dear life until the shockwave ceased.

The shockwave rolled under the landscape, and it rippled through the social relations converging in the pits. Some of the social artifacts thrust into the historical record by this tragic pulse have survived to the present and provide rare hints of antebellum social relations that otherwise have remained invisible. The most prominent such artifacts are the newspaper accounts of the disaster and its aftermath. The *Richmond Daily Dispatch*'s account was the most widely circulated and also the most revealing. Their initial report, "Terrible Coal Pit Explosion — Thirty-four Persons Killed, and Twelve others so Badly Burned that but few of them can Recover," began with a restating of the headline, specifying that the explosion had occurred at the Midlothian Coal Pits at five o'clock on Monday, two days before. It then immediately pivoted to claims suggesting how unusual and impossible to predict this explosion had been, that the pits had been perfectly safe, reporting that up "to the very moment of the accident, the superintendents and employees in the pits felt perfectly satisfied that there was not a particle of foul air afloat around them, and Mr. John Atkins, the agent, looked upon the pits as being so entirely free from danger, that he declared to us that he would not have hesitated to take his family into them to remain." Next came a guess as to the cause, some descriptions of the shockwave on the surface, and then a further reminder that the "Midlothian Pits have always been looked upon as free from danger, consequently the company found no difficulty in employing as many steady white miners as they desired."

Indeed, the article's primary purpose seemed to be addressing the concerns of white miners, noting that the toll could have been worse, that "if the explosion had taken place between the hours of 6 and 12 o'clock, we have no hesitation in saying that the loss of life would have been trebled, and the number of widows and orphans thereby created five times as great as that caused by the accident at the English Pits in May last; but fortunately, the men were not allowed to make over work, the supply of coal raised being greater than the demand, consequently, most of the white men had left the pits at 12 o'clock, and thereby saved their lives." The paper also highlighted what it saw as a brotherhood of white miners, not just from the Midlothian but from the neighboring English Pits (so named because the owners were from England), which was the sole Chesterfield mine to employ only white men and which had suffered a similar tragedy the previous year. The article went on to describe the rescue. "Mr. Job Atkins, the agent for the English Pits," and presumably a relation of John Atkins, the Midlothian agent, leaped into action "with a number of noble hearted volunteers" — those white survivors of the English Pits explosion together with the white Midlothian miners — who all "descended the Eastern shaft as soon as they could do so, and . . . immediately set about in search of such of the miners as they might find alive." Their bravery paid off, because before long they "succeeded in rescuing sixteen persons, more or less burned, four whites and twelve blacks, and took them to their houses and the hospitals, where they were immediately placed under medical treatment." Notably, this was the first mention that any black men worked in the pits. White miners had risked their lives to rescue twelve enslaved men, but the article was focused on celebrating white heroism, not on any transgressive solidarity that may have existed.

The remainder of the account centered on the sensational and graphic sufferings witnessed by and related to the writers. "Mr. Atkins describes the scene as heart rending in the extreme," the article reported. "Samuel Hunt, a small boy, who had been deprived of reason for the time, by the concussion, was calling loudly to the mule he had been driving to go along," but his mule was probably dead, one of the five killed in the explosion. The other survivors, "as soon as they heard the voices of their friends, begged earnestly not to be left, and then prayed loudly for a few drops of cold water to quench their burning thirst." Yet the true horror was in the hospitals; some "seven or eight negro men lay there, the skin burned from their faces, eyes, hands, arms and bodies, as if they had been roasted, and the groans that escaped from those who were conscious of their sufferings could not fail to pierce the hardest heart."

At the end of the article, the writers attached a list of the dead and injured. Whites were listed first, with names and brief obituaries. Meanwhile "Negroes Burned to Death" and "Negroes Injured" mentioned names only, concluding with the observation that a "large number of the above servants were owned by the Midlothian Company, and very few of them were insured." The damage to the mines, the article noted, was "serious, and it will cost a considerable outlay to get them fairly under way again"; but in the end, the greatest tragedy, and the final remark, was that this "accident has thrown a deep gloom over the neighborhood in which it occurred, and will be the means, no doubt, of driving many persons to seek other employment than that of mining." Fearing that free white miners would be driven from the mines, first by the disaster the year before at the nearby English Pits and now by their near-escape from the Midlothian, the *Dispatch* worried that labor relations in the region might be seriously disrupted.[9] This concern was likely one reason the *Dispatch* worked so hard to emphasize the heroic manliness of the white miners, to spin danger into opportunities for masculinity and brotherhood. The paper's fear, however, may have been overblown.

It was no accident that these mines had disproportionately taken black lives and spared white. The spaces and times of these Virginia coalfields were made and inhabited through a deeply racialized division of labor, life, and danger. First there was the wage relation between white laborers and white capitalists. There is no specific account of how free men in the Midlothian

Pits were paid, but if they followed the system developed in Britain and continued elsewhere in the United States, these white miners, most of whom seem to have been English and Scottish migrants, might have been paid by the ton of coal they raised rather than an hourly wage. According to many studies, this piece-rate system was the predominant method of payment in Pennsylvania and Ohio coalfields. But it probably worked differently in Virginia. At the neighboring Chesterfield pit of David Watkins & Co., account books show that payment was recorded according to days (or fractions of days) worked each month rather than by tonnage. Because many of the white miners in the nearby pit were also employed at the Midlothian Pits, it is likely that they were paid by the "day"—for their labor time rather than their labor products—in both mines. This wasn't merely a matter of accounting. How the colliers were paid (or not) had an enormous impact on how and when they worked, lived, and died in the pits. When they were paid simply for the production of a ton of coal, any time spent on anything other than directly getting coal—such as reinforcing timbers, testing for gas, or making sure all the ventilation doors were properly maintained—was a garnish off their wages, what miners revealingly called "dead work."[10]

Whatever the precise arrangement in the Chesterfield mines, when around 100 white miners left the Midlothian Pits at noon the day of the explosion, 40 enslaved black men remained behind. They remained to do the work that free miners would not willingly do: prepare, sustain, and extend the spaces of the mines themselves. Had there been no explosion, it might have been two o'clock in the morning before the dead-work crew finished propping the arches, preparing the timbers for the next day, boring and blasting new chambers, and caring for the mules.[11] Gouldin, at the direction of superintendent John Atkins, thus kept the enslaved colliers underground for hours to shoulder the risks and smooth the way for white miners to spend as little time as possible in the pits getting the most coal out at the least risk the following day. These black lives were risked for dead work.

Both white workers and white owners benefited from this division of labor in space and time. The free miners benefited by displacing some of the dangers of mining onto enslaved miners and then expropriating the slave labor that exposed and loosened the coal the white miners were paid to accumulate. And because this expropriation allowed white miners to produce more coal more quickly, mine owners who paid by the day could limit the total size of the industry's free labor force and the wage costs of that more expensive white labor. Many white miners listed in the Watkins account books as working full, half, or quarter days also appeared in articles documenting

Dungeons and Dragons and Gaslights

disasters at the Midlothian and Black Heath Pits (Richmond's two major collieries). This strongly suggests free colliers ranged over the Richmond coal basin as a class, quickly expropriating the labor of enslaved miners before moving on to the next "safe" mine. It also helps explain why the *Richmond Daily Dispatch* kept reminding its readers that the Midlothian Pits had been considered safe.[12] Crucially, this was a white working-class "safety" and wage produced through coercing enslaved black men into dangerous mine spaces and times. And through this coercion, this production of white safety, a firedamp explosion was violently and devastatingly sparked that evening in 1855 in the close, dark dungeon of the Midlothian Coal Pits.

I T would happen again. It had happened before. Neither the safety of white Virginians, the danger to enslaved black miners, nor mine explosions were born in 1855. Over the previous few decades, they had all been made together in this place. So, too, had the history of American gaslight. The subterranean fires of the Richmond coalfield, the oldest in America, traced their origins back at least a generation before the Midlothian firedamp explosion of 1855, when shaft depth, pumps, lamps, capital, and insurance combined to create fertile caverns for explosive life and industrial slavery. On March 18, 1839, Abraham S. Wooldridge's capitalized, incorporated dreams of coal were finally about to be realized, but perhaps not in the way he had imagined. Incorporated in 1835 with 200 acres and $150,000, the Midlothian Coal Mining Company directly owned between 140 and 187 enslaved people. The most heavily capitalized mining corporation in the Richmond coal basin, the Midlothian had been able to accumulate a reliable workforce of enslaved colliers, employing many in its own mines while hiring out the remainder to work in other Chesterfield pits. And as enslaved men neared completing the sinking of the first two workable shafts to the Midlothian coal seam some 800 feet belowground, a violent explosion suddenly upset the balance of power in the Richmond coalfields. It was a Monday in March, as it would be sixteen years and one day later. Two white superintendents kept forty enslaved colliers at work underground, in a situation almost identical to that in 1855. But the 1839 explosion was in the Black Heath Pits, not the Midlothian, and it erupted in the morning rather than in the evening.[13]

They knew the danger, but Virginians thought they had been prepared. All worked to avoid and prevent explosions, and mine owners relied on skilled superintendents to coordinate the safe and steady passage of workers, lights, and coal through tunnels potentially thick with firedamp. In the Richmond basin, more often than not these superintendents were seasoned Scottish or English miners. The knowledge, skill, and willingness to work underground that these men carried with them from the even deeper, older, and gassier mines of Britain put them in high demand in Chesterfield. And these qualities also earned them trust. John Rynard, a "Scotchman" and one of the two overseers below that Monday morning in 1839, "was a man of great skill in his profession, having been many years engaged in it in some of the most famous of the English mines." Considering Rynard's reputation, then, it was "hard to account for how *he* should have permitted the cause of the occurrence," the *Richmond Compiler* reflected with some incredulity.[14]

These British experts, however, probably did not arrive in Richmond with much experience as enslavers. Precisely what frictions existed in Virginia among American whites, British whites, black slaves, and free black men working underground is not entirely clear from the historical record, but some clues have survived. As in the later explosion, those shouldering the risk of Monday morning dead work "were all colored men," and the "superintendents above the shafts say that about forty were below." Interestingly, this was only a guess, as the superintendents "cannot speak with certainty." According to the *Compiler*, many of the enslaved "had gone to see their wives to distant plantations, and it was not known how many had returned. Those who had not, do not yet appear from terror at the news of the explosion, but forty is the *maximum*." While this uncertainty may have been evidence that black miners were considered so interchangeable that no one bothered recording who did what work when, some enslaved men, it seemed, moved regularly from mine to plantation, carefully negotiating the tangled social forces and terrains making them at once husbands, miners, and property. No doubt this "freedom" was severely circumscribed, but it may have been enough for some to slip through a regime of interchangeability that, at least after the explosion, left their very lives in question.

The explosion cast a cloud of uncertainty over the whole county, threatening to interrupt the critical movement of enslaved workers from plantations to pits. The Black Heath Pits, the property of the wealthy and well-regarded Heth family, had been mined for generations. By 1839, not only were the pits acknowledged to be "one of the richest and most extensive" mines in the region, but they were also reputed to be "the deepest in the Union: being more than 700 feet to its bottom."[15] But the Black Heath's very success proved its undoing. Miners had delved too deep, and awoken subterranean fire. It would be months before all the rubble and black damp could be cleared, timbers repaired, and the mine made once again useable. In the

Dungeons and Dragons and Gaslights

meantime, enslaved miners feared to return to the pits, while planters feared to part with them.

The 1839 disaster was a crisis, but also an opportunity. The Heths sold the Black Heath Pits to the English-owned Chesterfield Coal and Iron Mining Company, thereafter known as the English Coal Pits. The new owners' response to the crisis would eventually be to abandon leasing enslaved men altogether, instead employing English immigrants and free black miners. Eventually. In the aftermath of the explosion, the English Coal Pits were just dark, poisonous holes in the ground. Meanwhile, another vision for the coalfields, a vision of industrial slavery, was beginning to materialize immediately adjacent to the now-closed mine. Abraham S. Wooldridge, owner of the recently incorporated Midlothian Coal Mining Company, was poised to seize advantage with his heavily capitalized enslaved labor force and shaft-sinking engines. A few months after the Black Heath disaster, the *Richmond Enquirer* excitedly reported that the Midlothian colliers had "struck at the distance of 783 feet from the surface of the earth" what appeared to be a bed of "very rich" coal that "would be almost inexhaustible." More importantly, and more urgently, the paper had no "doubt Mr. Wooldridge will push this article into market, and thereby supply in some degree the deficit which is occasioned for the time, by the melancholy disaster which lately happened at one of the Pits in the neighborhood."[16] Already the name of Black Heath was being forgotten. This was going to be a Midlothian future.

First, however, Wooldridge would need to rebuild the relations of trust ruptured in the Black Heath explosion. No sooner had the Midlothian Pits struck coal than Wooldridge started circulating notices in the Richmond papers titled, "MID-LOTHIAN NOTICE. PIT HANDS WANTED." "In consequence of reports having been circulated of the insecurity of hirelings in the Mid-Lothian Mines," the ads began, "the Company deem it proper to make known, that the mines have recently been opened, and that not the slightest explosion from gas has occurred." If this absence of destruction was not particularly reassuring, Wooldridge wanted his readers to know that he had the best people in the business working to head off any danger. According to the ad, the Midlothian "workings were laid out and are now progressing, under a plan furnished by Messieurs Foster & Hall, two distinguished English Colliers, sent over from England to reclaim the Pits of the Black Heath Company after the recent explosion of gas — and the present under ground operations are now conducted under the management of one of the English foremen left here by these gentlemen, and supervised by Mr. Wm.

Hall."[17] It was, perhaps, ironic. The English capital and colliers pouring into a district still reeling from the Black Heath explosion might have established an archipelago of free labor capitalism in slave country. But redirected by men like Wooldridge, this potential archipelago shrank to the single island of the English Pits, while the British migrants, by working in slaveholding mines, stabilized, strengthened, and modernized the very industrial slavery they had thought to replace.

Having entangled English and Virginian capital, men, and expertise to project a safe and secure minescape, the Midlothian owners still needed to make their pitch to planters. "Owners of slaves at a distance from the Coal Mines, would do well to give some attention to the subject," the ad announced, and "the Company are now in want of some ten or fifteen additional able-bodied, active Pit-hands, on hire by the year, for which they will give the most liberal hire." Planters, pit owners argued, should be eager to get their field hands to the mines, for there "is no place in this country where slave labor commands as much, where their general health is better, and where the treatment and contentment of the slaves are surpassed." Of course it was true, Wooldridge conceded, "that within the last few years several disastrous accidents have occurred, but from the scientific and practical skill attracted to the mines, these accidents will be of rare occurrence, it is to be hoped." The ad was a plea for trust. When Wooldridge signed his name to the publication and included an additional assurance of safety signed by the "two distinguished English Colliers" Frank Foster and T. Y. Hall, he was putting his and their reputations on the line. He was trying to build, if not a shared vision, at least a shared interest in the industrial slavery of the coalfields. He was quite successful. By the Civil War, the two dozen largest companies in the Richmond-area pits hired or owned nearly 2,000 "hands," mostly enslaved, and they were raising 150,000 tons of coal each year.[18] When working properly, the system of pumps, furnaces, and safety lamps was supposed to minimize the risk of explosion or flooding by drawing off gas and water. But in the mines, nothing stayed the way it was supposed to for very long.

The pattern of an explosion followed rapidly by a sustained public reassurance continued tragically, predictably, in the Chesterfield coalfields through the Civil War. Reopening after an 1842 explosion had shuttered the smoldering mine for over six weeks, Wooldridge even invited reporters, ministers, and local ladies to take guided tours of the mines. As accounts of these tours were published, penned, or circulated by word of mouth, Wooldridge hoped to maintain the trust of investors, slaveholders, and free miners. Each time an explosion happened, newspaper writers expressed shock, surprise, and confidence that science would put an end to these deadly interruptions in mining. When investigations found that safety lamps may have been improperly used or that men may have unscrewed the tops to light pipes or for better illumination so that they could actually see the workface, company officers professed outrage and innocence.[19] But the explosions continued.

Focusing on how miners lived, and not only on how they died, further complicates the picture, but in useful and revealing ways. Finding information about miners' lives can be difficult. As was the case for enslaved turpentine workers, no slave narratives or interviews of men who were enslaved in coal mines have survived. But the existing sources do contain clues. Many of the enslaved miners may have considered working in the coal pits to be a relative improvement over their lives on plantations. They sometimes earned the chance to make some money. The ambiguities of runaway advertisements suggest that there was less surveillance of the enslaved at mines than on many plantations. The pits were also a space in which men from all over the region could exchange news, company, and experiences. In short, we should not presume to know the minds and goals of those already twice silenced, in the past and in the archive. Some of the enslaved men underground may have worked hard to be there. Neither should we ignore the impressive actions of the white miners in 1855 who risked their own lives to mount an underground rescue operation into poisonous air and possible additional explosions to save the lives of over a dozen black men, men they didn't own and whose owners probably did not see them as peers. Without a doubt, the white miners benefited from the racial division of labor and risk in the pits. But that did not mean it was impossible for friendships, alliances, information networks, or political solidarities to have crossed those same racialized boundaries.

The 1855 Midlothian explosion also revealed surprising frictions and alliances among capitalists. By the mid-1850s, when another flurry of explosions erupted under the Richmond basin, the most significant development in the disaster response was the growing presence of slave life insurance policies stabilizing relations between slaveholders. The social relations keeping the enslaved chained in and to coal dungeons were anchored far as well as near, sinking propertied hooks into their very mortality, hooks extending all the way to life insurance offices in Baltimore, New York, and Philadelphia. White miners were not the only ones concerned with and capable of displacing the persistent risks of coal mining. The Virginia planters and urban slaveholders who supplied the coal mines with a significant portion of the enslaved labor force were often reluctant to hire out the people they owned to such a notoriously dangerous industry, even when the mine owners were willing to pay higher rates. Coal companies like the Midlothian wanted to create strong and dependable spatial relations enabling planters, urban slavers, and pit operators to easily move (and work) enslaved people from field to city to mine. Overcoming the social relations holding enslaved people within plantations and cities presented a continual challenge to pit operators. Corporately owned enslaved miners alone could not meet the total regional labor demands. As the records produced in the aftermath of the 1855 explosion attest, of the 152 men at work underground that day, around 110 were free white workers, 18 were enslaved directly by the company, 8 were enslaved by the Wooldridge family (the company proprietors), and 16 were enslaved by others who had leased them to the company. And to persuade slaveholders to hire out their human property to work in the pits, boosters peddled the powerful technology of life insurance.[20]

Life insurance provided slaveholders with a tool of mastery that not only forced the enslaved into underground danger but could turn both their living labor and their working deaths to account. In the mines, even ghosts might not be free. Jordan and John were at the pits that fateful day because their enslaver, William Goode, had rented out the right to their labor to the Midlothian Coal Mining Company. Depending on the terms and duration of hire, Midlothian agents had agreed to pay from $1.20 a day to $90 for six months in order to be able to command and appropriate the products of Jordan's and John's labor. But this was not the only reason Jordan and John were underground with Gouldin and the others. William Goode had purchased life insurance policies on the two men through the United States Life Insurance Company, a Philadelphia-based corporation with an office in Richmond, contracting with local planters through the Valley Insurance Company. In this disaster-prone industry, pit operators increasingly relied on life insurance to force enslaved workers like Jordan and John underground. As an insurance agent noted in 1857, "Parties who apply for insurance of the Coal Pit Hands are very anxious to get the Policies as they are keeping the hands above ground and idle till they get them insured."[21]

In pooling capital and risk through a national insurance network, William Goode transformed the nature of his claim over John and Jordan. Now, not only could Goode sell and hire out John and Jordan in life, but he could recover most of their exchange value—their market prices—in death. The city of Richmond and its surroundings, where slaveholders hired out more than half of all the men and boys they enslaved, became rich recruiting grounds for regional industrialists and life insurance agents. The Baltimore

Dungeons and Dragons and Gaslights

Life Insurance Company's agents, 80 percent of whose policies covered an enslaved person, managed their own risk first by offering policies only to slaveholders they deemed "respectable" and "humane." In this way, life insurers offered slaveholders both security and a badge proving their paternalism. Second, slave life insurance policies only covered up to two-thirds of an enslaved person's "actual value," while charging $2 per $100 of insurance, twice the premiums they would charge on a white person's life. For dangerous work like coal mining, Baltimore Life charged an additional $3, but many slaveholders found these 5 percent premiums well worth the cost. All of Baltimore Life's competitors, meanwhile, charged even less. From 1846 to 1848, New York Life got in on the action, selling at least 488 policies on enslaved people, about one-third of all the policies the company sold. And one in six of those slave policies covered miners and coal pit employees.[22]

Through life insurance, enslavers and insurance agents conspired to create fictive slaves bearing perfectly stable futures. These knowable, ownable futures even became, to a certain extent, unmoored from the living people upon which they were based. Enslaved people never owned their own lives, but in certain circumstances they could use their chains to carefully shape and guide when, how, and where they worked and lived.[23] This limited but meaningful means of survival and control was sharply circumscribed by the necromancy of life insurance. As death — or at least certain kinds of death — no longer prevented enslavers from realizing most of the value of some of the people they enslaved, those safe, insured fictive futures pressed dangerously on enslaved people's real, lived presents. The future weighed heavily in the mines.

Indeed, a multiplicity of futures had converged to press John, Jordan, and thirty-eight other enslaved men into the mines at the moment of the explosion. When the black miners laboring in the Midlothian Pits to secure future safe and speedy passage for white miners triggered that catastrophic fiery assault, their own futures, or at least the parts of their futures to which others claimed ownership, were violently decoupled from their living selves. For eight of those killed in the explosion, the futures imagined for them by white enslavers and insurers were not permitted to perish along with their bodies. Their undead futures remained the property of planters and mine owners. And in this case, these were futures stored and coordinated entirely through the Philadelphia-based United States Life Insurance Company. Of the thirty-three enslaved men killed in the explosion, twelve were owned by the Midlothian Coal Mining Company and were uninsured. Colliers enslaved by corporations may have lacked or been denied insurance policies, but of the twenty-one remaining miners enslaved by individuals, including John and Jordan Goode and Stephen, Robert, and Orange, who belonged to the Midlothian-owning Wooldridge family, over a third were insured, and each through the United States Life office.[24]

As reported in the Virginia papers following the explosion, the "United States Life Insurance Office, under the Exchange Hotel (at the office of the Valley Insurance Company) loses by this accident an insurance on eight servants. The Agent, we see it stated in the Dispatch, is prepared to pay the insurance the moment the claimants make application." This was, they acknowledged, "comfortable" to slaveholders, and as for the losses to the insurance company, those "who insure lives, however, do it we presume with no expectation that insurance will save life." Saving life, of course, was never the point. Rather, life insurance made endangering enslaved people less risky for their enslavers. While John and Jordan lost their lives, both William Goode and the Midlothian Coal Mining Company were absolved of responsibility.[25] In this way, life insurance loosened the social relations keeping the enslaved "safely" close to owners, allowing planters, entrepreneurs, and capitalists to more easily shuffle such "liberated" enslaved men and women across and under southern landscapes, amplifying their exploitation, and multiplying their productivity at little risk to social and economic relations among southern elites. Insurance was a class-making institution. It was a coal-producing institution. It was, also, a race-making institution.

Through slave life insurance, white tidewater elites further secured the wealth appropriated from those they enslaved while wedging open space underground for white working-class miners to more safely pursue their trade. Insuring slaves thus gave whiteness tangible materiality, and this fact was eminently apparent to A. S. Wooldridge. He made his sentiments publicly known by publishing a letter he wrote in thanks *To the President and Directors of the United States Life Insurance, Annuity and Trust Co. of Philadelphia.* "Gentlemen," the letter began, "I avail myself of this method to express my appreciation and to commend the promptness of your company in the payment of your policies on lives lost in the Midlothian Coal Pits by the late explosion, (through your agent, O. F. Bresee, at office of 'Valley Ins. Co.' in this city,) without availing yourselves of the time allowed for settlement or deducting interest." The letter was not only an example of good business; it was an act of community building, one that Wooldridge hoped would protect the precious routes carrying enslaved men to the mines. "I can assure you," he concluded suggestively, "that such liberality will be appreciated by our citizens, and your Company will receive the encouragement and support by our community which it is pre-eminently entitled to, as well for its readiness to pay as its ability to do so." Read in context, the letter was also clearly a recruitment notice. Only a few days earlier, the Midlothian Pits had finally reopened, the damage wrought by the explosion having taken three weeks to repair, and Wooldridge needed to refill the mines with enslaved workers.[26]

In a sense, this cycle of fire and death presented a problem without a real solution. From 1854 to 1860, Baltimore Life and other insurers pursued industrial enslaved lives, while the continuing explosions created a halting, and deadly, cycle of enslaved underground movement. After the Midlothian explosion, Baltimore Life temporarily froze policies on coal miners for 1856, then again in the summer of 1857, finally restarting them only after raising premiums. But the work continued. So long as cities and factories valued coal, and especially gas-rich coal, there were power and profits to be had by convincing and coercing men to sneak rock-bound gas out of the dragon's den. Such would never end well.

Aboveground, in the gasworks and gaslights of antebellum America, the problems were both similar and different. But the struggle to know, control, contain, and harness flammable coal gas was every bit as desperate.

THE future had arrived in New Orleans, and everyone seemed to know it. In 1834, in the muggy heat of spring, reporters marveled as workmen dug up streets, laid more than two miles of footwide iron pipes, and began construction—a few blocks west of the city's bustling slave markets, near the Charity Hospital—of a massive cast-iron tank over fifty feet in diameter. This cistern, the observers claimed, was "the most extraordinary work ever seen in this country" and "will surprise all those unaccustomed to look at improvements upon a large scale." With a foundation of 100,000 bricks and a capacity of over 120,000 gallons, the remarkable container was meant to hold illuminating gas, a new material of the age, and by mid-June, the station where that gas was to be manufactured was drawing further attention. James H. Caldwell, the English immigrant, actor, and theater owner responsible for funding and building the awe-inspiring gasworks, had become a regional celebrity, to whom, local papers declared, "this city, and indeed the valley of the Mississippi is indebted, for the introduction of that beautiful, safe and economical light by gas."[27]

By early August, the public had gathered around Caldwell's theater to witness the first trial of his much-heralded gasworks. And this thespian knew how to put on a show. "The beautiful flame above the theatre, representing the Trident of Neptune, exceeded any thing of the kind that we have ever seen," reported newspapermen as they watched gas blaze forth into the symbol of a god. In the bright white flames issuing from these and countless other fixtures, boosters saw visions of a prosperous gaslit industrial region radiating from the great Mississippi. "Before many years," they proclaimed, "it is not difficult to foresee that every city, town, village or Hamlet, in the Western and South Western country, will be lighted with gas." Boosters claimed this coal-fired future was almost inevitable, "because the light is in every respect, so far superior over every other, and the materials of which it is composed, so plentiful and cheap." With perhaps some exaggeration, the report estimated that the "manufacture of gas for the light of this city alone, would consume, suppose the whole city to be lighted, 300,000 barrels of Pittsburg coal, which after making gas for the supply of 30,000 lamps, would leave 400,000 barrels of coke, to serve steam boats, locomotives, kitchens &c."[28] New Orleans, by draining the rocks and riches of all the waterways flowing into the Mississippi — by stockpiling the coal arriving on its wharfs from the mines of Pittsburgh, Kentucky, and western Virginia — imagined itself poised to light its way to regional and national prominence.

Strolling through the completed New Orleans gasworks, reporters from the *Picayune* were overwhelmed not only by the size and scale of the operations but by their intricate, immaculate precision. This was a new order: not just the quiet precision of a well-made pocket watch, but an industrial order of perpetual motion, heat, light, and sound — an order produced through mastered chaos. "Where huge furnaces are day and night and from year to year in an intense and perpetual glow," the *Picayune* reporters observed, "where ship loads of coal are constantly consuming — where lime kilns are burning, and pyramids of coke are piling up," a visitor would expect such a place to be one "of smoke and dust and sut and vallanous odors." Yet the *Picayune* claimed, incredibly, that on "the contrary, lady visitors may promenade the gas works, walk the fire rooms, the condensing rooms, the blacksmith shop, the carpenters, examine the gasometers, the scales, the pipes, the steam engine, the whole complicated apparatus of the place, and not soil a white satin slipper."[29]

The works themselves were arranged over an "immense square" between Perdido and Gravier Streets. "Mountains of coal are piled up in nice regularity, under substantial sheds erected for the purpose," one reporter described, while in one corner of the square a "vast limekiln is employed in the manufacture of lime from oyster-shells," lime being used to rid the raw gas of sulfur and soot. At the opposite end of this process, three "stupendous gasometers occupy positions within the square, and look like iron cages or caul-

Dungeons and Dragons and Gaslights

drons in which wicked genii are enchained." And the gas genies trapped in these industrial lamps were not the kind one wanted to awaken. If hydrogen gas ignited inside a gasometer, the explosion could level a city block. Gasworks were intended to be vast collective lamps where the flames were kept as far as possible from the font, or fuel reservoir. This meant the font could be made hundreds of thousands of times larger than that of an ordinary table lamp and be displaced, through a network of pipes, from domestic, commercial, and public spaces.[30]

In describing the process of gas manufacture, the *Picayune* painted a picture of passive, almost natural production. "After the generation of the gas," the paper reported, "it is passed through an immense number of iron pipes called gas washers.... After this process of cleansing with water the gas passes through what are called gas purifiers; they are large wrought iron boxes, in which are placed perforated iron plates, covered with lime." Once washed and purified, the "gas then passes through the grand metre, and thence to gas-holders or tanks, where it remains until conveyed through the pipes around the city." From generation to distribution, gas was simply supposed to make and measure itself, and nothing embodied this more than the station meter. "The metre is a very ingenious and beautiful piece of workmanship," the *Picayune* writer gushed, "telling with the accuracy of a clock the exact amount of gas made each day." Not only did it measure and control the outflow of gas with incredible precision, but the meter actually performed the work of record keeping on its own with "a contrivance attached to it, called a tell-tale, which, by means of a pencil and a sheet of paper attached, informs the superintendent, by looking at it, the precise rate of gas produced at any and all parts of the day."[31] Such industrial clockwork and clerkwork found further resonance for consumers in the self-operating individual meters admitting and measuring all the gas consumed in a given house or business.

But gasworks were also the sites where all that fuel was manufactured, and thus the human labor involved was impossible to completely ignore. At Perdido Street, reporters described "the workmen going through their variously apportioned employments, as the vast piles of coal disappear in the ovens, and the gas is generated! This is a spectacle of riveting interest, and is constantly going on. When the doors of the ovens are opened, the rapidity with which the coke is removed and fresh coal locked up in thrice seven times heated iron cages, is most remarkable." So remarkable, in fact, that another article suggested the "tremendous and continual fires kept up in the furnace, remind one of the hot place we read of, and the workmen who attend them must be almost salamanders, to endure the heat." "Sala-

"Making Gas." In this illustration of the Manhattan gasworks, in a room filled with coal dust, soot, and smoke, stokers loaded the retorts with bituminous coal. Furnace workers, meanwhile, fed the fires that heated the retorts, making sure they burned at the proper temperature. Despite the claims of boosters, neither were these men "salamanders" nor were gasworks immaculate, self-operating machines. Engraving in *Harper's New Monthly Magazine*, December 1862, 16.

manders," here, meant the mythical creatures of fire; boosters were painting these stokers as living manifestations of the elements, inhuman creatures of a mastered industrial hell. They cast the workmen who opened, emptied, and refilled fiery gas retorts (ovens that transformed coal into gas and coke) as unnatural native beings, both channeling and part of the tremendous forces flowing through the machine. Years later, *Harper's* similarly described a visit to the gasworks of the Manhattan Gaslight Company, where the fire under the lines of retorts "burns entirely around them with a fierce heat. Into these retorts the coal is put by gangs of stalwart men, who play about in the fire like salamanders, seeming really to enjoy the burning."[32]

But these stokers were not salamanders. They were not industrial forces like heat or steam to be engineered and harnessed. They were men, and at

Dungeons and Dragons and Gaslights

least in the New Orleans works they were most certainly not there "to enjoy the burning." One might never know it from reading the dozens of newspaper articles written about the antebellum New Orleans gasworks, but they were operated almost entirely by enslaved workers. "All manual labor was slave," according to one historian of the city, "whites being employed only in supervisory capacities," while the enslaved men and their families "lived in quarters inside the walls of the plant." When James Caldwell died in 1860, by then famous for establishing gasworks not only in New Orleans but in cities across the South, company books listed $53,000 in "live assets" and newspapers reported the public grief of his loving slaves. Whether this grieving was for Caldwell or for their own futures was mostly beside the point, for "in any case, Caldwell had erected a fifteen-foot brick wall and a first-class set of iron gates to insure the affections of his chattels."[33]

This was not simply some pre-capitalist labor force assembled to make do until free wage labor became available. In 1836, the company only "maintained a few slaves who were put to work about the station." And for the first decade, Caldwell and Dr. Rogers, his superintendent, tried to run these works mostly with hired workers. But in 1848, regarding white labor as too expensive and unreliable, the company explored, according to its minutes, "the practicability of substituting slave labor for white labor at the station and on the street main, as well as lighting the public lamps." It was a popular idea, and after discussing "the apparent economy of slave labor, from the estimate of the engineer, determined the board to authorize the engineer to make the experiment . . . to purchase from time to time . . . such slaves as will answer the purpose contemplated, either as stokers and laborers, as well as the necessary mechanics to supply the place of the white labor now of necessity employed at so large an expenditure."[34] The experiment was apparently so successful that the full transition to enslaved labor was almost immediate. This was a modernity explicitly dependent on slavery. Gaslight, the citizens and engineers of New Orleans believed, was the sine qua non of science, progress, and Western civilization. And slavery, they decided, was essential to it. But the decision to purchase an industrial labor force did more than provide the city of New Orleans with light more efficiently. It cast real lives into disruptive motion over considerable distances.

WHEN Davy discovered that his owner had sold him again, he could have had no idea that his life was being uprooted to feed the greedy dreams of the keepers of New Orleans's gasworks. Davy may not even have known what a gasworks was, since Caldwell had first fired his gas trident above the theater just over a year ago. The man who arrived in Mississippi to take Davy away from the chain gang might have let him know that he was headed for New Orleans, but that destination was hardly unusual. Thousands of enslaved people were sold to and from New Orleans every year. Court records show that a slave trader named Calvin Smith purchased Davy from the chain gang in Mississippi, marched him to New Orleans over a monthlong journey, and sold Davy to a New Orleans trader named George Botts for $500. According to the testimony, Botts had met earlier with Dr. Rogers, the superintendent of the gasworks, having "heard that the company wished to buy slaves," and told Rogers that although "he had not then on hand any that would suit them . . . a friend of his had gone to Mississippi to purchase prime slaves, and that if he could wait a few days, he would furnish him such a one as would suit." When Davy and Smith arrived in New Orleans, apparently Smith "represented" Davy to Botts "as a subject very hard to manage." The forces and decisions resulting in Davy first landing in a Mississippi chain gang and then being sold south to New Orleans were, while largely outside his control, a chance (however limited) to transform suffering and hardship into new possibilities. Both he and Botts knew that it was to Botts's advantage to hide or reinvent Davy's history. An unruly slave with a stint in a chain gang would be difficult to sell for much. Davy had just as much interest in leveraging Botts's greed into a clean slate for himself. Without the baggage of his history, Davy might be able to shape not only his passage through the market but the conditions and possibilities of his life after sale.

A few days after Davy arrived in New Orleans, Botts spotted Dr. Rogers "passing in his gig" and pulled him over to examine Davy. The ruse appeared successful for Botts. He resold Davy to Rogers for at least $1,050, more than twice what he initially paid. Whether Davy managed to shape his life effectively after sale is less clear. There are no records of his treatment, labor, or oversight in the gasworks, but two months after Rogers bought him, Davy ran away and successfully escaped his enslavement and his history. Without a laboring body, all that remained of the relationship between Botts and Rogers were debt and lies. Able to prove deceit, Rogers prevailed in court, and Botts was eventually forced to pay $1,050 in damages.[35]

The benefits of having property rights in labor, even when that labor stole itself, would not have been lost on Caldwell and Rogers. First of all, Davy almost certainly performed productive labor at the gasworks for two months, even if Rogers had to torture Davy to get it. And when Davy ran away, that labor became, for Rogers, better than free so long as he could

Dungeons and Dragons and Gaslights

make Botts repay him for not only Davy's past labor but his future, too. As Davy threaded his history, self, and work through the slave market and the New Orleans Gas Light and Banking Company, he left behind not only the products of his labor and his enslavement but the value of his past and future life in bondage. And most of the people enslaved at the works did not escape. While the works, which until 1845 was also a bank (capitalized at $6 million), kept expanding to provide more light, money, and credit to more of the city, the number of enslaved people living and laboring and storing capital within the walls of the yard only continued to grow.[36]

Company records at the moment of emancipation reflected the capitalist logic of the industrial enslavement of living, laboring, reproducing, aging people. In 1901, the *New York Times* reported that "after going over the books of the New Orleans Gas Light Company . . . the money losses sustained by the corporation through the emancipation proclamation of President Lincoln were $51,650. This amount represented the value of sixty-two slaves owned by the company at the time of the occupying of New Orleans by the Federal forces." Nat was enslaved at the gasworks from 1858 to 1863, but in only five years his recorded price had more than halved from $1,300 to $600, almost certainly due to a disabling injury. Company owners had enslaved a man named Crawford and his wife for the entire twenty-seven years the gasworks had operated. By 1863, the couple was, the *Times* report calculatingly noted, "too old to do any appreciable amount of labor for the company, but their investment was not a bad one. Crawford's wife bore six children who were given an aggregate value of $2900." Leonard, sold to the gasworks at the age of forty-three, had lost two-thirds of his price by the time he was emancipated fourteen years later, his life and labor and health having been burned into gaslight and profit. But capitalists could also force younger enslaved people to grow stronger and learn new skills, not only producing more gaslight and marketable goods for their owners but increasing the very capital contained within their price. For instance, the company bought twenty-nine-year-old Henry Barnes for $750, but by January 1, 1863, his value was listed at $1,000.[37]

Freedom was something the enslaved gasworkers would have to make for themselves. Shortly after Union general Benjamin Butler and his army of volunteers occupied the city in May 1862, he "placed the Gas Works under the Control of the Provost Marshall, from the insubordination prevailing among the slaves, who formed a large majority of the laborers." Not content with a mere change of flags, the more than forty enslaved gasworkers agitated for change, for freedom. But Union forces decided that gaslight was a "public good" and thus justified the continued enslavement of the gasworkers no matter the contradictions, no matter how strenuously the workers insisted otherwise. As officials began their counterinsurgent crackdown, W. Newton Mercer, the president of the Bank of Louisiana, wrote to Butler that soon after, "four who had run away were arrested at Carrollton," an upriver neighborhood in the city, "& sent back by Colonel French, — whereupon, believing an example indispensable, I directed them to be confined in irons — But a few days having produced the desired effect on two of them, their irons were stricken off, — the other two, who remained contumacious, were kept under punishment."[38]

By September 1862, Robert Harrison, Robert Morgans, and Joe Lewis had fled their continued enslavement at the gasworks, making their way four miles upriver to Camp Parapet, which was under the command of the Vermont abolitionist J. W. Phelps. Butler had ordered that all loyal slaveholders could keep their human property. Phelps wanted to make immediate abolition real. Frustrated with "the sluggish movement of the Lincoln administration, and Butler's accommodation with Louisiana slavery, Phelps began offering sanctuary to runaway slaves" and broadcast the message of freedom from Camp Parapet. The three men had already tried to run at least once before, as Phelps noted that "this is the second time they have sought the protection of the United States, in my camp." They had also likely run away several weeks before and been among the four men captured and tortured by W. Newton Mercer. Although unable to remove their irons, the fugitives carried news of their treatment to Phelps, who noted that "this time two of them were loaded with chains and one of them is badly maimed." Phelps was furious, and even more so that he was under orders to return these fugitives to a Mr. Kincalla, "the day watchman of the New Orleans Gas Company, to which these men belong." Phelps argued that returning these men to their enslavers would have violated recent war orders, but a responding missive instructed Phelps to return the men to Kincalla anyway, claiming that "they do not belong to Mr Kincalla but to the Gas Works which are now under military authority & we need them for public service." The Union army demanding slavery for the public good: little more clearly captured the contradictions of the politics of nineteenth-century gaslight. Rather than comply, Phelps resigned on the spot. Two weeks later, Lincoln issued his preliminary Emancipation Proclamation. Meanwhile, Robert Morgans, and likely all three fugitives, were returned to the gasworks, where they remained enslaved for at least several more months.[39]

Just as with turpentine, the antebellum years were boom times for industrial slavery as much as they were for industrial wage labor. And the two were clearly related. Northern operators shipped mountains of coal mined by (mostly) free laborers in western Pennsylvania, Ohio, Illinois, and Missouri downriver into southern landscapes made and sustained through the continual exploitation of enslaved human beings. Industrial visionaries like Caldwell wove waged coal and enslaved gas workers into fame, fortune, and an emerging Mississippi Valley industrial order. By 1841, Caldwell had won exclusive contracts to supply New Orleans, Louisville, Mobile, Havana, and Cincinnati with gas. By 1850, with the New Orleans works fully transitioned to slave labor, there were "consumed in the works, daily," more than twenty tons "of Pittsburg coal, 290 barrels being the most ever." Fed into retorts by about fifty enslaved mechanics and laborers working in shifts around the clock, this coal was transformed into approximately 200,000 cubic feet of gas over the course of each day.[40]

I n New York, a different set of visions of modernity was running up against the materiality of gaslight. Consumer advocates still had not achieved a reduction in the price of gas, which they hoped would bring gaslight and freedom to the masses and save them from the perils of camphene, but *Scientific American* continued to press their case. "The city of New York," the editors wrote in December 1852, "contains the most patient, suffering population in the world." Instead of serving as a force for freedom, gas had become a vehicle for robbing New Yorkers, while their "rulers, every public chartered company, every city contractor, and every speculator favored by these rulers, enjoy the most delectable privilege of getting the greatest amount of money out the 'dear people.' The taxes of New York City are much higher than those of any city in the world, and no city is so poorly served."[41]

The main problem, they felt, was that New York's gasworks were purchasing overpriced, inferior coal from Liverpool, when it was their "opinion that good cannel coal," the coal with the highest gas content, could "be obtained from Virginia for as low a price as $7 or $6 per ton." And if the Virginia gas coals weren't enough, or were running short, gasworks could at least find better British coals than what they were buying from Liverpool, "and if cannel coal was taken from Glasgow instead of purchasing the inferior Liverpool coal, a great saving in that quarter would be effected." If this were done, *Scientific American*'s editors were confident that the price could be cut by a third, and once "reduced in price, almost every private family would use it in place of oil, camphene, &c. We hope our gas companies will see to this; it would be the means of preventing many of the casualties which are constantly occurring from the use of volatile hydro-carbon fluids, and be a blessing to both rich and poor."[42]

Right away, it should be obvious that the way arguments for gas were being made in terms of safety, justice, and responsibility were just as problematic as the arguments for "safety lamps" and camphene. Gas would be a "blessing" to all as long as it was cheap, and so *Scientific American* specifically demanded either slave-mined coal from Virginia or that produced with the bonded and child labor of Scottish mines.[43] Part of the issue was that there was no easy way for private citizens to know where the city gasworks got its coal or how much it paid. Consumers were privy only to the price of gas; the rest of the story was hidden behind property rights and the commodity circuits that flattened and obscured the production of coal.

But then why ask specifically for the cruelest, most exploitative coals available? It was not as if the advocates of cheap gas were unable to at least imagine the realm of production. During a trial over the use of gas meters to charge customers, the lawyer for the Boston Gas Light Company went so far as to argue "what an incalculable saving [gas] has effected in human and animal life, by dispensing, in part, with the necessity for common oil — an article obtained only at the greatest risk of life, and at a very heavy expense."[44] Because whaling was so dangerous, they suggested, Americans should get their light from coal mining! Perhaps Americans associated coal with the foreign British, or they didn't care because white Americans weren't the ones mining in Virginia or because there were no coal-mining heroes to match the whalemen of American popular literature. Whatever the reason, when Americans discussed light, they readily connected whale oil to whalers but usually reduced coal to features of price, chemistry, and geology.

While northern consumers fought to secure the cheapest coals on the market, regardless of how and by whom they were mined, producers of gaslight built and maintained massive urban empires of infrastructure through armies of free laborers, clerks, and engineers. Based on his survey of seven major American gasworks from 1852 to 1854, none of which were farther south than Baltimore, one British commissioner reported to the House of Commons that the laborers were almost all foreign-born Irishmen, except in Philadelphia and Cincinnati, where both Germans and Irish worked side by side. Free gasworkers east of the Mississippi all made similar wages. In New York, firemen were paid $9.75 a week, while "labourers" and mechanics, respectively, made $1.12 and $2.00 a day. The same wages prevailed in Philadelphia and Baltimore. In Cincinnati, wages ranged from $30 to $45 per month, while in Chicago the workmen earned the least, only $1.00 per day.[45]

In Philadelphia, an 1856 report by the city Committee on Gas Works provided more detail on the work and wages of gas. To maintain and build the urban infrastructure of mains, pipes, meters, and lamps required considerable labor. The city paid men over $15,000 a year to clean lamps, and it paid for others to repair them. Philadelphia paid crews to haul new lamp posts and others to lay mains in the streets and still others for laying pipes and setting meters. The engineer's accounts listed $31,007.93 in wages paid to "laborers and fitters employed in erecting new works." In the gasworks themselves, engineers allied with master stokers to adapt and exploit an apprentice piece-rate system to increase productivity while keeping wages flat. "The very large change in the results of a unit of cost of labor," the committee noted approvingly, "results chiefly from the classification of the workmen, and payment by piece-work, whereby those having most skill can earn an increased amount proportionate to their ability, while those yet in pupilage obtain a per diem but little greater than ordinary day laborers." Increasing the rate of exploitation was of central concern to engineers when it came to stokers, as they represented the largest labor cost ($77,525.15) in the entire production and maintenance of gaslight in Philadelphia. But stokers weren't the only workmen at the retort houses. Carpenters, blacksmiths, bricklayers, fitters, and "laborers" were continually employed to repair and modify the various works. The white-collar labor of the clerks and meter inspectors, meanwhile, "employed in the inspection and Fittings and Meters, and taking Statements for Bills, Storing Coals and Coke at Works, measuring Coke, Tar, etc.," cost the city $47,737.60.

Although it too began as a privately chartered corporation in 1836, Philadelphia's gasworks were purchased by the city in 1841. The only major publicly owned works, Philadelphia's were also the nation's largest, which in 1855 held $3.4 million in assets and burned 771,533 bushels of coke and 830 cords of wood to transform 38,158 tons of coal and 438,558 pounds of the rosin distilled as a by-product of turpentine into 362,707,000 cubic feet of gas for 300,406 lights, including 3,349 streetlamps, delivered through 207 miles of mains and 63 miles of service pipes to 22,898 customers. Not only was it the largest gas system in the United States, but it also delivered gas at the lowest price. Philadelphia's rate of $2.00 per 1,000 feet of gas was approached only by Pittsburgh, at $2.10. Elsewhere, rates ranged from Boston's $2.50 per 1,000 feet to New Haven's $4.00 to New Orleans's $4.50.[46]

And this cheap, publicly owned, and publicly accountable gas depended in large part for its cheapness on access to the slave-mined coals of the Richmond basin. In 1853, the Philadelphia Committee on Gas Works agreed to purchase from the coal dealer Jacob H. Hill several thousand tons of high-quality gas coals from Chesterfield, Virginia, at $5.85 per ton, and they were pleased enough that after a few months they extended their contract for an additional 3,000 tons, but this time at $6.12 per ton. They came to believe, however, that Hill may have been cheating them and accused him of sending them the cheaper Clover Hill coal, from near Charlottesville, instead of the better Chesterfield gas coals while charging the higher rates. His indignant response mapped the economic geographies supplying gasworks like Philadelphia's with coal in the antebellum era. "Now, as far as the kind of coal sent you is concerned," he wrote in protest, "I sent you the same kind that you and your Superintendent of Gas knew as the Chesterfield coal. You had always received it from me as such," implying that he was a longtime supplier of the city gasworks and that they had long been using Chesterfield coal. This coal was, he continued, "the best quality coal; was mined in Chesterfield County, Virginia, and I bought it from the Clover Hill Railroad Company," perhaps explaining the confusion. And getting such valuable gas coal wasn't cheap, he complained, as "the price I paid for it was the highest price I ever paid for any coal in Virginia, viz.: $3 65 per ton; now add the price of freight, $1 50 per ton, 25 cents for carting, and weighing 5 cents, you will find that coal cost me $5 45 delivered at your works."[47]

As Jacob Hill's dispute with Philadelphia officials revealed, the urban politics through which systems of gas, iron, and flame emerged in antebellum America attached a new set of questions to light—questions of cost, quality, and public good. While issues of labor, justice, and individual rights were reframed as selfishly irrelevant, light also increasingly became a problem of government. In 1844, a committee was formed in Boston to investigate gaslight and came "without difficulty" to the conclusion "that *some* deduction should be made in the price of Gas." Not everyone in the city agreed. As the committee was quick to point out, "an active opposition is now carried on by persons who deal in camphine and various compounds, which can be afforded at half the present price of Gas." Camphene merchants wanted to protect their control over the swelling working-class demand for cheap domestic illumination. They wanted government to keep out of the market. Camphene, however, did not cooperate with its vendors. The reports of deadly and spectacular camphene explosions peppering newspapers across the country provided a moral opening for advocates of gas. "That most persons would prefer Gas to these compounds, there is little doubt," the committee wrote, and so the real "question then is, what shall that deduction be?"

The answer many proposed to this question of safety and markets lay in publicly chartered but privately held gas corporations. "The risks of the Company are so great," the report proclaimed, that "no insurance can be effected against fire, and any accident to a Gasometer would be attended with great expense." If gaslight were to rescue poor women and children from the violence of camphene, the real dangers and risks of gaslight, they argued, would be shifted not to miner or gasworker (or consumer), but to capital. And so the committee was "unanimous in the opinion that an annual dividend, of at least ten per cent, to cover all risks, should be realized by their stockholders, and it is presumed that a liberal public would not consider any thing less, an adequate compensation for so important and extensive an enterprise." A liberal public—a liberal city—required cheap gas, and cheap gas, explained the Boston Gas Light Company, required "an association, a clubbing together of purses and of minds. That was the way the Western Railroad was built—that was the way the Gas Company was established—that is the way that every heavy manufacture must be undertaken." In this vision, gas companies were progressive heroes, for the "Gas Companies do cheapen that article of prime necessity—LIGHT—and not only so, but they bring it within the means of poor and humble men, who, but for gas, would be deprived of a large portion of the light they may now enjoy." These monopolists claimed to be on the side of equality, for it "is a Corporation alone, 'monster' though it be, that can give us gas—and however hideous it may appear . . . it bears a very benignant aspect to the poor man whose midnight toil is rendered cheerful by its light."[48]

Still, gas monopolies had to be manufactured and tolerated in real space, not just in newspapers, trials, and reports. Among the most hotly contested material technologies of gaslight infrastructure were the meters admitting and measuring the flow of gas into every shop, home, and office served by the gas company.[49] Meters not only allowed the exchange of gas to be automatic but replaced a social interaction with one between individuals and things. With self-operating meters, customers received the gas they wanted when they unstopped and lit a fixture, while gas company employees visited each meter (located outside the premises) every few months and recorded the quantity of gas consumed. These records were then centralized by company clerks, and a bill would be sent to the customer.

This type of metering did not involve any human interaction between producers and consumers. For this system to work, in order to legitimize these bills, trust between consumer and company would have to be built to overcome gulfs in social time and space. It was a messy, contested process. In the winter of 1847, Boston tailor William Gault contracted with the Boston Gas Light Company to have gaslight provided to his shop on Washington Street. A gas fitter hired by the company began the installation by drawing tin piping from the gas main running under the street into Gault's tailoring shop. This piping was arranged to run through a gas meter placed in the cellar under the shop, where a company employee could inspect it occasionally. Each evening around half-past four to five o'clock, Otis Foster, one of Gault's apprentice cutters, lit up the six gas burners fitted around the store and kept them lit for six hours before extinguishing the flames at ten to half-past ten. "There were six burners in the store," Foster later testified, "and I think that they were never all lighted at one time. We sometimes lighted three and sometimes four. We had two in the window, two in the centre, and two at the back of the shop. Sometimes we lighted none at the back end. When business required a light in the cutting room, we lighted a fourth." Carefully regulating light as people moved and worked through the shop, Foster was responsible for efficiently and economically deploying illumination over processes of labor and exchange. "It was my duty to light and take care of the gas," Foster explained. He kept an eye on the adjustable height of the flame in the globe lamps, as his "attention was directed to it by Mr. Gault, and I took particular pains to economize, and to see how much gas was used." After a few months, a bill was sent to Gault for the gas used from October 1, 1847, to January 1, 1848, which the company expected him to pay. He had different ideas. Gault believed he had been overcharged, that his meter was inherently suspect. And if what the meter said he owed might not be the case, well then, Gault felt there was no good reason he should have to pay anything at all. He said as much. Understandably, the gas company disagreed.

In the ensuing trial, the company marshaled testimony from gas experts, scientists, and a host of employees involved in the making, delivering, and bill-collecting of Gault's gas. George Slater took account of the meters in Gault's neighborhood, at a salary of $12 a week, then transferred his account books to the company clerk, who entered the meter accounts for all customers. Henry Davis then delivered a bill to Gault and tried to collect. Nathaniel Turner was the gas fitter. Richard Hodson was the meter maker, and William Lawler was in charge of proving the meters. Finally, John Blake superintended seventy Irish workmen tending dozens of benches, loading scores of retorts with Nova Scotian and English coals, to manufacture the gas delivered to Gault and hundreds of other Bostonians.

Gault's defense attorney, meanwhile, paraded an array of small and medium-sized business owners — theater managers, hatters, jewelers, machinists — to challenge the notion that the gas company, despite its scientific pretensions, could ever fairly or accurately deliver gas and charge its customers. This, the attorney argued, was "a question of immeasurable value to my client, to all consumers of gas in the city, to you, gentlemen, and to the citizens and public in general." Few were ready to cede to gas manufacturers the sole authority and power to charge for gas. This trial, Gault's attorney argued, was nothing more than an attempt by gasmen to cloak their unreliable meters in the guise of science.[50]

The company eventually won and then published the proceedings in triumph. Although Gault and similarly aggrieved customers lost, the trial provided a glimpse of a middle class self-consciously organizing itself around gaslight. As small business owners tried to grapple with the problems of metered gas, they formed a faction challenging the authority of corporations to determine the relations of exchange.

While business owners articulated a middle-class politics around gas meters, small property holders began to organize around a different aspect of the geography of coal gaslight: that of the gasworks and gasholders. In 1852, the mayor and aldermen of the city of Boston agreed to hear complaints against the Boston Gas Light Company from propertied residents concerning planned and existing gasholders and retort houses. Residents claimed that the manufacture and storage of real and imagined gas was physically poisoning their neighborhoods, marring them with soot, and lowering property values. Those close to the gasworks made the additional (for some, the principal) claim that the Irish laborers employed to work the coal yards and retort houses were destroying their communities and threatening the American character of the city. "No Gas manufactory can be established in any way and not be a nuisance," they complained, noting that in the South End, "the population has changed, the American families leaving and the lower class of Irish coming in and taking their places." The gasworks of the company, meanwhile, "must throw out its noxious effluvia and its smoke, and, as we can show, be productive of great danger to the passers-by."[51]

The city government responded to these private complaints by pitting them against a vaguely defined public interest. Not only should a gas corporation be granted special privileges, the aldermen argued, but the needs of these systems of gas, iron, and flame should take priority over private property rights — and the biological rights of individual citizens living near the gasworks not to inhale soot, fumes, and poisons — if the gasworks could be made to better and more broadly serve all the "people" of Boston. To that end, the city clarified, "reservoirs of gas are indispensably necessary, and the

Gasworks and gasometers. The massive structures producing and storing gas stood as monuments to the modern age, their scale and operations celebrated by publications like this one. Engraving in *Harper's New Monthly Magazine*, December 1862, 14–15.

interests of the consumers of gas are united with those of the manufacturers, in having those reservoirs at the points most suitable for distribution." A truly liberal public city would be made, then, by enshrining a relationship between a corporation and the people, unimpeded by selfish individual businessmen or property holders. And unlike in Philadelphia, politicians in Boston and every other major city in the United States chose to keep that corporation in private hands and were content with delegating administration of a public good to capitalists who were in it for profit. This choice sometimes required justification. "It is not therefore with an eye to their interests alone, but with a comprehensive regard to the accommodation of the public," the report thus concluded, that there should be "a large and increasing body of persons who depend on the supply of gas for lighting their houses, stores and shops."[52]

Middle-class petitioners, city officials, and corporation owners had ulti-

Dungeons and Dragons and Gaslights

mately converged on largely common ground. They disagreed about the particulars of exactly when, where, and how gas would be made, delivered, and charged, but all agreed that these systems should be built, that governments should play some regulatory role, and that fair terms of measurement and exchange should be established. But it was not just aboveground and in the marketplace that Boston's gasworks endangered the citizens they were supposedly meant to serve. Nor were the fears of citizens restricted to the slow degrading of their neighborhoods through industrial waste and the character of gasworkers. At least as acutely, the threat of sudden fiery violence haunted the cities where gaslights were made.

In Boston, on the morning of February 21, 1852, "the clerk in Simon M. Gove's furnishing store, 8 Hanover street, discovered that the gas in the cellar was escaping rapidly, filling the shop with its odor." There was little he could do, and to make decisions even more difficult, by the time he noticed the leak, the gas had likely already affected his thinking. Hoping to locate and maybe even stop the leak in the cellar, the clerk "immediately descended with a lighted lamp . . . when the gas took fire and exploded with a very loud report, completely shattering the large bow window in the front of the store, and to some extent the wood work around it." Miraculously, the clerk survived as the "cellar was instantly enveloped in flame," although his "eyebrows were burnt off, also some of his hair, and his face and hands considerably scorched." Not surprisingly, "the explosion nearly stunned him, besides the danger to which he was exposed from the stifling atmosphere." Poisoned, burned, but somehow still alive, the unnamed clerk was lucky to escape. As fire brigades quickly extinguished the fire without further damage, the store owner and his neighbors must have counted themselves lucky as well. But that didn't mean they were naive. The *Evening Transcript* pointedly suggested that the "frequent repetition of similar accidents, which we have to notice, leads to the question whether there is not some dangerous defect in the prevailing mode of introducing gas fixtures into buildings, which ought to be inquired into." The explosion on Hanover Street was explicitly cited in the formal petitions made later that year.[53]

This danger was an issue beyond Boston. The unintended and catastrophic sparking of gas fires, which continued to ravage and enforce discipline in the mines, became violently entangled in parallel processes in cities across the United States. That same year, a gas main running under the street of a former New York City alderman burst, pouring gas into the ground all around it and, unknown to him, filling the ex-alderman's coal cellar. He decided he wanted some coal brought up, so "he requested the servant girl to go to the vault, and fill the scuttle." He may have suspected some danger, and letting the "servant girl" lead, he "followed her to the door with a lighted camphene lamp, and the instant the unfortunate woman opened the door, the gas ignited from the blaze of the lamp, and she fell upon the flagging." The girl died from her wounds, while the alderman, despite his attempt to distance himself from danger, survived with disfiguring burns. Two years later, "a little apprentice boy" arrived at the building of a New York book and stationery dealer to "the strong smell of gas." As with the clerk and the servant, it was the apprentice boy's job to proceed "with a lighted lamp into the basement, and upon bringing the light near the end of an inch-pipe which had been carelessly left uncapped, a terrible explosion occurred, the shock and report being felt and heard at a distance of two blocks." Immediately, the stairway leading back to ground level "was entirely demolished, and portions of the floorings and ceilings of the first and second stories torn away and scattered in every direction." Windows were shattered and fire ensued,

Dungeons and Dragons and Gaslights

but again, remarkably, no one died. Soon after, the same occurred at a New York paper warehouse while a gas fitter was inspecting a leak with "a lighted candle." The resulting explosion "blew up the first floor, tore down the ceiling of the second story, broke the windows, shattered the rear office wall," while a young employee "was blown through a window and landed on the sidewalk, but escaped unhurt."[54]

As people experienced these explosions directly and indirectly through nationally circulated news reports, they were violently reminded of the price of living with the gasworks colonizing their cities. And though many suffered, it should be noted that in each case it was the servant, the clerk, the apprentice, or the laborer who was forced to assume the most proximate and terrible risks of gaslight. Combined with the gendered displacement of camphene's dangers, the violence of gas lighting made the antebellum work of seeing through dark urban spaces as much into a vehicle for hierarchy as into a rallying cry for public regulation.

This gaslit class struggle extended to shop floors as well. Large establishments with considerable illumination needs often preferred to manufacture gas themselves, rather than rely on the city gasworks. "We burn on average 100 Batwing burners, costing us less than one half cent per hour, for each burner," the owner of a machine shop wrote approvingly to the makers of a self-contained gas system. If companies with lighting requirements on the scale of factories could afford the investment in their own gas system, not only was it far cheaper in the long run than using oil or camphene lamps, but the "day watchman makes the gas without assistance and with no addition to his regular pay." During the day, a single worker could easily (and at no additional cost) make all the gas the machine shop would need to fuel 100 lights for a few hours in the evening. These systems were also popular in cotton mills. Brighter than oil lamps, the light from gas burners was easier on the eyes of mill workers in carding, weaving, dressing, and spinning rooms. And because the gas fixtures could be permanently and expertly placed and wouldn't spill and start fires, they afforded managers a far greater degree of control over the lighting and dangers of the mills by distancing laborers from the work and means of light.[55]

Together with the gaslights installed in the shops and offices described in the Gault trial, use of gas solidified employers' control over the length of the workday and so forced servants, clerks, apprentices, and factory hands to relax and replenish themselves at later and darker times; employers used gas to push the exhausted working class into the deadly domestic embrace of piney light. To protect their homes, property, and selves from the poor and uncouth, and to create a night space for respectable (and lucrative) night activities like shopping, dining, and attending theater performances, middle-class urbanites relied on bright streets like Broadway (the "Great White Way") in New York.[56] In antebellum cities both north and south, gas was the means of law, order, modernity, and commerce. Without gas, city elites feared rioting and a halt to work, exchange, and profit.

New York writer George Foster described the front lines of this contested geography of light and darkness, portraying gas lamps as bourgeois forts radiating liberal power into hostile frontiers. As with sperm-oil lamps, the policing power of gaslight was greatly exaggerated, but that didn't stop middle-class Americans from demanding more or projecting their hopes and anxieties onto gas streetlamps. Foster painted a midnight scene in the center of the Five Points, New York's most infamous urban frontier, where over "our heads is a large gas-lamp, which throws a strong light for some distance around, over the scene where once complete darkness furnished almost absolute security and escape to the pursued thief and felon, familiar with every step and knowing the exits and entrances to every house." Before that lamp had breached the night, he claimed, in "those days an officer, even with the best intentions, was often baffled at the very moment when he thought he had his victim most secure. Some unexpected cellar-door, or some silent-sliding panel, would suddenly receive the fugitive and thwart the keenest pursuit." But proponents believed that gaslights would break open the Five Points to the forces of law and order, and so "the large lamp is kept constantly lighted, and a policeman stands ever sentinel to see that it is not extinguished. The existence of this single lamp has greatly improved the character of the whole location and increased the safety of going through the Points at night."[57]

Elsewhere, the liberal city was ready to sacrifice the laboring city to save its gasworks. In May 1849, the city of New Orleans suffered a terrible flood, threatening not only the lives of thousands of poor and working families but the city gasworks. "The water is pouring over the right bank of the canal in one almost unbroken sheet from the Basin to the toll gate," the *Daily Picayune* reported, panicking "the residents in rear of the 7th Ward, who are seriously threatened by the encroachments of the flood." This unstoppable rising tide had "already inundated Gravier, Common and some other streets in rear of the Gas Works, and as it rises with considerable rapidity, we are fearful that much damage will be done in this vicinity." As the flood continued, *Picayune* writers marveled that there was "a fair chance that we are not to be left entirely in the dark, although we should be inundated. Col. Campbell is

Dungeons and Dragons and Gaslights

protecting the Gas Works by levees, and, although now surrounded with the waters, has rigged a steam pump which would keep the yard free, even if his levee were broken and the yard submerged." Streets were flooded, but city officials were "in contemplation to establish a line of packets between the Gas Works and high-water mark in Canal street"; the energy and hopes of New Orleans's elites were desperately aimed at trying to keep their gasworks from being swallowed in water even as residents drowned or were driven from their homes.

This campaign to protect the gasworks, and only the gasworks, from flooding was, as even the *Picayune* acknowledged, more a matter of priorities than possibilities. "A similar levee not more than five times as long," they surmised, "extending from the New Basin, would have protected many long streets from overflow and had the levee of the New Canal been strengthened in time, the loss from inundation throughout the city" would have been negligible "compared with what it now is." And such would have been entirely feasible, as "the Gas Works levee has employed the labor of less than fifty men," with the result that "[we] have much faith now that we are not to be deprived of gas light." With pumps furiously keeping the gasworks dry though completely surrounded by rising waters, the "danger from excessive rains is so feared at the Gas Works that another inside levee is to be constructed around the retort-house to insure as far as possible the requisite supply of gas." But should the rains and waters continue, "it is not beyond the possibilities that we shall yet be left completely in the dark, and the oil and candle dealers succeed in their much talked of speculation."

Apparently Col. Campbell and the enslaved gasworkers kept the flood at bay. Over three weeks later, the *Picayune* writers visited the still-islanded gasworks. Traveling by barge through the flooded sections of the city, they arrived at the gasworks, which had "the appearance of an entrenched castle, an embankment of earth of about four feet having been thrown up all around it." This was, they observed, "the only dry spot we know in the whole inundated district. It presents a singular aspect, the grounds of the work being about three feet below the surface of the water." After describing the heroic efforts of steam pumps, ditches, and trenches, the writers cheekily concluded, "We have thrown all the light we can upon the subject, and think our citizens need have no fears of being 'left in the dark.'" The city's poor neighborhoods, in good times blackened and poisoned by the incessant manufacture of gas, had now been left to drown while the gasworks lighting the wealthy quarters had been saved. Rallied to by New Orleans's leaders, the gasworks continued to enslave, it continued to make gas, and it continued to keep the city from falling into "that very ancient state of things 'when darkness was upon the face of the deep.'"[58]

G ASLIGHT was a process of displacement, arrest, and pursuit. In this tangle of push and pull, capitalists and middle classes made and negotiated new alliances in the gaslit cores of cities by first displacing labor, violence, and accumulation out of class, out of homes, and out of cities, and underground into enslaving coal dungeons. Meanwhile, through these coal dungeons and in southern gasworks, slaveholders were constructing an alternative vision of social and technological progress founded in industrial slavery. These internally related but divergent gaslit projects also enabled the rigidly controlled division of space and time into labor and leisure (and unpaid work) in both "free" and slave cities. Nor was the easy compatibility of slavery and engineering only a southern phenomenon. In Boston, a City Council committee tasked with investigating alleged overcharges by the local gas monopoly relied on the ledgers of a prison gasworks—worked for free by prisoners—to establish the "true" cost of manufacturing gas.[59]

As cotton spindles spun sperm oil away from lamps and into lubrication in the decades before the Civil War, there occurred a much forgotten invasion of the continent, an invasion of explosive lights. Much of antebellum American history stands as an unsung monument to this invasion, to the attempts to feed and control the lights manufactured from coal mines and turpentine camps, settling in cities as camphene lamps and massive gasworks. These lights were unlike the kinds of lamps, candles, and fires that for centuries had coevolved with the largely wooden environments of American cities. Where a candle might tip and set fire to a curtain, a camphene lamp might explode like a grenade, drenching anything or anyone nearby in liquid fire. When a gas main or gasholder exploded, it was more like the detonation of a powder magazine or a direct assault by a battalion of cannons. Yet these lights' very instability was as much a source of power as it was a rallying cry for their regulation or eradication. The struggles among producers and consumers, enslaved woodsmen and camp operators, miners and bosses, and laboring and propertied interests were struggles over the meanings of modern lights. Their attempts to navigate the explosive and illuminated (and the explosively illuminated) spaces of mines, camps, lamps, and gasworks materially circumscribed the politics of making cities and defining the "public good" in antebellum America.

But the politics of antebellum lights drew and challenged cleavages among more than just humans. The new chemistry of sulfuric acid opened vast new frontiers of fat, bone, and guano to the valuation of capital and the manufacturing of modern candles and phosphorus matches. As human struggles among children and capitalists, enslavers and enslaved, and farmers and manufacturers became entangled in this new chemistry, they also conscripted millions of captive hogs in the Ohio Valley, cattle in the Pampas, and birds on guano islands into battles over life, death, work, and profit.

Chapter Four

LARD LIGHTS AND THE PIGPEN ARCHIPELAGO

"Hark to the haste of pattering feet, / That splash through the mud of the slippery street." The annual march of hundreds of thousands of hogs from Ohio Valley farms to Cincinnati, the world's leader in pork packing, was enough to inspire poetry. "Here—gathered from the fruitful cornfields of Ohio, Indiana, and Kentucky, where their lives have hitherto passed in blissful ease," one Cincinnati observer rhapsodized in 1841, "comes a drove, staggering under the weight of their accumulations, to shed, like true patriots, their blood for the good of their country."[1] From November to February, a mere ninety-day naturally refrigerated window, farmers and drovers in a 300-mile radius forced up to half a million hogs into Cincinnati by rail, steam, and hoof. There, the hogs were rapidly consumed in the city's astonishing system of mass death, now known widely as the disassembly line, emerging as lard and pork.

In most places and times that lard would have been of limited value, but by the 1840s Cincinnati disgorged so much lard so fast in such concentrations that those who could get their hands on enough of it could realize economies of scale that made by-product industries like candles not only possible but enormously profitable. What set Cincinnati apart from other nineteenth-century slaughter zones such as Buenos Aires and the Russian steppe was the fact that, instead of just rendering hogs into mobile producers of pork, people marched them into Cincinnati deathworks as captive producers of the means of light. Their living labor and forced concentration made possible the creation of profitable pens, slaughterhouses, packinghouses, and steam-powered factories for rendering lard oil and candles, making Cincinnati one of the world's leading centers transforming fat into the means of illumination, producing in 1851 more than twice as many pounds of candles as Nantucket did at its peak.[2]

Like scores of travelers and scholars, Frederick Law Olmsted was drawn to this spectacle of carnage. "We entered an immense low-ceiled room and followed a vista of dead swine, upon their backs, their paws stretching mutely toward heaven," those fallen "patriots" who had, by this point, already shed their blood. Beyond the "vanishing point" of dead hogs lay what Olmsted could only describe as "a sort of human chopping machine." This was what made those deaths into something meaningful, something valuable. "A plank table, two men to lift and turn, two to wield the cleavers, were its component parts," Olmsted wrote, impressed by the simple, mechanical elegance of the human violence: "No iron cog-wheels could work with more regular motion. Plump falls the hog upon the table, chop, chop; chop, chop; chop, chop, fall the cleaver. All is over. But, before you can say so, plump, chop, chop; chop, chop; chop, chop, sounds again. There is no pause for admiration." Each "human chopping machine" could consume, in a single working day, up to 850 hogs, or 170,000 pounds of pork.[3] And Cincinnati had scores of these "machines."

But the processes and struggles that made wintertime Cincinnati into one of the world's most productive dealers of death and light were never wholly, or even primarily, human dramas. The hogs were more than simply objects of human labor. They were *themselves* living labor. It is not enough to say that pork was an important commodity in greater Cincinnati. Living hogs and hogwork—that is, hogs as both product and labor—were inextricably entangled in human struggles in the Ohio River Valley over property, class, white supremacy, slavery, and the power to determine when, where, and how to turn those living hogs into dead pork, and that dead pork into oil and candles.

In and around fields, markets, hog trails, and hog pens, men and boys managed and oversaw the lifework of countless hogs, making the captive and semicaptive animals immediately responsible for the work processes of, first, turning corn and acorn mast into pork, lard, and manure; second, making more hogs; and, third, moving themselves on foot over field, stream, and road. The spectacular rise of the Cincinnati pork packers and candle manufacturers rested on a newly organized rural system of exploitation, one by which the owners and managers of living hogs squeezed the most biological work in the shortest possible time from their porcine property. This was a story of industrialization, but as with the production of whale oil, camphene, and even coal gas, the industrialization of lights made from fat happened in places not usually associated with industry. Cities captured most of the glory and most of the profits, too, but the industrialization of hog death and hog lights was a process dependent on changes as much in the countryside as in the city. Each year, farmers and drovers forcibly drew millions of piglets into adulthood across thousands of square miles of seasonally shifting rural terrain toward just such urban convergence points in time and space and death. Meatpacking centers like Cincinnati and Chicago were but the most visible nodes of these vast webs of work and energy, webs that the hogs and swineherds spun.

Recent scholarship has shown that the industrial revolution — usually understood to mean the rapid ascent of coal, steam, mills, iron, and rail — was just as much an animal-powered transformation as a mineral one. These studies have focused on horses, whose numbers and uses, from cities to roads to rail depots to warfare, expanded exponentially for over 100 years before peaking in the early twentieth century. The history of horsepower and steam power was not a case of past succumbing to future or of two competing systems, but of a complementary relationship. The application of animal power in the movement of goods and people increased over the nineteenth century not in spite of the expansion of rail, but because of it.[4]

What, then, about animals that were themselves sites of production, both producers and products? What should we make of the millions of hogs transformed into food and lights? In one sense, what many antebellum agricultural reformers said was true: farmers were manufacturers of pork. But looked at from another perspective, farmers and drovers were merely overseers of the real work of making pork and lard, which was done by the hogs themselves. It might be counterintuitive, but to see the full range of the human geography of pork and candles requires first centering hogs in the story as actors. Farmers built and moved pens to forcibly contain unruly hogs, but the captive hogs turned over and fertilized the penned-in fields. Farmers let hogs run loose to find food for themselves in forest commons or they purchased the raw materials of production (corn, feed, or range), but either way only the hogs transformed those materials into muscle, lard, and manure. As they passed through shifting seasonal landscapes, hogs made themselves and were made into everything from mobile pork factories to forced migrants to property; they might be the embodiment of centuries of breeding labor or a package of raw materials for production; hogs were trespassers and runaways, thieves and violent territorial bands; and sometimes they were all these things at once.[5]

As the hogs changed shape and meaning, they were put into new relations of time and space by the expanding construction of steamboats, canals, and rail lines reconfiguring the Ohio Valley from the 1820s. The new geography of transportation helped create the extraordinary reach and compression of humans and hogs in wintertime Cincinnati (a process then replicated in St. Louis and Chicago). And it was this space-time compression that so revolutionized the work of death and disassembly awing the onlookers, like Frederick Law Olmsted, who gathered in Cincinnati to bear witness to the birth of the human chopping machine. Not only did the death march of the hogs concentrate enormous stocks of life into a narrow window of time and space, but the hogs were made to do much of the work of moving themselves to market and, by staying alive, all of the work of preserving their flesh from decay (both tasks could have been accomplished by the paid labor of butchery, packing, salting, hauling, and carting). The drive to compress the products of hogs' biological work, accumulated over thousands of square miles of Ohio Valley land over twelve to eighteen months, into a three-month moment in the confines of Cincinnati made it possible for both capitalists and free male workers to plunder the hogs' living energy and matter for surplus value much more thoroughly than if their deaths had been stretched only over the times and spaces of their rural lives. Cincinnati capitalists captured so much of the surplus life and labor of the hogs that they were willing to pay farmers 7 to 10 cents more for a hog than packers in other cities. And the laborers who worked for the farmers and packers managed to siphon off some of that surplus too, in the form of higher wages.[6]

One of the most important formations to emerge in this extraordinarily productive (and destructive) geography of life and death and light was what I call the pigpen archipelago. Hogs in the antebellum Ohio Valley were born, raised, and marched toward death through spaces their captors increasingly circumscribed by constructing chains of wood-enclosed islands. From breeding pens to field pens to fattening pens to the pens on the ferries and railroads, at the "hog hotels" where droves rested and refueled, and in the massive pens surrounding slaughterhouses, the always contested movements of the hogs within and between the pens transformed the region. They made hog lights possible.[7] The pigpen archipelago was the outcome of a struggle among farmers, hogs, and herders in which wealthier farmers fighting to better realize and defend a fixed private property regime triumphed over a more migratory regime of swineherds and commons. With fence and violence, farmers transformed (parts of) a geography of hog roaming into an archipelago of forced labor pens extending from rural nurseries to urban killing floors. And it began in the countryside.

E ACH year, the negotiated, contested movement of hogs from farms, plantations, and woods stitched together this patchwork process of life across an oscillating social landscape. As in any agricultural environment organized around private property, Ohio Valley farmers faced continual challenges to raising and then transporting corn and hogs, organisms that, despite thousands of years of domestication, only ever imperfectly respected the boundaries and allegiances of social institutions like private property. The social lives of hogs were woven through and around four different cycles: their own biological cycles and that of corn; the seasonal cycles of rains, mud, freezes, and thaws; the price cycles of corn, whiskey, pork, and lard; and the cyclical movements of enslaved and free farmhands from fieldwork to postharvest urban employment. Ensuring that cycles of biology, ecology, transportation, and climate in the Ohio Valley all aligned in the right places and times with cycles of economy remained an ongoing challenge that demanded new configurations.[8] But while the precise shape and pattern of this annual patchwork was continually shifting, two poles remained firmly anchored and separated in the process, their bridging forming the challenge and substance of each year's hog trade. These were the spaces of life and of death.

At the ecological center of this geography of human and nonhuman labor lay a deceptively simple material relationship between soil and the gastrointestinal tracts of hogs. Everything else revolved around it. It cycled land into pork and shit, and shit into more land. It was the mushy, rooting, smelly labor process white farmers and herdsmen exploited to pursue a politics of white rural independence. It herded hogs through life and death and markets so that black enslaved cotton pickers could be fed, white workingmen employed, white women and families "provided for," cheap lights massproduced, and all while keeping industrial city joined to but separate from white yeoman country joined to but separate from slave plantation.

It was neither practical nor possible to fatten hogs all year or in all places. But nearly every farmer agreed that hogs should be kept continuously growing from birth to death. This meant that during the summer, farmers either allowed hogs to forage in forests and streams or had them feed on clover or grass in mobile pigpens. Fattening was confined to the autumn, usually a period no longer than six to eight weeks, and was almost exclusively an affair of corn (on the stalk, raw, cooked, mashed, distilled, or otherwise). It was a race against cold. Farmers strove to get their hogs to weight (between 250 and 300 pounds) by ten or twelve months of age. Threading a hog's life from sow to slaughterhouse without having to winter the animal meant cir-

Lard Lights and the Pigpen Archipelago

cumventing fat-sapping freezes, higher costs of fodder, and slowed rates of growth. Just as critical, the extra human labor necessary for accelerating, scaling up, and industrializing the fattening of hogs was not readily available until during and after harvest. Thus enterprising hog farmers felt even more pressure to make fattening time count.

First, however, the hogs had to be made. Hogs had large litters, twice a year, and gained weight faster and with less food than cattle. The work of sex and gestation was something the hogs did mostly on their own. Farmers could try to organize reproduction by penning animals together or by assisting during birth, but it was not work they could be directly involved in. Artificial insemination was generations in the future. In the nineteenth century, reproduction was work that farmers expected livestock to provide for free. What farmers could do was cull, assemble, and discipline a productive hog labor force from birth. According to one farmer, a hog raiser should start with piglets of "the right breed, and then pick out the good-natured ones from the litter; I can't afford to feed a cross critter; I sell them when they are pigs." Pigs that fought were "cross." Pigs that ate were "good-natured." For farmers, describing a pig as cross was an expression of the limits of mastery, where individual pigs rubbed up against farmers' expectations and demands. "Crossness" had everything to do with the core metabolic processes of the pigs that farmers hoped to exploit: "'How can you judge?' said I. 'Well, if you watch them when they are feeding, you will find that some pigs are allers fighting about their victuals, and some go in for eating. There is as much difference in pigs as there is in folks.'"[9]

When farmers did try to directly assert control over hog reproduction, they focused their efforts on sows, attempting to align the behavior of sows with the interests of their owners. Sows, at least those living in captivity, frequently killed and ate their young, a problem common enough that the agricultural reformer Solon Robinson publicized and endorsed "an easy and sure prevention, 'to give the sow about half a pint of good rum or gin, which soon produces intoxication, and the drunken mother becomes entirely harmless toward her young.'" The main purpose of disciplining sows by getting them drunk was to keep them from interrupting the chain of nutrients that farmers hoped to usher through their property, from soil into corn into hogs (then back into soil as manure) over the hogs' lifetimes. And when hogs were still piglets, the only way to maintain this chain was to pass it first through a sow in the form of milk. The importance attached to this critical stage in the reproduction of hogs meant that farmers erected special structures for nursing, separating it spatially from the rest of the lifework.[10]

Starting with "good-natured critters" and keeping them alive was a critical first step, but as any farmer knew, productive hogs were not just born. They were trained and made. And so were hog overseers. "My own training in the business was of course progressive," the mid-nineteenth-century Tennessee hog farmer Edmund Cody Burnett recalled; it "began when, as a child, I gleefully watched the little pigs get their meals from mammy sow." If the piglets survived weaning, farmers were quick to transform them into private property. The young Burnett first witnessed this violent process "when a terrible squealing at the barn drew my curiosity thither, where, through a crack, I perceived that something fearful was being done to the pigs with knives and needles." He watched as men sheared the young pigs from their tails, probably castrated them to reduce "crossness," and then inscribed property relations into their flesh: "With hogs belonging to so many different people running loose, it was necessary that they be marked. My father had, I think, the simplest mark in our entire valley, which was a smooth crop off the right ear." To keep track of their property, hog overseers had to learn an entire visual grammar of mutilation. As Burnett recalled, other "marks that I became acquainted with were: half-crop, swallow-fork, underbit, overbit, hole, slit, in one ear or both ears, singly or doubled, in almost any possible combination."[11]

Farmers began raising in the spring, after weaning, and while practices differed across regions and from farmer to farmer, raising often meant they expected the hogs to gather, consume, and excrete their food on their own in either field or forest. This saved labor and even helped to close the metabolic loop between soil and animals, leaving pigs to fertilize the land that nourished them. As *Harper's Weekly* described in 1860, hogs were "raised, as the term is in that part of the world, all over the Western States; though there are many, in several convenient localities, who make the business a specialty, and breed the animal in large quantities. In some places they run at large in the woods, feeding abundantly, and fattening rapidly, on 'mast,' the beech-nuts, hickory-nuts, and acorns that abound in the forests of the luxuriant West. Thousands are confined in pens and yards and fed on fodder and corn; and immense numbers crowd the pens attached to the larger distilleries, fattening and corrupting, as is too often the case, on the warm slops from the stills and mash-tubs."[12] Even the distillery-fed hogs were basically fed for "free," as distillers had to do something with the waste from making corn whiskey. Such hogs acted as organic recycling machines, turning that waste into manure (the better to fertilize more corn), pork, and lard.

It may not have been obvious, but the bodies of hogs were being shaped by the new chemistry of candle making. Formerly, hog lard had been fairly useless to either candle or soap manufacturers, who had relied overwhelmingly on cattle and sheep tallow. Lard wasn't hard enough, it smoked and stunk when it burned, and it easily became rancid. The only valuable lard had been the high-quality edible kind. But in the 1840s, candle and soap manufacturers began processing a wide range of fats with steam, lye, and sulfuric acid to produce a hard, clean, stable, uniform substance they called stearine. The industrial chemistry of sulfuric acid made a hog's fat newly valuable, and it was the availability and quantity of the fat, no longer its quality, that mattered most. Fattening hogs on the waste from whiskey distillers suddenly became a common practice that only fully made sense in the new internal relations of commodification that the chemistry of stearine formed between hogs and candles. Such mash fattening almost certainly made the hogs' edible meat worth less (if not worthless), but a sickly, fat pig fed on "corrupting" slop was, precisely because it was cheaper, at least as attractive to a candlemaker—who might just throw a whole hog into a steam vat—as a tasty, healthy, meaty hog. This same chemistry also explained why, half a world away, tallow produced on the Russian steppe that had to travel months slowly spoiling before even reaching St. Petersburg was suddenly so sought-after by British soap and candle manufacturers. The cheaper and lower-quality the fat the better, so long as there was lots of it, because once processed, stearine was stearine no matter the source.[13]

Although most hog *death* in the region ultimately converged in Cincinnati and other cities along the Ohio River, the spatial politics of hog *life* differed based on what side of the river that life began on. Most hog raisers south of the Ohio River let their young pigs run free through common fields, woods, and creeks to feed on grasses, nut mast, and crawfish. Plantations always raised some hogs, but the vast majority of hogs in the South were raised by yeoman farmers. They were the "poor" whites who enslaved no more than a few (if any) people, imagined themselves culturally distinct from both large planters and Ohio free-soilers, measured their freedom and superiority against the domination of black slaves and "wage slaves," and secured their economic "independence" by owning and selling goods and services to the plantation economy. As such, their hog commons were implicitly racialized. "Unfenced territory was free to all comers—provided they came on four legs," the Tennessee hog raiser Edmund Cody Burnett wrote ominously. Hogs may have been free to range over white southern commons, but strangers and black people were another matter.[14]

In the agricultural regions north of the Ohio River, especially in the corn and livestock centers along the Miami and Scioto Rivers, a more complete and enclosed private property regime led to different spatial politics and demanded other forms of hog discipline. "As a general rule, our domestic animals are never unruly, except when taught to be so," the agricultural reformer Solon Robinson contended. Complaining that too many lazy farmers, when "turning stock from one field to another, only let down a few of the top rails or bars and force the animals to jump over," Robinson feared that hogs were being taught to disregard, and even destroy, the barriers and divisions of private property, the fenced landscapes that farmers relied on to control and exploit their hogs. To discipline and educate animals that had to be free to find fodder, but not so free as to escape their condition as private property, one "writer says his practice has always been to teach his . . . hogs to go through or under, rather than over, the bars or fences, always leaving a rail or bar up at the top. Taught this way, they never think of jumping, and he has never been troubled with unruly animals, even when his fences were low."[15]

But many farmers went a step further and sought to multiply their gains by directing hog lifework into transforming their land with more than manure. One enslaver, frustrated by the relentless advance of "the despised wire or joint grass" over his cotton fields, sought to deploy his hogs against this counterplantation pest. With the wire grass's thick root clusters grabbing hold of soil meant for cotton and choking out any new plantings, he "concluded to fasten hogs up in the field without any other food, to see if they could live upon it, and in some degree destroy it, or at least thin it, so as to render the land fit for cultivation. The hogs were put in in February, 1840, when very poor. Result, in four weeks: they were in order fit for pork, and had rooted the field where the grass grew." Using railings and fences to contain and concentrate the lifework of his hogs onto particular parts of his land, this enslaver was able to keep land suitable for cotton and raise his hogs on food he didn't have to force people to plant.[16] Put another way, he turned his hogs' will to life into labor he neither had to pay for nor coerce from the people he enslaved: "free" labor that was procotton, proslavery, and propork all at once.

Moveable fences were critical technologies for guiding this hog work over land and season. Once the ears of field corn had ripened sufficiently, if "a moveable fence is provided to confine hogs to a small quantity, little is lost by field feeding," one farmer noted. The moveable pigpen assembled hogs, soil, and plants into some of the most effective ecological agents of agricultural space in the Ohio Valley. Advising his readers of ways to "make the pig-pen valuable," Solon Robinson suggested using hogs to de-grub and improve grassland: "Fence off a piece, and shut your swine in upon it for a few days without feed, and if they leave a sod unturned or grub uneaten it will be a wonder. It is the best preparation of such a spot for a hoed crop, or for sowing again in grass, that can be given. There is no good reason why the pig should be always kept in idleness and mischief. Let him be trained to be useful in his life as well as at his death."[17]

Yet neither hogs nor their owners came easily to respect fences as absolute barriers. Enclosure in the West was a gradual process, with some fiercely resisting the elimination of the hog commons, while others pressed hard against letting any captive animals run "at large." "I hope the voters of Scott county come up to the polls in April, and vote for the law prohibiting hogs from running at large," wrote one Iowa farmer who hoped to legislate the region into modern political economy, noting, "It is now the law of Clinton county, and will be that of Cedar and Jones after next July. It is the law of more than one-half of the State of Pennsylvania, why not then be the law here?"[18]

In the prairies and pastures west of the Appalachians, where wood and free labor were relatively scarce and settlements more spread out, hogs seemed especially empowered to subvert fenced-in landscapes. Describing his training in the hog raising business in Tennessee, Edmund Cody Burnett recalled that "when nimbleness of feet and legs was called for, as when the hogs had broken into a forbidden cornfield, it was I, the small boy, who was all too often assigned the task of chasing them out. Then I began to dislike hogs." With corn increasingly colonizing the landscape, hogs had even less cause for staying close to their captors in order to find food. And their captors had equally little incentive to control them, undoubtedly recognizing that if their hogs escaped their pens and ate their neighbor's corn, it was, for the owner of the hogs, a subsidy akin to that of the commons. Controlling access of one's hogs to corn, both within one's own land and across property boundaries, emerged as a critical problem in the hog-corn belt. Forcing hogs to adhere to the rules and expectations of this spatial regime was part of what farmers meant when they said "hogs should be kept as gentle and tame as possible."[19]

But keeping hogs gentle was no easy task. It was "customary with some farmers, if a hog don't exactly please them," one reformer claimed, to "set a dog on them, and . . . literally amputate their ears; but this, in general is a very bad practice." Instead of torturing and terrorizing hogs, enlightened commentators like this farmer recommended an olfactory discipline, one designed to confuse, disgust, and haunt hogs: when they "become troublesome about getting into the corn field, and waste corn, a good method to keep them out is literally to soak" the hogs' sides "with a mixture of bran shorts, clabber [spoiled, curdled milk] and buttermilk, this applied to them daily in the above manner, will soon enable any farmer to keep them out of mischief." Mischief, it seemed, was the word of choice. "Hogs running at large are always in mischief," wrote another reformer, who was "satisfied that, for the last five years, there has been more destroyed by hogs than all the exports would amount to of pork from Scott county for the same length of time."[20]

THE race to weight accelerated during the summer, propelling hundreds of thousands of hogs each year into migratory movement through the uneven geography of life in the Ohio Valley. The movement of hogs through these archipelagic chains of pens produced a spatial and class division between two types of farmers called, respectively, "growers" and "fatteners." Sometimes growers were merely the smaller, poorer neighbors of large farmers who controlled access to better land and equipment, but just as often, growers operated many miles away from the nearest fattener. Fatteners, on the other hand, tended to cluster near the rivers, canals, and railroads that fed into the pork-packing centers. What all fatteners had was corn. And fall was the time for fattening.[21]

The majority of fatteners arranged the work and spaces of fattening in large fenced-in pens called feedlots. Feedlot owners corralled hundreds of hogs in muddy, tramped-down fields, where workers prepared and carted out vegetables and dried corn periodically to feed to the animals. Feeding required skill and training on the part of both humans and hogs. A visitor to one feedlot south of Columbus watched as farmers continued to train and discipline their hogs in the fattening process, particularly in the sonic conditioning that would be so critical for maintaining control over the hogs during the final drive to market. "In wheels to the hog pasture, a great heavy Dutch wagon with four stout horses," the amused traveler wrote, "the driver astride on the near hind one, coolly whistling some animating air and keeping time with the flourishing of his whip in loud pistol cracks, while another genius, standing on top of the load, commences pitching it to the right and left, stopping and standing up now and then to give the long drawn roll-call, at the top of his voice, of whoo-oo-hoo, or perhaps more poetically from a horn slung at his side, he draws forth a clear tremulous blast that rouses the whole grunting field from their recumbent positions and sets them on the move."[22] Feedlots, the most common mode of fattening in the antebellum Ohio Valley, were labor-intensive projects designed to produce conditioned hogs and

Lard Lights and the Pigpen Archipelago

a managed chaos within spaces circumscribed by pens, whips, and sounds. But many farmers thought there were better ways to fatten their animals.

According to reformers, the ideal fattening camp would function somewhere between a factory and a plantation; they hoped to industrialize the pigpen. The reformer and publisher Thomas Affleck, in his mission to empower, enrich, and modernize western farmers (before his later mission to empower, enrich, and modernize southern slaveholders), routinely visited and published descriptions of hog farms near Cincinnati that he felt conformed to the ideal. "Mr. M. has gone to work in the right way," Affleck wrote, "beginning with a good barn, good fences and good roads—his barn and stables, hog pens, &c., are rather close to the dwelling house to please the taste of many, but not too much so where the farmer intends that every thing shall be well attended to, under his own eye." Mr. M. had arranged his hog camp within lines of sight, but separate from the sounds and smells of his human domain. And as Mr. M's hogs worked to transform corn into flesh, not only could he easily surveil them, but "every hog can be put in a separate pen, if necessary . . . with a passage along the whole front of them direct from the cutting and steaming house, in which are two large set boilers, with hogsheads for souring food for the hogs, cooling troughs, &c." Here was an industrialized practice and vision of hog farming: corn carefully prepared for a captive hog labor force made to focus all its energies on the singular task of eating. "Hogs to fatten best should not know what liberty is," the *Prairie Farmer* advised, and "they should have a warm dry bed—their feed at regular hours, and in sufficient quantities. As soon as the meal is over they then lie down and rest until the next feeding time comes round."[23]

Food alone was not enough. To press their hogs into rapidly producing weight, farmers tried to construct pens that allowed them to better manipulate the hogs' passage through season and weather. "In the first place, *there must be a good piggery*," wrote one exasperated farmer for the *Prairie Farmer*. Too many farmers, reformers believed, thought only of fodder. They forgot that their hogs were living animals vulnerable to the elements, and "many have hogs that are continually *scolding and crying*; not so much on account of being scantily fed, as for the want of a comfortable piggery." The tension was less between cruelty and kindness than in the contradiction between working hogs in life and preparing and accumulating value in hogs for death. As one farmer noted in a particularly clarifying example of the violence of economistic attitudes toward life, "An idle hog will make 12 pounds of pork as easily as it will make 8 pounds if the animal is allowed to exercise his natural propensity to root. In this we entirely agree, and have often contended that when a hog is shut up to fatten, if he was confined in a slip so narrow that he could not turn round, having one side of his narrow prison made so as to be moved out as he increased in bulk, he would fatten faster than in any other position." It was a prescient antebellum vision of a future that has become our present, a world in which the logic of holding unlimited property rights in the bodies of living labor was played out to its awful conclusion.[24]

But many fatteners, like raisers, spurned both feedlot and industrial pigpen. Instead of hiring men to harvest corn and paying a miller to grind it, they set their hogs loose into a cornfield to harvest and consume it on-site. Using moveable fencing to guide this work, farmers aligned several biological, seasonal, and economic cycles by field fattening, or "hogging down a cornfield." Remembering summer days spent overseeing hogs as they fattened themselves within the chestnut rails, Edmund Cody Burnett described how a big hog "would stride up to a standing stalk, put his nose against it, give it a gentle shake, and would know at once whether the stalk carried a light ear or none at all. If he felt weight up there, he would push hard with his nose, and if that did not bring down the stalk, he would rear up on his hind legs, put his fore feet against the stalk and push and shake it till it came down." Never far away, the smaller hogs followed the large as they shook and knocked down stalks. Burnett liked to imagine "that the little hog, unable to knock down a big heavy stalk by himself, felt grateful to his big brother, who having got his own bellyful, moseyed off to the shade, leaving parts of ears unconsumed." Within the fences, cornstalk by cornstalk, ear by ear, hogs worked with and against one another in a struggle that rapidly transferred the energy of corn into the portable commodity form of living pork.[25]

It was a labor process that produced pure surplus value for the farmers who owned the means of lifework, and many sought to precisely measure this gain. In one "experiment," a moveable fence was used to confine 189 hogs, weighing an initial 19,600 pounds, "to an area sufficient to afford feed for two or three days." Just a few days later, the hogs weighed 30,340 pounds, had doubled in price, and had "improved" the land, "making the actual gain per acre $16 64, equal to 40 cents per bushel standing in the field."[26] As farmers made their hogs fatten themselves in the field, the hogs reproduced the very conditions for their captivity. Producing value both in their own flesh and in the land, these fenced-in hogs ate, shat, and maintained the means of their and future hogs' captivity in life, while drawing themselves, pound by pound, ever closer to the exchange that would end in their mass slaughter.

MOVING the hogs from farms and fatteners to slaughterhouses, from the spaces that had made them in life to the spaces that would unmake them in death, was a massive undertaking. The death march of the hogs meant the pigpen archipelago would have to travel. As the autumn harvest drew to a close in the antebellum Ohio Valley, hogs, men, and market rumors were set annually into motion across the landscape. "When we read that about 18,000 hogs were driven from Marion County in 1845, about 40,000 from the Chillicothe vicinity in 1847, and similar numbers from other defined areas," one historian noted, "we realize that getting the animals from place to place and from owner to owner must have furnished employment for a small army of drovers and helpers." Finished with the harvest, farmhands — enslaved and free — looked to follow the hogs they had helped raise on their journey to Cincinnati. In 1856, one Kentucky author described how each year "hundreds of hogs were purchased miles away, and taken to Porkopolis in droves, on foot. This was practiced to such an extent that hog driving in the Fall of the year became a regular business, and many were the farm hands who annually calculated on buying a new 'rig out' for the Winter, with the receipts of a 'drive' to Cincinnati behind a lot of porkers."[27]

Moving so many hogs required people to drive them, called drovers and drivers, and in the interstices of hog movement, some enslaved Kentuckians used the forced march of hogs from life to death to secure their own freedom. In 1856, a group of enslaved women and men from Boone County, Kentucky, were tried as fugitives in Cincinnati. They had been led by an enslaved man named Simon, who had, for several years, helped drive hogs into the city (and across the important political boundary between slavery and freedom of the Ohio River) with his enslaver. Over the course of his drives, Simon had made contacts with members of the sizeable free black community living in Cincinnati, some of whom he was connected to by family. In the winter of 1856, using slave passes previously issued to them and the death march of the hogs as a pretense for entering the city, Simon and six other enslaved Kentuckians escaped. Their trial became a flash point for the politics over the Fugitive Slave Law and the geography of freedom and slavery in Cincinnati, a city right on the border of the Ohio River and through which the enslaved, fugitives, free black people, and the products of both enslaved and waged labor regularly passed.[28]

Some followed hogs to freedom and others followed them to money, but first the hogs had to be concentrated into droves. Droves were assembled in the hog hinterlands of Cincinnati by two main processes. As the *Cincinnati Gazette* observed in 1843, in "Kentucky, the drovers frequently buy the hogs alive of the farmers by gross weight, as is sometimes the case in Ohio and Indiana. But generally the farmers club together (each one having his hogs marked) and drive them to market themselves in droves of 500 to 1,000, and seldom less than 500, except in the immediate vicinity of the city." Sometimes based on prior contracts, sometimes on speculation, drovers fanned out across the countryside purchasing or collecting hogs from smaller farmers (those with up to 100 hogs for sale), until the droves reached anywhere between 500 and 3,000 hogs. They also had to recruit enough free or enslaved men and boys to help on the drive, and this was rarely possible until the completion of the harvest released farmhands to find new jobs. As drovers made contracts with slaveholders and free laborers and finished gathering their droves, which could take a few weeks, they often constructed large pens in rented fields in the area. This way, drovers could afford to assemble their droves piecemeal, purchasing a supply of recently harvested corn to maintain already-fattened hogs at around 200 to 300 pounds until they had accumulated sufficient numbers, or until a change in the weather or market or labor availability made a drive more attractive. Another strategy was to combine droving with fattening. Some drovers bought up stock hogs of only 100 to 125 pounds and then "bought in some convenient locality in the district, a standing field of Corn, into which all the stock hogs purchased in the neighborhood are driven." Then "the hogs tear down the stalks and Corn and are thus self-fattened."[29]

With hundreds of thousands of hogs marching toward Cincinnati each year from a radius of 300 miles, drovers, hogs, and farmers came to produce a highly specialized geography of hog transportation. Like the better-documented hog and cattle trails passing east over the mountains, the Cincinnati trails were reproduced each autumn over a network of roads, ferries, and penned-in feed stands known as "hog hotels." The march itself was long, arduous, and unsure. A drove took many days, sometimes weeks. And as the pens where each day's marches began and ended were neither impermeable nor continuous, this migration formed the most vulnerable and contested branch of the pigpen archipelago.

For the hogs, the transition into a moving drove was abrupt and exhausting. Some of the "wilder ones" fought so hard to resist that drovers would stitch their eyelids shut before starting on the road. And even once the drove got marching, hogs continued to cause trouble by refusing to move or trying to wander off. "It is no light job to trudge over a muddy road, day after day," one Cincinnati writer recounted, "urging on the hogs continually with a whip or a switch, yelling 'so-boy' constantly at the top of one's voice, now running

Lard Lights and the Pigpen Archipelago

like all fury to head a spry hog, which has taken a notion to go the wrong road, and again helping a lazy fellow along, by wrapping his tail around one's hand, and giving him a boost with the knee and arm together."[30]

Through a combination of pain, whip cracks, and conditioned vocal commands, drovers attempted to manipulate the hogs' memories, training, fears, and instincts into a system of control over their movements from pen to pen. It was a struggle that could be heard for miles. "As soon as a drove came through the gap in the ridge about a half mile distant across the river," Edmund Cody Burnett recalled, "we could hear the 'ho-o-o-yuh! ho-o-o-yuh!' of the drivers, and sometimes we could hear the crack of their whips." One reason these disciplinary practices and relations relied so heavily on sound was because they had to extend past sundown, when drovers' visual power reached its limits. Forcing hogs from pen to pen was not only a day-time struggle; it was often a dash through the dark. Burnett remembered one night "when a drove was overtaken by darkness 2 miles or more from the ferry," and he "could hear an unending stream of 'ho-o-o-yuh! ho-o-o-yuh! ho-o-o-yuh!' mingled with the resounding crack of the whips, as the drivers sought to prod the weary hogs a little farther, a little farther, to where they could be lotted and fed." Pushing hogs through day to night, from station to station, from meal to meal, hog drivers knew that whether with voice or whip, in the passages between the pens, their work and power were products of sound. And while not "every driver took pride in his voice," according to Burnett, "every driver seemed to be proud of his whip." As he observed, a "skillful manipulator of the whip could not only make it talk with a resound-ing thwack but could produce a pretty good imitation of a clap of thunder. There is no room to doubt that a combination of the driver's voice and whip impelled a lazy sluggish hog to quicken his pace."[31]

"Lazy" hogs seemed to be the bane of a drover's existence. "That black, spotted critter raised by Squire Sidebottom, was so lazy that we had to wollop him every step. He's a plaguey ill-mannered hog, and I came mighty near being the death of him afore we got half way thar," one drover com-plained. With whip, prod, and shout, drovers prided themselves on getting the hogs up to speed. The first few days were the hardest, according to the *Cincinnati Gazette*, during which "the hogs cannot well travel more than four to six miles; but after that they travel eight and sometimes ten miles per day, depending upon the condition of the roads. The Yorkshire are said to be the best travellers." But conditioning hogs into better travelers also made them faster runners, and as droves sped up, they commonly lost hogs, sometimes to death, sometimes to injury, and sometimes to escape.[32]

At the stands where hogs congregated to prepare for and recuperate from the day's march, Burnett's childhood friend and hog drover had "known ten to twelve droves, containing from 300 to one or two thousand stop over night and feed at one of these stands or hotels." Knowing where hogs and drovers wanted or needed to move, and buying and building up the land, cornfields, pens, and human accommodations that made up a hog hotel, could bring a farmer a considerable and predictable stream of additional revenue. Burnett himself grew up on land purchased for that very reason, his grandfather having "acquired a considerable amount of land about the mouth of Big Creek some fifty-odd miles down the river from Asheville" in 1834 for the purpose of setting up a hog stand. By controlling the fuel by which hogs and drovers moved themselves from farm to slaughterhouse, hog stand owners used corn and pens to secure their own position and power in a geography transitioning hogs from life to death.

Because hogs could only travel four to ten miles in a day, these hog stands had to be carefully and densely staggered over drovers' routes. Hog stands were often also agricultural craft and trade centers, where farmers, travelers, craftsmen, and merchants would congregate and exchange company, work, and money. One of the more famous hog stands along the French Broad River contained "'a hotel, store, tanyard, shoe-shop, harness-shop, farm, blacksmith-shop, waggon-factory, grist mill, saw mill, ferry, and bridge.'" Such enormous hog stands were capable of housing and feeding thousands of hogs and dozens of men at a time. Burnett related a friend's story of how once, when he, his father, and their hogs stopped at one major stand, "there were as many as 10 separate droves there for that night. 'This,' he remarks, 'was about 4,000 hogs, with one man to a hundred hogs—40 men, with a manager for each drove of hogs—making a total of 50 men to find beds for.'" Fifty beds, and fifty human stomachs to fill. And it "'took a lot of food to feed 50 hungry men.'" Accounts were silent on who actually did the work of feeding all these men, but most of it was likely performed by the enslaved, standkeepers' wives and children, and other domestic workers the standkeepers claimed as dependents or property.

A steady and sizeable amount of food, labor, and money was needed to feed all the men passing through the hog hotels. Vastly more was required to feed the thousands of hogs passing over the trail during the autumn droving months. Applying Burnett's figures for the 150,000 hogs passing through Asheville to the 2 million hogs typically arriving for slaughter in cities along the Ohio River, Ohio Valley farmers would have had to raise around 60,000 acres of corn, or up to 300 million ears, simply to move the hogs from lifescapes to deathscapes. This was a lot of corn, and a lot of money. Not only were drovers dependable and desperate buyers for their corn, but the stand-keepers could capture some of the surplus value of the hogs' lifework by becoming corn merchants and overcharging drovers. Many standkeepers weren't even farmers. Rather, they set up in the center of a town along a popular road and assumed positions as middlemen between corn growers and hog drovers. They were both standkeepers and "storekeepers who for the most part obtained their supplies of corn by providing the farmers with goods. The price paid the farmer for corn was almost invariably 50 cents a bushel; the price charged the drovers was 75 cents."[33]

The established trails were also the product of decades of pursuing and fighting over food in an uneven but energy-rich terrain. In Kentucky, hog droves chased food across the landscape, just as hogs foraged in commons and across property lines during raising. "In Madison and other counties, mast and acorns are very scarce. It abounds, however, in the county of Estill," one article explained. Chasing this "free" energy, herdsmen from Madison drove many of their hogs through Estill County, "which the Estill people considered an infringement on their rights." Local citizen councils were con-vened to discuss this outrage and many plans were debated, until finally "one was adopted. It seems that hogs have great fears of bears. Accordingly the skin of a bear was procured, and a large sow was caught from one of the droves. She was covered with the bear skin and then let loose. She immedi-ately returned among the droves, but on her approach all the hogs took flight, pursued by the sow with the bear skin. It is stated that since this experiment not a hog has crossed the confines of Estill County."[34] The consolidation of hog trails around corn and feedlots was, to be sure, largely a process of geog-raphy and market relations. But it was also the result of insurgent marginal farmers, resistant hogs, and bearskin-clad terrorist sows.

If hog stands and corn circuits were the major energy bottlenecks in the geography of droving, ferries constituted some of the most important choke points in the migratory pigpen archipelago. Burnett, who grew up near such a ferry, vividly described the risky but unavoidable process of moving a drove of several hundred hogs across a river. Because of the considerable difference in size between a drove and most nineteenth-century ferryboats, this pro-cess was almost always further complicated by a requisite disassembly of the drove into groups of no more than fifty hogs and then its careful reassembly on the other side of the river.

Maintaining and projecting the same power relations between humans and hogs that farmers and drovers worked to construct in the terrestrial pens and on the trail was an especially pressing concern over open water. "For ferrying hogs or cattle movable railings were set up at each side of the ferry-boat with gates at each end, making an enclosed pen" and forming a floating extension of the pigpen archipelago. According to Burnett, the hogs were usually "perfectly content to remain quietly in the pen until the river was crossed, but now and then a hog, who had not been wholly subdued by his fattening-lot schooling or who was not sufficiently restrained by the load of fat he carried, would plunge over or through the railing into the river." Canoes were kept alongside the ferries or landings to capture such fugitives, and "two men would jump into the canoe and go after that hog with all possible speed. One man would grab him by an ear and hold his head above water, while the other managed the canoe. Once in a while a hog would be drowned before he could be rescued. Then a hurried butchering followed, and some of us would have backbones or spareribs for supper."[35]

Eventually, the droves arrived at the outer reaches of the urban land-scapes of death. When Frederick Law Olmsted departed south from Cincinnati, he and his fellow travelers had to pass through the tide of tens of thousands of hogs converging on the city from Ohio, Kentucky, and Indiana. "Our progress was much impeded by droves of hogs," he wrote of the herds "grunting their obstinate way towards Cincinnati and a market. Many of the droves were very extensive, filling the road from side to side for a long distance. Through this brute mass, our horses were obliged to wade slowly, assisted by lash and yells."[36] A diffuse geography of life was compressing itself in the time and space of the winter city, where an assemblage of ferries, depots, pens, and markets waited to absorb the droves and transition them into the spaces of death.

Railroads financed and dominated by Cincinnatians came to dominate the final leg of transportation for hogs moving through Indiana and Ohio during the 1840s and 1850s. These railroads constituted a geographic triumph that helped Cincinnati to capture a plurality of the more than 1 million hogs killed and disassembled by packers every year in the Ohio Valley over the 1840s. It was a spatial strategy so successful that, whether or not the total number of hogs living in the region increased over the 1850s, the number killed in packing centers doubled to an average of nearly 2.5 million. With the aid of railroads, industrial winter deathscapes were capturing exponentially more hog life.[37]

The thousands of hogs delivered to Cincinnati every day by rail made the depots into spaces of frenetic activity. "The arrival at one of the principal

Lard Lights and the Pigpen Archipelago

depôts of one of these hog-trains, as they are appropriately called," *Harper's Weekly* reported, "is the signal for the commencement of a scene of uproar and confusion as interesting and peculiar as one would wish to see. From the crates, the pigs, as a temporary disposition, are driven into pens, arranged, with convenient gateways, along the side of the track." The same chaotic entrance was true for hogs reaching Cincinnati by steamer.[38]

Getting the hogs into the unloading pens was only the first step. They still had to reach the slaughterhouses, most of which were located at the outskirts of the city. This final drove, usually through two miles of city streets from the rail depots, was, for hogs, "a different portion of their journey, which they are forced to accomplish on their own feet." Even with railroads dominating the final approach to the city, drovers continued to be indispensable agents in this last movement of hogs from life to death. As *Harper's Weekly* wrote, the "direction and management of this transit is undertaken by drovers experienced in the business, who engage for the occasion the assistance of a suitable number of boys; scores of whom, of every age, color, and nation, are generally collected about the depôt when a hog-train is expected, clamorous for an engagement." Here, German, Irish, and American-born whites and free and enslaved black men and boys all competed for work. "Whoever succeeds in securing the job by contract with the owner of the hogs is instantly beset by dozens of these boys, vociferously eager to be employed in the enterprise," the magazine explained, and the "shouting and screaming of the boys, gabbling in several languages at once, the quarreling and tussling of the unruly among them, the angry and peremptory exclamations of the men, combined with the squealing of the hungry pigs, produces an exciting scene of tumult and contention."

The final droves from depot pens to slaughterhouse pens were enormous street battles, from 200 to 1,000 hogs struggling against a drove manager and a team of hired boys who harried them with whips and shouts straight through the city. "A drove once started on its journey," *Harper's Weekly* insisted, "is bound, at all hazards and against all obstacles, to go through." Such obstacles were "a consideration that troubles in no degree the heads of the contractor and his yelling and slashing gang of vagabonds. You may be splashed and run into, delayed and otherwise offended, upset." During these final droves, the city belonged to the hogs and to the drover, for whom, "with his boot-tops over his pantaloons, a coon-skin cap on his head, red flannel sleeves on his arms, and a cracking whip in his hand, it is all one. He will heed you not at all. The main business of his life at this moment, mark you, is to 'land those pigs on the other side of Jordan.'"[39] Given the very real power of the hogs, this single-minded intensity was something that no drover could afford to lose.

A likely fictional-but-informed account by "Bill Jenkins," a young drover from Kentucky on his first visit to Cincinnati, recounted how runaway hogs could lead to confusion over the boundaries and identities of droves and property. "Directly I seed two hogs goin' off from the drove, an' as a matter of course I goes arter 'em," Jenkins recalled. He chased the two fugitives up a lane, turned them around, and was driving them back to the drove when a man said, "'Hey, you, what you going to do with them 'ar hogs?' 'Take 'em to the drove,' sez I. 'Drove,' sez he, 'them's my hogs.'" Words were exchanged, a brawl began, drawing the city watchmen, and worse might have ensued had not "just then, the boss of the drove cum up, an' sez he, 'Bill, them aren't my hogs!' 'Nough said,' sez I, as I walked off to the drove, might glad the boss had cum up." Considering the confusion, violence, and police attention that could erupt over a fugitive hog, Bill Jenkins "was monsus careful all the way" to the pens "not to run after any more hogs, because, thinks I to myself, the boss may do it." For a hired hand like Jenkins, the risk of chasing after a hog and potentially getting into street fights over property claims was simply not worth taking.[40]

Many hogs pursued this small but real chance for escape, some with more enduring success than others. On the journey to the slaughterhouses, some "hogs invariably became separated from the droves and made free use of the city's streets, alleys, and even the sidewalks where they competed, unattended and uninhibited, with citizens for their use. Many pigs roamed unrestricted into, around, and under the homes of tolerant Cincinnatians." Still, most hogs remained trapped within the force fields of whips, commands, and street boundaries until they were more securely contained inside the slaughterhouse pens.[41] Only then were hired drovers like Jenkins or the boys seeking work at the hog train depots finally paid.

THE spatial and temporal division of hog life and death in the Ohio Valley concentrated enormous quantities of hog flesh in winter cities like Cincinnati, and it did so at a time when workingmen were suddenly cut free from the commercial and manufacturing relations employing them during the warmer months of the year. Arriving from cities pushed out of global and regional trade networks by frozen rivers and reduced steam traffic, and from countrysides finished with labor-intensive harvests, a seasonally constituted reserve army of labor made winter in Cincinnati an industrialist's paradise. "The value of these manufacturing operations to Cincinnati," Charles Cist wrote of the hog killing and processing industries, "consists in the vast amount of labor they require and create, and the circumstance that the great mass of that labor furnishes employment to thousands, at precisely the very season when their regular avocations cannot be pursued." Around 1,500 coopers from city and country were hired to make lard kegs and pork barrels "at a period when they are not needed on stock barrels and other cooperage, and the country coopers, whose main occupation is farming, during a season when the farms require no labor at their hands." And it was not only from the auxiliary container industries that employers recruited workers into the political economy of hog death. Manufacturers also recruited from the city's seasonally unemployed winter workers, and so considering "that the slaughtering, the wagoning, the pork-house labor, the rendering grease and lard oil, the stearin and soap factories . . . supply abundant occupation to men, who, in the spring, are engaged in" brickmaking, bricklaying, and street paving, "employments, which in their very nature, cease on the approach of winter, we can readily appreciate the importance of a business, which supplies labor to the industry of, probably, ten thousand individuals, who, but for its existence, would be earning little or nothing, one-third of the year."[42]

Slaughterhouse and candle manufactory owners used the wealth they wrested from control over the work of death to build ostentatious Georgian mansions on nearby hills and fashionable brick townhouses downtown. These structures were in sharp contrast, and contrast was the whole point, to what one German visitor described as the "wretched houses where foreigners live in miserable poverty." Cincinnati was the kind of place that Laia Odo would have immediately recognized, where the "shocked decent people" spent fortunes of plundered flesh and fat to convince themselves they were, as she'd put it, better than mud. But capitalists weren't the only ones profiting from hog plunder. In the 1820s, Cincinnati workers had commanded wages considerably higher than those of their eastern counterparts. During the antebellum years, despite the growth of trade unionism, employers were able to force wages closer to relative parity with those in other regions in the United States—closer, but still not the same. In 1851, the ten slaughterhouses of Cincinnati employed among them as many as 1,000 men, "selected for this business, which requires a degree of strength and activity, that always commands high wages." Through combinations of striking, organizing, and political action, candlemakers and slaughterers, like many other workers in Cincinnati, managed to slow the capitalist erosion of their power and continue to capture some of the surplus value of hog lifework.[43]

By the eve of the Civil War, the "human chopping machines" of Cincinnati that had so overawed Frederick Law Olmsted were slaughtering nearly half a million hogs each winter. For a mere three-month window, the city was transformed in time and space to become a massive, industrialized deathscape where the lives of 450,000 hogs were violently disassembled by 10,000 wage workers into pork, soap, lights, and enormous amounts of money. Slaughterhouse workers were a "motley crowd" of free and freed black men, Irish and German immigrants, and American-born whites, and in the slaughterhouses of Louisville and Covington on the Kentucky side of the river, some were enslaved. As the *Cincinnati Gazette* described one slaughterhouse, "In the uproar of a Babel of confusion, worked the laughing negro, mirth-loving Irishman, and the sedate, hard working German, side by side."[44]

The work of death began in the slaughterhouses, the final islands of the pigpen archipelago. "Arrived at the slaughter-houses, the way-worn pigs, with waled backs and bleeding feet, are deposited in pens and fed to restore their condition," *Harper's Weekly* recounted, adding that these slaughterhouse "pens are, in most cases, connected with the killing-sheds by inclined plank ways, up which the pigs are driven as fast as they may be wanted by the butchers." Kept organized, disciplined, and imprisoned by teams of well-paid jailers overseeing a vast system of pens encircling the slaughterhouses, the hogs entered the terminal reaches of a geography of captive life everywhere walled in by whips and fences. Here was the end of the pigpen archipelago, where the "hogs for slaughter are allotted, as they are owned, to different pens regularly numbered, all of which communicate to one leading in to the upper end of the slaughter house."[45]

The slaughterhouses, situated at the critical juncture between the worlds of hog life and death, were flexible technologies designed to navigate unpredictable changes. The Cincinnati editor Charles Cist described the slaughterhouses as typically "fifty by one hundred and thirty feet each in extent, the frames being boarded up with movable lattice-work at the sides, which is kept open to admit air, in the ordinary temperature, but is shut up during the intense cold, which, occasionally, attends the packing season, so that hogs shall not be frozen so stiff that they cannot be cut up to advantage." Writing in 1851 of the enormous integrated slaughter-, packing-, and lard-rendering house of Milward & Oldershaw, situated on the opposite side of the river from Cincinnati in Covington, Kentucky, Cist informed his readers that the "slaughter-house, which will contain four thousand hogs, is on the upper floor, and the hog-pens are on the roof, the hogs being driven up an inclined plane." The slaughterhouse, which measured "three hundred and sixty feet front, and runs back one hundred and sixty feet," was, according to Cist, "doubtless the largest building for the purpose in the United States."[46]

Driven by men with whips and by the momentum of the animals behind them, the hogs were forced up the inclined planks into the killing pens at the beginning of the long slaughterhouse buildings. This would be, as *Harper's Weekly* observed, their last act of living labor, made to "raise themselves to the second story of the building by the use of their own muscular power" against the force of gravity and into a structure designed to take advantage of that gravitational gradient to pull the accumulated biomass of the hogs down through the descending slope of slaughter. "A pen selected for slaughter is open[ed]," the *Cincinnati Gazette* began, "the hogs driven up thereto, and some twenty admitted into one of two Knock Down Pens." Men drove twenty hogs at a time into what *Harper's Weekly* called "the death chamber," "where they are crowded as thick as they can stand," when the "door of this room is then closed, and on the backs of the hogs crowded in this narrow pen, walks Tom Broadman with a double hammer, constructed for the purpose, weighing from 1½ to 2 pounds, and with one blow on the head generally fells to the floor each one of the hogs." Once the hogs were knocked unconscious or dead, two other men immediately seized them, hauled them "out a few feet on to a platform where their throats are cut, the blood escaping through a lattice floor, and sometimes saved for use, and the doors opened to admit other twenties in succession during the working hours of the day, or until the supply of hogs are terminated."[47]

What happened next was a rapid flurry of death work. From the moment the hammer man knocked a hog unconscious, the animal was "immediately seized by the butchers inside, and stabbed, bled, scalded, scraped, and cleaned out before he has a very distinct or satisfactory impression of what has happened to him. He is converted into pork [and lard] in about three minutes from the time his tail glides unsuspectingly beneath the insidious trap that slips down at last between him and the trials and comforts of the outer world forever." From the "death chamber," men rolled the bled hogs off the bleeding platform into tubs of scalding hot water heated by steam. The *Cincinnati Gazette* described the scalding tub of a Deer Creek Valley slaughterhouse as "a wooden tub, 16 feet long, 4½ feet wide and 4 feet deep, filled with cold water and heated by steam, and kept heated uniformly by a furnace and boiler adjacent." The steam was almost deafeningly loud but still not loud enough "to drown the squeal of an unskillfully felled porker by the murderous blow of Broadman's 'double knocker.'" Amidst the shrieks

"The Death Chamber." In what *Harper's* called "the death chamber," hogs finally reached the deadly end of the pigpen archipelago. Cincinnati slaughterhouses like the one illustrated here pioneered a new division of labor later named the "disassembly line" after the Fordist assembly lines it helped inspire. Engraving in *Harper's Weekly*, February 4, 1860, 72.

of the animals and the roar of the escaping steam, men rolled a continuous avalanche of dead hogs into the boiling-hot tubs. At the other, lower end of the tub, two men watched to make sure the hogs rolling toward them had been sufficiently scalded. Satisfied, they would "press a lever, and thereby remove two hogs that are fit for the bench, by an apparatus called a rack, the top of which is a series of rollers that is constantly throwing out the scalded animal upon a long platform, even, smooth, and gradually inclined to the lower end."[48]

Powered by steam, gears, human muscles, and the hogs' final counter-gravitational labor, the carcasses moved onto and down the inclined bench. There, upon the benches, "one of the busiest scenes may be witnessed of the manoevres, of the twelve or fifteen men stationed up and down each side of the bench, on six scalded hogs undergoing the process of being scraped, shaved and ham-strung." While three pairs of men facing each other across the bench sequentially de-bristled, scraped the hair off, and strung up each pair of hogs by the legs with a wooden bar "while the hog is reeking with steam," additional men slid and rolled the carcasses down the incline of the bench from station to station. Others were employed sharpening knives or

Lard Lights and the Pigpen Archipelago

pouring water over the scalded hogs to keep their flesh wet (so the workmen would not too badly burn their hands).[49]

The next stage was organized around a machine called the wheel. Using the bar placed between the hind legs of each hog, a team of three men hung each carcass on one of eight iron hooks attached to a circular wooden framework. As the wheel revolved, the suspended hogs were passed through several more work stations. First, a washer threw a bucket of water over each hog to clean off any remaining hair or blood. Second were the gutters, "two experienced men, one of whom cuts open the hog and the other cuts and cleans out the offal with wonderful rapidity and skill." Washers then doused the hogs again while others cleaned them out thoroughly, and finally the wheel conveyed the carcasses to a group of three or four men, wearing oil-cloth pants and coats, who expertly used the mass and momentum of the hogs to swing them off the wheel onto their own shoulders and carry them off to the drying room. The speed and scale of this death work was extraordinary. In 1843, the *Cincinnati Gazette* reported on one Mill Creek slaughterhouse, noting that it was a typical establishment and employed forty men whose "greatest achievement was killing 827 hogs in one day out of a little over 8 hours," while "at another time they killed in three days 2385 — and at another in four days 2809. Thus the thing has been repeatedly done of killing and completely dressing more than one hog in a minute through the day."[50]

From this point on, the organic material involved in the production of hog lights diverged temporarily before recombining in lard-rendering establishments. The gutters who ripped open the hogs also collected huge amounts of offal and gut fat. In most times and places, this gut fat would simply be waste, but in Cincinnati, where slaughterers produced millions of pounds of it, gut fat was gold. At the start of the 1840s, when Cincinnati was beginning its ascent as the pork capital of the world, "slaughterers formerly got the gut fat for the whole of the labor thus described, wagoning the hogs more than a mile to the pork houses, free of expense to the owners. Every year, however, enhances the value of the perquisites," especially the organs and fat that could be sold to soap, lard-oil, and candle manufacturers. By 1850, "from ten to twenty-five cents per hog have been paid as a bonus for the privilege of killing." By 1854, slaughterhouses were each employing half a dozen men in cleaning off and washing the gut fat tossed away by the gutters, gut fat that had become worth 30 to 40 cents a hog.[51] In other words, because of the by-product industries transforming pork waste into candles, oil, and soap, slaughterers actually paid hog owners for the rights to kill the hogs and claim the fat. They paid in order to gain access to portions of hog lifework that could only be transformed into exchange value, into capital, where the scale of death of reached Cincinnati levels.

The other path through which lard became lights followed the gutted carcasses. After hanging in the drying rooms, the hogs were carted from the slaughterhouses to the pork-packing houses. There men transformed the carcasses into ham, shoulders, bacon, sides, and tender loins, while their kidney fat was "then torn out, and every piece distributed with exactness and regularity of machinery, to its appropriate pile," and when the price of lard was high, it "tempts the pork packer to trim very close, and indeed, to render the entire shoulder into lard." As *Harper's Weekly* noted, "Every scrap, even the apparently most worthless, is saved and turned to account, either as an article of food or for use in the arts. Its flesh is converted into hams and pickled pork; its lean scraps into sausage meat, and its fat scraps into lard, stearine candles, and lard oil."[52]

The factories mass-producing candles and lard oil from the reclaimed waste of the pork industry seized full advantage of this ecology of living human labor and dead hogs. By 1859, Charles Cist boasted that there were twenty-five manufactories producing lard-oil or stearine candles in Cincinnati, "and mostly on a large scale," so that the light-making industry in the "aggregate forms our heaviest manufacturing department, except that of clothing, which it, however, exceeds in importance, the raw material, in these articles, being entirely of home growth." These lard light manufacturers were the recycling centers where every shred of hog flesh, every iota of embodied hog work, was transformed into value. As the *Cincinnati Gazette* reported of "the establishment of Koeble & Miller, near the Brighton House, but one of several establishments in the city, we learned to what extent even the smallest and most inconsiderable portion of the hog was used." Situated on the banks of the Mill Creek, the "rendering apartments are built on the banks of a ravine which carries all the waste matter off to Mill Creek. On the side of the building next to the ravine is a row of twelve large wooden tanks with tops, which are raised when required, to admit the stock, of which black and white grease are made." Into these rendering vats, men dumped gut fat, leaf lard, and other hog parts. They also purchased whole hogs such as those "that have smothered to death, or such as were scalded by the recent explosion, worth 2½c. a pound," and threw them "into these tanks, whole, and with the big entrails are boiled and steamed, the grease at proper times being scummed off and the bones and refuse matter let down into the ravine by touching a lever which opens a trap door."[53]

As workers threw the grease into vats to boil it with lye, the stearic acids were bound into soap. To make candles, workmen then reversed the chemistry by decomposing the soap in sulfuric acid, leaving solid cakes of nearly pure stearine. Now clear and white in color, the cakes were subjected to a process almost identical to that which spermaceti manufacturers used to process sperm oil. After wrapping the cakes in linen, men placed the packages between thick wooden boards or, "in some establishments, under steam or hydrostatic pressure, and others in presses, arranged in great numbers up and down a long apartment, regulated by cogs, levers and weights, by which a gradually increasing pressure causes the lard oil to exude, which runs down into vats in the lower story to be barrelled, the residum in the cloths, being the article of commerce which is called stearine, of which Star candles are made."[54]

Most refiners relied primarily on gut fat and leaf lard, but some specialized in consuming entire hog carcasses, a practice that became even more common as the value of lard began to outpace that of any part of the hogs but hams. One of these hog-dissolving factories consumed "in one season, as high as thirty-six thousand hogs. It has seven large circular tanks — six of capacity to hold each fifteen thousand pounds, and one to hold six thousand pounds." The owners of this factory, a combined packinghouse and lard-oil manufacturer, first hired and arranged men into packing teams to cut away the hams from slaughtered, dressed hogs but, unlike other packers, left the rest of the hog intact. Other men then dumped "the entire carcass, with the exception of the hams," 600 carcasses per day, into the enormous circular tanks, "and the mass is subjected to steam process, under a pressure of seventy pounds to the square inch; the effect of which operation is to reduce the whole to one consistence, and every bone to powder," while the fat was then drawn off. This factory was a recycling plant for the entire city: "Beside the hogs which reach this factory in entire carcasses, the great mass of heads, ribs, back-bones, feet, and other trimmings of the hogs, cut up at different pork-houses, are subjected to the same process, in order to extract every particle of grease. This concern alone turned out, the season referred to, three millions six hundred thousand pounds lard." By 1851, lard-oil and stearine manufacturers were producing 3 million pounds of stearine and over 1 million gallons of lard oil a year, almost all destined for use as illumination. In 1850, Cincinnati exported 67,447 boxes of candles. By 1861, candle exports had risen to 138,234 boxes, the candles classified as "exports" forming but a portion of the millions of candles manufactured from the fat of the hundreds of thousands of hogs slaughtered annually in Cincinnati.[55]

Because the fracturing of hog carcasses into illuminants also produced the ingredients for soap, many candle manufacturers were also soap makers. The most famous of these joint operations was Cincinnati-based Procter & Gamble. Their factory was, in 1859, "probably engaged more extensively in manufacturing operations, than any other establishment in our city." Using hundreds of tons of lye and sulfuric acid, vast steam-powered screw presses, and mold and wick machines, Procter & Gamble transformed the lard from around 100,000 hogs into over $1 million worth of value, all with only eighty wage workers.[56]

Evidence of working conditions in the candle factories of Cincinnati is difficult to find, but if they were like those in New York, this increased scale of hog destruction would also have likely witnessed an increased exploitation of factory workers. In 1853, the Operative Tallow Chandlers of the city of New York held a series of meetings to protest both the low wages ($1.25 a day) and the round-the-clock twelve-hour shifts. "The hours of work are from 7 A.M. to 7 P.M., one hour being allowed off, and from 7 P.M. to 7 A.M.," one article reported, and the "day and night work is taken by the men alternately. The operatives demand that the working hours be reduced to ten, and their wages increased to $1.50 per day." If candle workers in Cincinnati were able to restrict the working day to daylight, as workers in the slaughtering and packing industries had done, it would have represented a tremendous struggle and a considerable triumph. But the evidence indicates that night work was the norm. In 1851, two devastating candle factory fires, one in Cleveland and one in Cincinnati, erupted at night, the Cleveland fire starting at four in the morning.[57] It seems likely that in the highly capitalized candle and lard-oil factories of Cincinnati, employers ran the machinery as continuously as they could. The final, furious, creative destruction through which hogs were passed on their way into and out of Cincinnati as light, then, was a space of nearly pure profit for capital, extracted from disassembled hogs and the labor of workingmen with the aid of industrial technology.

Stearine candles competed most directly with spermaceti candles, but they also had a wider market. As white, as hard, and as clean-burning as spermaceti, but with even higher melting points, candles made of stearine also cost one-third less than those made from whales. This meant that high-quality stearine candles were accessible to middle-class and even working-class Americans. They were also the preferred light of workers in hard rock mines, where temperatures well above 100 degrees were commonplace. Indeed, brightness was never the only quality Americans looked for in their lights. Gaslights were associated with commerce, industry, and public space. And because gaslights were too bright to look at directly, manufacturers

Lard Lights and the Pigpen Archipelago

began to sell thick glass globes, not only to shield consumers from the heat of burning gas and to cover up some of its unadorned industrial appearance, but to block and diffuse some of the too-bright light. Bourgeois households certainly embraced gaslights, but they excluded them from the centers of their homes. Few mid-nineteenth-century drawing rooms would have had gaslights. Gas was for entryways, corridors, kitchens, and cellars, and sometimes for use in dining rooms or servants' quarters. When people wanted or needed smaller, more intimate lights that they could look at, could gather near, and could move, both rich and poor Americans chose to use candles and lamps burning whale oil or camphene. A camphene or oil lamp was brighter than a candle, but sometimes a candle or two was all that was needed to read or work or eat by. And with the parallel proliferation of incredibly cheap phosphorus matches, once-complicated maneuvers like lighting, snuffing, and relighting candles and lamps became simple acts, making small, portable lights not only more useful, but more economical. Lastly, in the summer, the fact that candle flames weren't as hot as camphene lamps, let alone gaslights, may have been a compelling reason to choose candles.[58]

The new chemistry using sulfuric acid, which could transform even the poorest-quality fats into high-quality candles capable of competing with spermaceti candles at a fraction of the price, made many fats interchangeable. And this interchangeability put lard centers like Cincinnati into new relationships with fat-producing and fat-processing centers around the world. Sulfuric acid and stearine meant that Procter & Gamble had to compete with massive British candlemakers and their products, such as Price's "palmatine candles," which meant that Ohio Valley hog raisers had to compete with West African palm-oil plantations. A quick look at the geography of fat crisscrossing the globe and converging in London's soap and candle industries shows that Cincinnati was but one node in an emerging constellation of mass-produced fat and death. In 1860, the United Kingdom imported 1,430,108 hundredweights of tallow and exported £238,622 worth of stearine candles, including to the United States. Of that imported tallow, £204,060 worth came from Uruguay, while £197,423 came from neighboring Buenos Aires. From the United States, Britain imported £388,004 worth of lard and £347,345 of tallow. In 1850, Australia had been a source of fat comparable to the United States, but by 1860 Australia exported little.[59]

The main story of Victorian British fat, however, was the story of Russia and West Africa. In 1860, the tallow from cattle slaughtered by nomads on the Russian steppe constituted the second most valuable export from all of Russia, and almost all of it went to Britain. First, it had traveled thousands of kilometers from the southwestern steppe up the Volga River to St. Petersburg. The £2,759,493 worth of tallow exported to Britain from St. Petersburg and the £281,504 from Odessa made Russia, by far, the largest exporter of tallow in the world. This tallow, rendered by steppe nomads from herds of cattle numbering in the millions, on the millions of hectares of land that Russia claimed but the nomads controlled, was making the nomads rich and the St. Petersburg merchants organizing the trade even richer. This geography of life, labor, and power even made the ethnic Russian settler farmers and serfs in the steppe economically dependent, at least sometimes, on ungoverned pastoralists. This was not what was supposed to happen. The settlers were supposed to make the nomads dependent on them, not the other way around. But either by choice or as tribute, settlers paid nomads for the privilege of fattening Russian pigs on the waste from the slaughter camps. And this tallow, produced only in the loosest sense by Russian subjects, formed Russia's most (or in some years after wheat, second most) valuable export. It accounted for three-quarters of all the tallow imported into England in 1860, even during what was called in England the "Great Russian Tallow Speculation" of 1859 to 1861.[60]

Also incredibly significant was the £1,684,532 worth of palm oil imported from West Africa in 1860. Palm oil had originated in the "legitimate trade" movement spurred by British abolitionists attempting to encourage West African nations to export goods rather than people. Opening and operating new palm plantations in West Africa may have somewhat reduced the incentive for selling people into an illegal overseas slave trade, but it actually invigorated the need for and use of plantation slave labor in Africa itself, introducing and strengthening plantation slavery on African coasts where formerly no such form of slavery had existed.[61]

The chemistry of stearine was transforming the internal relations of candles, throwing Cincinnati hog drovers, West African palm-oil slave drivers, and steppe nomads into economic relations open to further exploitation. What made Cincinnati so unusual in this constellation of fat and slaughter was how tightly it integrated not only the geographies of life and death but the rendering of the fat with the secondary industries of candles and soap. The specificities of particular geographies mattered. Cincinnati never competed with the major centers of cattle death to supply markets with bones and never became a source for the phosphorus of lucifer matches, but it did outstrip most of the cattle camps in the production of candles. That Cincinnati capitalists combined as much of the work and spaces of life and death and candles as they did — displacing risks onto farmers and hogs while controlling and reaping the rewards for themselves—made the city unusual and made some residents very rich. Here then, was a vanguard of capitalist light on the eve of the Civil War. In the Ohio Valley, struggles over slavery and freedom, capital and labor, and owner and livestock drew bright lines in the adipose tissue of hogs. This was what capitalism—always omnivorous by choice—looked like when both capitalists and workingmen sought to exploit hogs, workers that were unable to claim the rights to life that had been won by (some) humans. Not only extinguishing the living labor of hogs and men in the production process but extinguishing life itself, the continual remaking of lard lights through the pigpen archipelago plundered the hogs' pasts and futures absolutely, letting not a drop of work or life escape valuation. Or, as one visitor to the Cincinnati packinghouses concluded lightly, "The pig was used up."[62]

Chapter Five

LUCIFER MATCHES AND THE GLOBAL VIOLENCE OF PHOSPHORUS

The knock on the door came at three, as it did every morning, as the watch-man made his rounds. It was too early for even the light of dawn to begin its dim journey through smoke and soot and shadow into his Manchester tene-ment, but fifteen-year-old Richard Toye never woke in complete darkness. Like hundreds of other boys and girls in Manchester, England, and thou-sands more in cities across Europe and the United States, Toye glowed in the dark. He'd done so since he was nine years old. Many children began glow-ing as young as six. The wisps of softly luminous smoke rising from his skin and clothes were always there, if mostly invisible save at night and in dark corners. As he'd tell the man from the government later that day, "If you are in the dark and rub your breeches a bit you can see them 'shine like a cat's eye.'" But even when people couldn't see the glow, they could smell it. Toye didn't much mind the smell anymore, unless his clothes were wet. Then he could barely stand himself, as his damp, glowing clothes "'stink him out the room pretty nearly.'" Across town, eleven-year-old John Stafford lay in his bed, where he could "see his shirt shining and smoking and smells it." Only on Monday, after laundry, did his shirt not reek and glow. "'Mother says I stinks her out of the house,'" Stafford would confess to the man interviewing children at the factory. "'I tell her I can't help it.'"

By quarter-past three, Richard Toye had left for work. Joining him on his more than two-mile walk to the match manufactory outside the city, the largest in the world, were a dozen other boys his age. Most would be glowing; some might be missing parts of their jaws. Three-quarters of an hour later, this first wave of glowing youth arrived at the Newton Heath factory complex of Dixon, Son, and Evans. Just as Toye reached Newton Heath, thirteen-year-old Michael Johnstone's mother woke him, a "'knocker up'" having earlier come to the door and woken her. Like most mornings, Johnstone felt sick. He'd been losing his appetite, and sometimes he vomited. He knew all this was tied to his match work, that the work might be killing him. Still, his mother was counting on his wages, and he had a job to get to, so he hurriedly left the tenement. The same scene repeated itself all across Manchester, and soon Johnstone joined a second, considerably larger wave of over 100 stinking, luminous girls and boys marching to Newton Heath.[1]

Match manufacturers in cities across Europe and North America purchased the labor of thousands of children and hundreds of adults, employing them to combine phosphorus, potash, sulfur, and American and Norwegian pine into billions of what were most commonly called lucifer matches. Importing sticks of purified phosphorus from Birmingham and Lyon, American match factory owners hired children to melt the sticks into a dipping paste and employed thousands of other girls and boys to make around 40 million dipped matches a day by the 1850s. As the American reformer Virginia Penny observed, the "making and selling of matches have furnished employment for hundreds and thousands of boys and girls in all our large cities."[2] Most of them would have glowed.

The daily and nightly march of glowing children to and from Manchester may have been the most eerily spectacular, but they were hardly the only specters haunting nineteenth-century cities with their toxic, disfiguring light. "At night," Annie Brown of London, who had glowed since she was nine, lost her jaw at thirteen, and long ago become used to the stench, "could see that her clothes were glowing on the chair where she had put them; her hands and arms were glowing also." Emilia Block recounted how, though she and the other girls at the factory in Aberdeen frequently washed, the "'compo' sticks to their hands, and has a bad smell. Noticed it when she first came to work here. Mother said she could not touch the dishes after she (witness) or any one from the place had touched them." In Belfast, nine-year-old Patrick Morgan matter-of-factly recounted how he "'went a-fire in the street one day,'" attributing his spontaneous combustion to the "sparks" he saw "about his clothes and legs at night," luminescent sparks that were not only strange but terrifyingly dangerous. Richard Chidleigh fled the deadly stench of one London factory for a better-ventilated one, testifying, "'It aint such dirty work here.' It was the carrying in to dry that he did not like. There was such a nasty smell in his mouth from the steam. 'Yes,' the smell was in his mouth." He claimed carrying the frames of dipped matches into the vapor-filled drying room "did not make him ill," but it "did those that had been at it long. They had the 'flute.' That means their jaw swelled, and they had it cut out. He might have had it" too, he noted, if he had not moved. In Lyon, observers described workers whose very breath not only stank but glowed. When they were "in the dark, the gases they expel from the stomach by belching, become luminous, so that they seem to make flames by the mouth, and they make a real game of this remarkable phenomenon."[3]

While the stinking glow was isolating, identifying, and strange, it was more a marker of danger than the danger itself. The problem was the purified, highly reactive phosphorus. As the phosphorus slowly oxidized, it transformed some of its potential energy into eerie light and smell. But the two other pathways for that potential energy were what embedded phosphorus in a global ecology and politics of violence. Whether in match factories or in the hands of match users, that phosphorus, like a coiled spring, would eventually release its energy. It could do so quickly in a sudden burst of flame, as when a match was struck, or the phosphorus could react more slowly with organic molecules, transforming into toxic organophosphates. How and where and to whom these transformations, fast or slow, would happen was never certain and could be, within material limits, prolonged, paused, or displaced onto others. This was what made the nature of phosphorus political. The most permanently visible reminders of this inescapable ecology were the many match workers missing pieces of their jaws; sometimes the "gums [were] entirely gone from the upper jaw, leaving the bare bone grinning out, a living death's head." This was the fate, the "flute," that Richard Chidleigh had tried to flee. Maybe after a few months, but more often after many years of working in match factories absorbing phosphorus vapor, a relentless pain began in the jaw that not even laudanum could dull. The exact cause of the disease was a mystery at the time, and there is still no medical consensus. The phosphoric acid may have directly destroyed the jaw, or the vapor molecules may have transformed into toxic chemicals in children's mouths, lodged themselves in their gums, and killed critical cells responsible for the normal turnover of bone. According to one survivor of the "jaw disease," who found relief from sleepless agony only after surgeons removed his whole lower jaw, "'No one can describe it if they don't know it; it's like everlasting pain.'"[4]

Phosphorus could be both the stuff of life and the stuff of death. It could also be, in the interstices between, the stuff of light. One of the most revolutionary technologies of the century, phosphorus matches were also among the smallest, cheapest, and seemingly simplest things. But those matches, incredibly cheap, found their way into the hands of the highest and the lowest, everyone now a strike or a jostle away from producing instant, useful, but also potentially devastating flames. And each American lucifer match was the culmination of world-spanning pilgrimages of bones, shit, wood, sulfur, and glowing and poisoned child laborers, journeys of work and energy and struggle that crossed, unsettled, and reinscribed boundaries between life and death, enslaved and free, capitalist and laborer, adult and child, and even human and animal. Indeed, the owners of a match factory in Wallingford, Connecticut, told the local newspaper in 1864 that their "Phosphorus is mainly imported . . . from France," the rest coming from England, and "it is now, we understand, mainly obtained from the bones of animals and human beings."[5]

Beginning in the 1830s and 1840s, at factories in Lyon and Birmingham, French and English workingmen directed sulfuric acid and massive flows of heat to transform phosphate-rich bones and guanos into pure elemental phosphorus, all of which was for use in matches. European manufacturers reached first for locally available bones; then they scoured the world searching for more, settling especially on the millions of pounds of bones thrown out annually as waste from the coastal slaughterhouses, or *saladeros*, of Buenos Aires and Uruguay, where free, enslaved, and freed men transformed the massive cattle herds of the Pampas into hides, meat, and tallow. They may not have lived and labored in the United States, but they were all bringers of American light. And as the Connecticut match manufacturer well knew, the phosphorus that made its way into lucifer matches and fertilizers came from more than the bones of animals. "The battle-fields of Europe have even, in some instances, been dug up," *Scientific American* reported, "and their long pent treasures sent to the bone mills to be converted into 'superphosphate.'" Following nineteenth-century chemical revolutions in cities like Lyon, the bone rush led even to the plundering of hospital bone deposits, as "from then on, the bones were much sought after; they had a value which increased and doubled in a few years; we passed, to obtain them, markets with hospitals; to procure them, men searched the rubbish, and this debris became the basis of a very considerable commerce. These are the masses of dry bones that feed the beautiful [phosphorus] factory in Barraban."[6] By the late 1850s, English manufacturers began to supplement these bones with phosphate-rich rock guano mined on West Indian guano islands like Sombrero.

Making lucifers was an extraordinarily, kaleidoscopically political process. The workplaces of phosphorus were never simply sites in which capital and labor confronted and mutually constituted each other. As the history of the production of phosphorus matches reveals, "labor" was always a shifting constellation of nested hierarchies: enslaved South American slaughterers and contracted Afro–West Indian guano miners struggling with cattle and birds and white overseers who worked for white owners; children struggling with one another and the adults who claimed them; masters and apprentices working alongside "unskilled" laborers paid directly by the owners of the means of production; younger children working under older siblings; and girls working for women working for owners unbound by any ties of kinship.

None other than Karl Marx found himself drawn into the politics of lucifers, an occasion that would prove crucial to his theory of living labor. As he noted in *Capital* in his chapter on the working day, the "manufacture of matches, on account of its unhealthiness and unpleasantness, has such a bad reputation that only the most miserable part of the working class, half-starved widows and so forth, deliver up their children to it." And not only were these sacrificed workers overwhelmingly minors, but many were truly young. As Marx noted, of the witnesses interviewed by John White, the commissioner of an 1863 report to Parliament on child labor, "270 were under 18, fifty under 10, ten only 8, and five only 6 years old. With a working day ranging from 12 to 14 or 15 hours, night-labour, irregular meal-times, and meals mostly taken in the workrooms themselves, pestilent with phosphorus, Dante would have found the worst horrors in his Inferno surpassed in this industry." Here Marx worked out his theory of capital as the consumption of life through the consumption of time, only pages before penning his famous formulation that "capital is dead labour which, vampire-like, lives only by sucking living labour, and lives the more, the more labour it sucks. The time during which the worker works is the time during which the capitalist consumes the labour-power he has bought from him."[7]

The luminous wisps of smoke that could be seen rising from the forms of thousands of children were the visible, material remains of this necromantic violence. But to make the phosphoric fangs the vampire capitalists of lucifer matches and their collaborators used to suck child life, phosphorus first had to be wrenched out of an organic ecology of DNA and ATP, of soil, sea, and bone, and pushed up the slippery faces of potential energy precipices into an inorganic industrial chemistry. The story of nineteenth-century phosphorus was the story of coercing people to continually roll phosphorus up steep mountains of potential energy while those on the downslope scrambled to direct it into the means of flame, profit, and power — or merely to survive it—before the phosphorus inevitably tumbled back down into organic chemical relations.

It was a Sisyphean struggle, and it began with bones and shit. In the cattle-killing fields of the Río de la Plata and on West Indian guano islands, entrepreneurs and workers struggled with cattle and birds through racialized structures of power to make lives, make money, and make massive amounts of phosphates by unmaking animals and islands. The struggle then recrossed the Atlantic to the English and French chemical manufacturers in Birmingham and Lyon, who, diverting some of these South American bones and Caribbean guano phosphates, negotiated with well-paid chemical workmen to transform the phosphates into pure inorganic phosphorus. From that point on, whether in the American match factories where children dipped splints into that phosphorus or in the hands of match-wielding seamstresses, tobacco smokers, bored children, or insurgent slaves, the struggle of phosphorus was the struggle of living in its inescapable ecology of oxidization. There, people could, at best, try to choose among glowing, stinking, poisoning, or fire. But most of the time those choices were made for them. The violent ecology of phosphorus provided real opportunities for even the most powerless, but it was a slippery power that might turn in its purported owner's hand, and was ever escaping control.

SEVEN thousand miles from the glowing children marching to and from Manchester, a different sort of march was taking place. From the Argentine Pampas the cattle came, hundreds of thousands every summer, to die. Some came from nearby, others from fifty or sixty leagues, the drivers, or *receros,* "pushing, and goading, and hallooing with might and main" to drive hundreds at a time nine to fifteen miles each day. Beginning in the late eighteenth century, but accelerating in the nineteenth, ranchers had commanded generations of enslaved workers and hired peons to dig out of the Pampas hundreds of *estancias,* deep circular ditches surrounding complexes a couple hundred yards across, each capable of enclosing tens of thousands of cattle. This vast geography of captive cattle life had formed over two centuries as Spanish and then postcolonial armies in the provinces of Buenos Aires, Entre Ríos, and Bando Oriental (modern-day Uruguay) pushed outward from the Paraná, Uruguay, and Plata Rivers to protect settlers carving estancias out of indigenous land. Meanwhile, the powerful equestrian Mapuche Indians of the southern Pampas and the Guaykurú horse nations of the Gran Chaco to the north kept the estancia empire hemmed in through hunting, raiding, slaving, diplomacy, trade, and war well past the middle of the nineteenth century. Still, by the 1850s, the estancias spanned twelve degrees of latitude, from twenty-eight degrees in the north to forty degrees in the south, enclosing millions of animals. It was a geography at once deeply similar to and deeply different from the pigpen archipelago. The Pampas were plains, not prairies, and the needs of the thirstier cattle meant access to water shaped the limits and contours of this semiarid geography more than for the hogs of the Ohio Valley. And not only were cattle far larger and stronger than hogs, but these South American longhorns had enormous deadly weapons growing out of their heads. Descended from cows that had become wild after Spanish settlement and multiplied on the plains, these longhorns had complex herds and hierarchies that further separated them from hogs. These herd behaviors could make leading and moving large numbers easier for receros, but they could also lead to uncontrollable stampedes.[8]

As the cows approached their destination, the signs of their future became harder to miss. On the road to Buenos Aires were "dead horses and oxen everywhere," one traveler noted, and so many bones that the "road is repaired by filling up the holes with them, and in some places you see hedges made of them. I have seen one or two *corrals* . . . surrounded by fences made entirely of the bones that form the cores of the horns of oxen, and so close and thick that you cannot see light through them." The nearer to death the cattle got, the more they walked through a landscape made of the deaths that had come before them. After the cattle crossed the bridge over the Matanza River, about three miles from Buenos Aires, "the stench was too awful: there were ditches filled with blood *instead* of water, actually in all stages of putrefaction; *miles and miles* of fences, three or four feet high and two thick, dividing off the different establishments skirting the road, and forming *corrals*, made entirely of the bones of the bullocks' horns." Here they were met by the continual "death bellow" of dying cows, "a noise more expressive of fierce agony," Charles Darwin recorded in 1833, "than any I know. I have often distinguished it from a long distance, and have always known that the struggle was then drawing to a close." They had reached the saladeros, as the slaughterhouses of the Plata region were called. The sounds, sights, smells, and remnants of mass death were as haunting as they were impossible to ignore. "The whole sight is horrible and revolting," Darwin wrote, "the ground is almost made of bones; and the horses and riders are drenched with gore."[9]

When these drives began two centuries earlier, men had killed the cattle mainly for their hides, loading slave ships with animal skins after they unloaded their human cargoes. The rest of the skinned animals they had either left to local economies or recycled into the landscape itself, bone roads and bone fences steadily accreting year after year, decade after decade of mass slaughter. At the end of the colonial period in the late eighteenth century, saladeros and merchants began to increase the scale of death while digging deeper under cows' skin for new commodities. First a burgeoning trade in Río de la Plata salted beef grew into a critical protein supply line for the slave plantations of Cuba, where the sugar industry was recentering in the aftershocks of the Haitian Revolution. Following soon after, tallow exports from the Río de la Plata soared into the nineteenth century to meet the demands of European soap and candle manufacturers. No longer were skinned carcasses left along roadsides to decay. The market had reached cows' muscles and fat. When Darwin penned his 1833 description of the saladeros, capitalist markets had not yet penetrated bone. But that was about to change, and lucifer matches were at the center of the story.[10]

An 1851 visitor to the saladero of Santamaria, Llambi, y Cambaceres, who wrote to Charles Dickens's magazine *Household Words* about his experiences, watched as the drovers herded their cattle through a gateway on the road, entering a wide, oblong corral in a large yard surrounded by buildings on three sides and packed with animals. Nested inside the enclosure, was "a smaller *corral* or *vreté*, entirely round, and paved with wood," both pens "formed by closely-wedged fences of the trunks of poplar trees." Once trapped in the larger corral, standing "knee-deep in mud and manure, the animals awaited slaughter for days, without water or food."[11]

But sooner or later, the time to die would come, and the work of death employed thousands. "It is three o'clock in the morning of a South American summer," observed the visitor, "and a bell has already summoned the workmen to their various avocations." A large saladero like Santamaria's, which, from three in the morning to four in the afternoon, slaughtered and processed 400 to 500 animals a day during peak season, employed several hundred men at a time and thousands of seasonal laborers over the course of the summer. It was a geography of labor that had been constructed principally through slavery. By 1841, across the river from Buenos Aires, Montevideo's saladeros had amassed so large a labor force of enslaved Africans that the Buenos Aires press termed the Montevideo saladero district the Villa Angola. But Montevideo saladero operators also contracted Spanish colonists — Basques, Galicians, and Canary Islanders — as *siervos temporales*. Employers relied on city police and public institutions to enforce their private power over both enslaved and indentured contract laborers. Just as the press denounced and advertised "escaped" (*huidores*) slaves, so too did it warn of "escaped" Canary Islanders who fled to the countryside.[12]

Following the wars of independence in the 1810s, warfare both within and between Spain's former colonies racked the Río de la Plata into mid-century, when the nation-states of Argentina and Uruguay largely solidified their modern territories. Only then did the thousands of enslaved people (and their descendants) who fought in these wars, and against the proslavery armies of Brazil, finally force both Uruguay's and Argentina's governments to officially abolish slavery; official emancipation in Argentina didn't come until 1853, a few years after the *Household Words* account was written. In the upheaval, the Río de la Plata saladero workforce became increasingly composed of European migrants, but people of African descent remained critically important. Workers' persistent, sometimes illegal practices of mobility won and maintained a real degree of worker freedom, even in the midst of war. One saladero owner "even complained to the chief of police that his workers regularly and audaciously insulted him," while determining their own hours and holidays. Defying vagrancy laws by moving from job to job in search of higher wages, Afro-Argentinians and Afro-Uruguayans "found opportunities galore, but at the bottom of the rural social ladder. They worked in the *saladeros*, and on ranches, doing the shearing, harvesting, and branding; driving cattle; and conducting oxcarts."[13]

The bloodstained workmen who arrived hours before dawn to begin the work of death at Santamaria's saladero, therefore, likely spoke dozens of languages and had crossed oceans; had strained or broken chains of enslavement and contract; had worked, fought, and lived through wars; and had challenged, navigated, and been circumscribed by a shifting order of racialized power. The men were seeking their own freedoms: from enslavement, against reenslavement, to navigate coercive contracts, and to find better pay in a relatively mobile geography of labor. "The work people are paid wages which would astonish the European operative," noted one English visitor, as even "boys gain from four to five shillings a-day. While the more skilful workmen can net as much as from six to seven pounds sterling per week."[14] Such human freedoms, however, were inextricably bound to the captivity and mass destruction of cattle, to violently constructing and maintaining the boundary between man and animal.

This is not an indictment, at least not specifically, of these workmen. It would be a mistake to simply insist that all life was equally valuable and leave it at that. Cattle were not equal to slaughterhouse workers, or match-making children to match-factory-owning adults, or the enslaved to enslavers. These categories were descriptions of power relations, which meant they were descriptions of inequality. And neither racism nor child abuse nor slavery

Lucifer Matches and the Violence of Phosphorus

was ever going to be ended by getting everyone on the right side of oppression, by making everyone into white people or adults or slaveholders. Or "humans." Trying to stay on, or get to, the dominators' side of a wall, rather than dismantling it altogether, never has saved the world. But it has often saved (some) lives, and we need not condone domination to empathize with those struggling to navigate the worlds domination has made. A full accounting of the work, struggle, and energy of nineteenth-century lights forces us to confront the ways that the weak have sought to escape, or survive, the crushing weight of power structures by displacing that weight onto others, or by plundering strength from the even weaker in order to resist plunder from the stronger.[15] And perhaps no lights demand this confrontation so clearly as lucifer matches.

The work of death began by moving cattle from one enclosure to another. At Santamaria's, receros first closed the gate to the road, preventing escape attempts. Then they pressed some of the cattle in the larger, now-locked pen through the thick, slippery morass of mud and manure into "a narrow lane, consisting of two rows of enormous stakes—whole trunks of trees in fact—driven into the earth," leading them into the smaller reinforced pen, enclosed on one end by a great barred doorway that could open into the slaughterhouse. Such solidly massive structures of nested pens of bone and tree trunks and hardwood fences were tools of domination, but they were also a measure of animal struggle. Against the cattle's mass and will to live, the workmen's need to counter a collective breakout required skillfully constructing and using landscapes of captivity to their advantage. The receros and their highly trained horses coordinated to shout, prod, body-slam, lasso, drag, and round up cattle that were beginning to panic. Many attempted to escape, some succeeded, and others gored horses and drivers with their giant horns. "Occasionally, indeed frequently, a point of two, three, or half-a-dozen animals will break way from the lot," noted one observer, who watched as the receros spurred their horses in pursuit to recapture them one by one. Unable to turn such a fast-moving massive animal, the recero instead expertly lassoed one of the cow's enormous horns, careful not to let the animal gore either him or his horse, while his mount "gallops to one side and comes suddenly to a stop; the animal is swung round, a second lasso is thrown by another horseman; one takes one side, the other the other; the enraged animal rushes to and fro but he is checked by the lassoes; and so, by the dragging of the lasso-men and his own mad rushes, he is worked on to the remainder of the herd." This was an inter- and trans-species battlefield of humans and horses against cattle, of clashing, contradicting freedoms and oppressions, fought out at one of the many bottoms of a global power structure of capital and empire and the means of light.

Many cows tried to escape. Some succeeded, if only briefly, but most did not. At the appropriate time, the *Capataz de los Coralles* and a half-dozen mounted boys began galloping around the larger enclosure in all directions, "making the most hideous noises their lungs are capable of . . . till they have frightened from eighty to a hundred of the beasts into the lesser enclosure." With the animals finally packed into the inner cage, the portcullis door was lifted and the "bovine victims rush in; but the moment they enter, they encounter sights and smells portentous of their coming fate, which impel them to make a sudden retreat. Alas! the instant the last tail has passed under the opening, down falls the door to oppose all egress; and the unhappy oxen find themselves as completely imprisoned as rats in a trap." Waiting inside the trap were scores of the bloodstained men, each with a pair of long knives stuck into their sashes. At one end of the killing floor, right at the entrance, was a small wheel wound with a thick-roped noose connected at the other end to a pair of horses. It would be the first thing the cows would see, and the last.

Next to the trapped cattle by the wheel, a man stood upon a railed stage above the throng. The writer for *Household Words* looked on as the man, "with unerring aim throws the noose of his lasso over the horns of the nearest animal, and catches it. He then gives the word 'dele!' (go on), to the horses harnessed to the other end of the lasso; they move rapidly on, the lasso travels round the wheel till the ox's head is pressed so tightly against it, that he is powerless, and forced into a position most convenient to be slaughtered." The man then stabbed the pinned ox at the base of the skull. "Death is instantaneous." After each death blow, the workman lifted a bar, letting the wheel drop the carcass into a tram car running back and forth into the interior, drawn by other horses unwillingly habituated to the terrifying smell of blood. These horses spent their workdays delivering carcasses to the skinners and returning to the wheel for fresh victims. In the interior, meanwhile, teams of skinners and butchers rapidly disassembled carcass after carcass into skin, meat, offal, and bones.

Here the animal matter diverged onto two separate paths: in one direction went the hides and salt beef; in the other went the tallow and bones. It is the latter path that concerns us here. "While the meat is being salted and piled, the bones, fat, and intestines are hurried to another part of the yard," the visitor wrote. Two tall chimneys marked the melting house, continually spewing forth ash and odors that settled over the saladeros and drifted to the city. Inside, enormous boilers drove jets of steam into eight wooden vats. Each steam vat could hold more than 100 carcasses, and as the work of slaughter continued, teams of men spent hours loading the hundreds of dead, disassembled animals killed each day into the wooden tanks. Once the tanks were full or no more carcasses were forthcoming, "the steam is turned on and the whole is steamed incessantly from forty-eight to seventy-two hours." Returning after the days of steaming, workers removed what remained of the dried, whitened large bones — shanks, shins, skulls — for shipment and drew off the valuable tallow, "purified in flat vessels and packed in barrels for shipment," destined for stearine candles and soap. Left in the vats was the stuff of lucifers: tons of residue of small bones and animal fiber that were "heaped up in immense piles; and . . . set light to and left to burn till it is reduced to ashes." European phosphorus manufacturers began to import these bones and bone ash, formerly no more than problematic waste but now made increasingly valuable through matches and fertilizer.[16]

The Río de la Plata was by far the world's largest exporter of bones to Britain, where manufacturers turned them into fertilizer, glue, and most importantly for this story, the phosphorus for lucifer friction matches. Together with manufacturers in France, these British phosphorus makers were the sole sources for the American match industry. From 1857 to 1861, the Plata states of Buenos Aires, Entre Ríos, and Uruguay combined to kill a total of 4.371 million animals, nearly 1 million a year. The total volume of bones imported into the United Kingdom tripled from 20,000 tons in 1851 to over 62,000 in 1860, the product of the skeletons of nearly 1 million animals, one-third of which came from the Río de la Plata. As *Scientific American* noted, "Bones are collected along with old rags in every country in the world, but the largest supplies are obtained from South America, where an immense number of cattle are annually slaughtered for the sake of their hides and fat."[17]

Mammalian bones were not, however, the only source for phosphorus. Manufacturers also made it from guano. "In place of bones," the phosphorus historian Richard Threlfall wrote, "mineral calcium phosphates could be used," but the best was "the Caribbean *Sombrero* phosphate (the best, because soft and easy to grind)." Guano islands formed from the accumulated waste of billions of birds excreting fish skeletons over geological timescales. Sombrero, a coral island covered in such mineralized bird shit, lay between the Virgin and Leeward Islands, forty miles from St. Martin, the "nearest inhabited land." Boston entrepreneurs first claimed Sombrero in 1856 and then promptly sold their interest to a New York firm, but the U.S. government never asserted sovereignty or recognized the legitimacy of the mining operations. Yet mining went on under American management anyway until 1863, when the British took formal control of the island. According to a 1932 U.S. Department of State report, Sombrero rose out of "a small coral bank" and was less than a quarter square mile in size, "1800 by 400 yards square." After decades of guano mining, the island had been reduced to "jagged points of coral rock rising to a height of 40 feet in the center, and covered with a few prickly pear bushes. The sides are precipitous and rocky, and there is no good harbor, but there is an anchorage on the west side of the island."[18]

In the summer of 1860, either shortly before or shortly after the English Quaker firm of Albright & Wilson, one of the world's two largest manufacturers of phosphorus, began importing "sombrerite" guano to supplement the bone ash they shipped from South America, New Englander Myrick Snow wrote a letter to the *New York Herald* after assuming the position of acting superintendent, boasting that the "island of Sombrero and its appurtenances are of no small magnitude." Besides the officers and overseers, the more than 200 workers on the island were all free or freed black men contracted from nearby Danish and English islands, promised $12 a month to dig and load vessels with guano. All of the owners and officers of the mining operations were white Americans: the owners were New Yorkers, Snow was a Free Mason from Maine, the company agent was a proslavery secessionist from Baltimore, and none of the twelve other white employees was listed as a "laborer."

The operations consisted of houses for officers, overseers, and laborers. There were stores for supplies, and a separate powder magazine. "Substantially built" stone and cement cisterns held the precious fresh water that had to be continually brought in by ships. A yacht had "just arrived from Guadeloupe, with a cargo of water (3,000 gallons), the stock on hand being three days supply only." The deep dependence on these irregular supply lines was central to the spatial politics of life and labor on Sombrero, circumscribing the possibilities and targets of both domination and resistance. The port, which Snow claimed had seen 110 ships loaded with guano in the previous three years, consisted of an assemblage of docks, lighters, tenders, and iron buoys and moorings. The island also housed stables for the horses, mules, cars, carts, and derricks used to transport tools, guano, and men over treacherous rock and the island's double-track railroad. Hogs, sheep, goats, and "fowls of nearly all kinds," most of which were likely the seabirds whose endless cycles of shitting and tunneling and nesting had created the island itself, served as food for the "average of about 200 men to feed and pay daily."[19]

A month and a half later, it seemed those 200 men had had enough of

Lucifer Matches and the Violence of Phosphorus

Myrick Snow. According to a flurry of hysterical and racist American news-paper reports, a group of four, or maybe one, or maybe all of the black men on the island had decided Snow needed to die. "Snow was a man of energy and determination," the *Petersburg Express* explained, "and being long ha-bituated to the exercise of a vigorous command on shipboard, did not change his system of discipline on the island." The politics of slavery and antislavery, the *Express* suggested, were at the heart of the dispute, and the writer stated that the "negroes, who were principally English and Danish, were obtained from the neighboring islands of the West Indies belonging to those nations, and were consequently always hostile to Americans, on account of the latter's slaveholding proclivities, grew dissatisfied with their new superintendent, and meditated every variety of revenge." The original report claimed that "only four of them, however, seemed to have brought their designs to a concerted measure, and on the 24th of July, one of them was delegated to murder him, while the others were near to see that it was effectually done." Whether the deed was discussed and planned weeks be-fore or an accident not planned at all, on July 24, 1860, Myrick Snow's head was crushed. American reports, which appear to be based solely on the testi-mony of white sailors, claimed the blow occurred when the four men "were engaged in loading a car, on this day, and the chosen murderer being on the top of it, took advantage of the foreman, as he was stooping on the ground below, and hurled a tremendous lump of guano at his head, crushed his skull with the blow, and left him for dead on the ground." But if the plan was to murder Snow, why didn't they finish job? Notably, these miners appeared to have used the tools, materials, and landscape of their labor to both fight back and provide plausible deniability. Guano miners were injured and crushed all the time. Maybe they thought the owners would believe it had been an acci-dent. Maybe it was an accident.[20]

As guano islands went, Sombrero Island was probably not the worst place to be. But that was an incredibly low bar. "No nineteenth-century job," according to one historian of the guano rush, "was as difficult, dangerous, or demeaning as shoveling either feces or phosphates on guano islands."[21] Simply moving across a guano island like Sombrero was dangerous work. Samuel Rinder wrote of his visit to a guano island for *Household Words* in 1852 describing how he waded knee-deep through stratified layers of com-pressing guano as he attempted to traverse a treacherous landscape of shit that tripped, cut, and scraped anyone forced to cross the ground. The surface was crusted in a light-colored layer of fresh guano honeycombed with holes scratched out by nesting birds, and "you can scarcely put a foot on any part of the islands without sinking to the knee and being tickled with the sense of a hard beak digging into your unprotected ankles." In a landscape continually renewed through the remains of marine life deposited and excreted again and again by the nesting birds, the friction of the terrain was incredible. Crossing was possible only with "the loss of sundry inches of skin from our legs." Miners built the work camp by excavating a pocket of reduced friction "on a small space cleared of guano"; it consisted "of twenty or thirty miserable shanties." On Sombrero, with three times the miners, the camp was considerably larger.[22]

At all guano islands with jutting cliffs and steep harbors, such as Sombrero, miners delivered guano to ships using gravity and canvas chutes. While some men risked falling to their deaths to pour hundreds of tons of guano down chutes into ship holds during the day, others hauled eighty-pound loads of guano down to the cliff-side enclosure on their backs at night. But before the ship could safely get close enough to be filled through the large chute, it needed sufficient ballast. And captains typically ordered their crews to spend several days and nights, sometimes weeks, perilously rowing back and forth through narrow channels to jockey with crews of other ships for the privilege of filling their boat under one of the smaller chutes. There the crew tried to keep their boat in place while cascades of guano rained down on their heads, plastering them with shit dust, only to then have to row back to the ship and load it up with guano ballast. Then they repeated the process over and over, one boatload at a time. At the time that Myrick Snow's head was crushed, there were at least two ships moored at Sombrero, the *Emma Tuttle* of New Haven and the *Warren* of Baltimore. The Sombrero workmen had just finished loading the hold of the *Emma Tuttle* with guano. The *Warren* was waiting its turn.

When the time had finally come for the *Emma Tuttle* to receive its full cargo of guano from the larger chute, one of the white men in the camp piloted the ship into position. Accompanying him was a team of six or so black men from the English and Danish Virgin Islands. In Peru, where Samuel Rinder's ship prepared for its turn under the larger chute, such workers were "Indians." Nowhere were they white. These teams of men were tasked with the unbelievably awful work of "trimming," or evenly distributing, the guano as it poured out of the chute into each new ship's hold. "The hatchways are quickly choked up, and the atmosphere becomes a mere mass of floating guano," Rinder observed, likely from a safe distance. In attempting to navigate the nature of the work, crews "retreated into the rigging to escape as much of the odious, airborne powder as possible" during loading. The trimmers had no such option. Working naked, "the only article of dress with some of them being a bunch of oakum tied firmly over the mouth and nostrils, so as to admit air and exclude the dust," the trimmers tenaciously strove to keep the nature of the work outside themselves, especially their airways, with these improvised respirators. The trimmers sought to regulate the materiality of trimming spatially with masks and temporally by dividing "themselves into two parties, one relieving the other every twenty minutes." Rinder, watching the work unfold, was amazed at how hard and skillfully the Indian trimmers "toil . . . coming on deck, when relieved, thoroughly exhausted and streaming with perspiration." Naked, choking, and sweating alarmingly, the men in the relieved team of trimmers desperately sought shade and drink until it was their time to return.

No one and nothing on the ship could truly escape the reeking dust. The ship's crew could retreat to the rigging, but they, too, were "compelled to wear the oakum defences, for the clouds of dust rising from the hold are stifling. . . . Not a cranny escapes; the very rats are set a-sneezing, and the old craft is converted into one huge snuff-box. The infliction, however, does not last long, three days being generally sufficient for the loading of a large ship." As soon as the ship was loaded, the crew eagerly set sail toward the nearest port, glad to be gone and desperate to wash. From there they carried guano to market; hundreds of tons of it made its way to Birmingham's phosphorus manufacturers. But for the trimmers, the infliction never ended, as they were forced again and again with each new ship into the thickest clouds of shit. They did their best to navigate the spaces and times of this work; but it was ultimately a losing battle, and trimmers commonly suffered nosebleeds, temporary blindness, and worse.[23]

By 1854, just two years after Samuel Rinder's account, the geography of labor in the Peruvian guano islands had changed dramatically, and because of an integrated global market in guano, such changes may have influenced practices of recruiting and disciplining labor as far away as Sombrero Island. From 1854 to 1855, more than 25,000 enslaved black Peruvians forced the government to formalize their emancipation, while indigenous Peruvians and their allies succeeded in abolishing the Indian head tax and tribute system. No longer able to draw on such sources of coerced labor, Pacific guano island entrepreneurs, most of whom were Europeans and Americans, turned to China and the Pacific Islands, building a vast network of enslaved islanders, Chinese "coolies," debt peons, convicts, and "contract" workers, even as laws were passed banning such trades and practices. The West Indian and Peruvian emancipations of the 1830s and 1850s, then, meant guano work was not so much deracialized as reracialized.[24]

In 1854, American newspapers reported that contractors held nearly 600 Chinese laborers on the Peruvian islands. They were forced to work on five-year contracts and were paid at the officially stated rate of $48 per year. Worked seven days a week all year, the men, many of whom had been misled into believing they were signing up to go to California, "commence work in the morning as soon as they can see to work. They have five tons of guano to dig and wheel to a distance of over one-eight of a mile. It is all or nearly all, so hard that it has to be picked up; and if they do not accomplish the five tons by five o'clock, P. M.," white overseers whipped any stragglers bloody. Not only were the miners—thousands of miles from home and no-where they had ever expected to be—tortured into filling five-ton quotas by men speaking strange languages; their bodies were breaking down from starvation and malnutrition. Observers described the Chinese captives as suffering from swollen legs and arms, with "bad sores on their legs, feet and hands." With little prospect of escape or relief from the pain and drudgery, many apparently chose suicide by "leaping from the rocks one hundred feet high, cutting their throats and burying themselves alive. This last has actually been the case to my knowledge. One morning three were found who had so buried themselves; two were dead and one alive. The last recovered to pro-long his miserable existence for a short time."[25] Whether the men had delib-erately buried themselves or been crushed accidentally was unclear, but on the most brutal guano islands, so too were the lines between suicide, murder, accident, and just giving up.

The work of mining guano on barren islands under coercive labor regimes assaulted miners from within and without. Chute-loading caused blindness as the dust ripped through noses, mouths, and lungs, causing miners to faint, cough and spit up blood, and choke from ammonia vapor. They almost certainly contracted fungal infections caused by breathing in spores found in guano. In their guts, men suffered from cramps, diarrhea, vomiting, and infections from contaminated food and water. Eventually, even the very connective tissues in miners' bodies began to break down, re-sulting in swollen limbs, burst blood vessels, and scurvy. Overseers set "half-starved dogs employed to instill fear" onto flagging workers and whipped them with cat-o'-nine-tails. But even without the direct violence from over-seers and dogs, the work itself broke the miners. Trenches caved in, on some islands burying miners alive. Even under "the best of circumstances, their hands cracked open from constantly wielding picks and shovels, from push-ing heavy wheelbarrows, and from mucking about in cargo holds—sores, ulcers, and other open wounds frequently infecting but rarely healing." On islands where miners had most completely lost the struggle with employers, men worked until their hands failed, then overseers yoked them to carts and forced them to haul. And when they could no longer walk, they were forced to kneel and pick up guano as they crawled, literally scraped and ground to death as overseers tortured them into the stuff of fertilizer and light.[26]

On Sombrero, the workmen may have decided to act before Myrick Snow could make Sombrero like the Peruvian islands, could grind them into slaves, into dust. But we don't really know what any of the workmen were thinking or if they planned an attack on Snow at all. What seems far more certain was that none of the white Americans on the island or aboard the waiting ships cared to ask these questions, their fears and expectations of black revolt driving them into what seemed like a rehearsed counterinsur-gent response. "Considerable commotion followed among the whites," the *Petersburg Express* reported, "upon the discovery of the attempted murder, and the negro who committed the deed was immediately arrested by Capt. Burnell, of the Emma Tuttle," whose crew delivered (along with a full cargo of guano) the first news of the revolt when they arrived in Petersburg, Vir-ginia, a few weeks later. Back on Sombrero, "Capt. Birdsell, of the barque Warren, of Baltimore, placed" the arrested suspect in irons and hauled him "on board the latter vessel. It was their determination to take him into a United States port for trial." It was not clear that this was legal. The arrested suspect, later identified as Joseph Sahara (or Lahara, or Sarasa) was neither a citizen nor a subject of the United States. Moreover, the Department of State had repeatedly refused to recognize any American claims to Sombrero Island, explicitly and publicly denying petitions under the Guano Islands Act of 1856 to even offer the protection of the American navy.

If there had ever been a plan, it had begun and ended with murder, to remove a brutal overseer and protest American racism and slavery. After Sa-hara's imprisonment, the unrest turned into something bigger. It seems that some or many of the free black miners, feeling mistreated and determined to both challenge what they feared might be their own enslavement and make a broader statement against American slavery, may have decided to use the tools they had—mining carts, mined guano, and their own strength—to strike down an oppressor. Now, outraged that one of their compatriots had been kidnapped by Americans without inquiry or trial, the free black miners took control of the island and staged a labor strike.

Reports were clear that the miners secured neutrality from the twelve white employees, who "were unharmed, since they remained non-committal, being too much intimidated to attempt to suppress the rage of two hundred half savage negroes, and at the same time afraid to join with them against the vessel, knowing that punishment was sure to follow from the United States authorities." The miners gathered by the wharf, and when the captain of the *Warren*, where Sahara was being held captive in the empty hold, tried to return ashore, they picked up pieces of the guano they had mined and "stoned him until he was forced to take refuge back upon his barque." When the captain of the *Emma Tuttle* tried to force his own way onshore, he "was met with savage demonstrations from the insurgents, who were armed in their right hands with huge clubs, and in their left with heavy blocks of guano." He wanted to reach the company agent A. C. Elliott, but before they would let him pass, the miners made the captain "surrender all the English money he had in his possession." Elliott was furious. Barricaded in his office, he threatened the strikers by claiming that fifty Irish fighters were about to arrive with "as many revolvers and other instruments of government and civilization." The miners called his bluff. Enjoying some of the limited food and water they could easily claim to have earned, they stopped working and relaxed. The rumored Irishmen, meanwhile, seem never to have arrived. What eventually broke the several-week strike was the impending threat of starvation and dehydration. Remarkably, there is no evidence that a single person died, with the possible exception of Joseph Sahara.[27]

The day after the miners struck, the *Emma Tuttle* departed for Petersburg, Virginia, with news of the revolt and guano for sale. The *Warren* remained another two weeks, and then it departed for the United States, still with no guano, still with Joseph Sahara on board, and with the strike still unresolved. When the *Warren* arrived in Virginia, the crew handed Sahara over to U.S. marshals and brought additional news. Myrick Snow, to everyone's surprise, had lived and recovered. The *Norfolk Argus* also reported that after the *Tuttle* had left, A. C. Elliott had taken "a bold stand, exhibited great coolness and presence of mind, threatened to shoot the whole party, locked up the water and provisions, and showed them plainly that he was not afraid of them." Clearly. "Having seen some remarkable instances of his sharp shooting," the report continued wryly, "they were rather desirous to keep out of the way of the balls from his revolver. Assembling at night in front of his residence, and howling like a drove of hungry wolves, Elliot made his appearance, and immediately told them in a stern voice to form a line and range themselves so that he could kill as many as possible." The men did not do this. But they did laugh, "loudly and heartily, and the bold and fearless course [Elliott] pursued compelled the hungry fellows finally to submit and go to work again, without further trouble." Elliott controlled firearms, but the more important tools of domination in the end were thirst and hunger. Starvation and laughter may have broken the strike, but not before the men got rid of Snow and earned themselves a break. They may even have negotiated passage off the island, as a year later only forty-five workers remained.[28]

These were the worlds producing the means of match flames. But before these phosphates could become the phosphorus tipping American lucifers, they first had to cross the Atlantic, to where European manufacturers combined coal-fired heat, sulfuric acid, and wage labor to transform these flows of bones and shit into phosphorus and riches.

I n 1846, the workman Pierre-Napoléon Laurent spent his days and many nights distilling phosphorus. He and five other men had the job of shoveling coal and tending eight ever-burning furnaces, fires heating 108 sixty-liter retorts filled with acid-decomposed bones. The Lyon workshop of Coignet et Fils was almost always filled with thick clouds of sulfuric acid and white phosphorus vapor, reeking of something like garlic, that took a great deal of practice to breathe. Laurent was proud that he no longer coughed from the fumes. Not everyone could handle it. That man Seraphin, unable to stop coughing, had had to leave for another job in the factory.

It took twelve to fifteen hours of heating before any of the pure phosphorus vapor began to collect in the receiver vats, and nearly sixty hours before the process was done. Fortunately, once the charcoal hearths and the water containers were set, most of the men could leave and breathe fresh air, returning only to keep up the fires. But one of the men had to stay. In order to keep the necks of the receivers from stopping up, he had to remain inside, frequently clearing the necks out with an iron ladle or a spatula made from a spare bone. This was dangerous work. "In performing this manipulation," one writer advised, "the hand should be covered with a long glove of chamois leather well moistened with water," to protect hands and arms from scalding and chemical burns. The "same precaution," the writer added, "should be taken in forcing below the surface of the water, the portions of phosphorus which first come over and float at the top, being buoyed up by the gases mechanically enclosed within their substance."[29]

In a different workshop, free of the stench of sulfur but still brimming with the reek of phosphorus, Benoît Gagne and a partner molded the glowing phosphorus into sticks. From morning till night, the two men stood with a basin of hot water between them. Taking a mass of phosphorus, Gagne dropped it into the basin to melt it and plunged "into the liquefied phosphorus a tube 1 centimeter in diameter and 40 to 50 centimeters long." Sucking on the tube, Gagne drew up some phosphorus, and when the tube was full, he covered the end with his finger and carried it to a tank of cold water next to him, plunging the tube and his hands into the frigid liquid, where the stick of phosphorus solidified. Able to mold about sixty to eighty kilograms of phosphorus in a day, Gagne would keep at this work for at least the next ten years. And so, behind every match was a man who spent all day pressing his face into phosphorus at the very moment it reached its unsteady peak of potential energy, mouth-pipetting the means of lucifers. Unsurprisingly, molding was "sometimes attended with terrible accidents, in consequence of unskilful workmen drawing up the phosphorus into their mouths, where it immediately takes fire."[30]

In 1845, the Health Council of the Rhône recommended that Coignet's phosphorus factory cease stocking fresh bones on the premises and reduce its output of smoke, claiming "their workshops belong to the first class of unhealthy establishments." The factory was located on an open plain far from any houses, but the chimneys spewed forth such "a great quantity of smoke and very unpleasant emanations" that notwithstanding the distance from the city, many "complaints have been made against them: they have been wrongly accused of exerting a deleterious effect on the vegetation, and, with foundation, of spreading in the atmosphere a very disagreeable foul odor." The claim about vegetation was revealing. The waste spewed out from phosphorus factories themselves may not have killed surrounding flora, but they were almost always located in barren landscapes made toxic by the sulfuric acid manufactories next to which phosphorus makers tended to locate.

When John & Edmund Sturge, the largest phosphorus manufacturers in England, decided to relocate in 1851 to expand their production, they chose Oldbury "chiefly so that the works should be near the South Staffordshire coalfield and Chances' alkali works. This had been established by Chance Brothers in 1837 (against strong local opposition)." The locals were right to worry. So corrosive were the gases emitted that "'even grass and the hardiest of plants failed not to succumb. Not by incessant labour could Oldbury housewives keep fire-irons or other utensils of bright steel in any desirable condition of cleanliness; metal tarnished in a single night.'" The Sturges moved to Oldbury because they could capitalize on others' hard-fought work of establishing an industry-dominated toxic landscape, but they also moved to be close to coalfields and the alkali works, since sulfuric acid was a

Lucifer Matches and the Violence of Phosphorus

key material in phosphorus production and heating was the largest expense. There was literally a pipe delivering sulfuric acid from the Chances' works into the new phosphorus factory, with only a wall between the two.[31]

In 1850, before relocating, the Sturges were manufacturing 26.5 tons of phosphorus a year in 120 clay retorts from 265 tons of bone ash and 232 tons of sulfuric acid. By 1863, Albright & Wilson, as they were renamed, produced over five times as much in 648 retorts. Their main competitor was the French firm of Coignet et Fils. Most of the French and English exports went to the enormous match industries in Germany and Austria, but the manufacturers also competed over exports to the United States. Before 1851, American match makers had been buyers of the Sturges' phosphorus, but American demand grew faster than the Sturges' ability to supply it. As the manager noted one year, "'We have lost . . . one order for a ton of phosphorus from not being able to get it ready; on the other hand we hear what was sent of ours to America sold well and the buyer wishes to have more.'" American demand kept increasing, but no serious phosphorus production was established in the United States until at least the 1880s. "An Englishman who had had some experience of this manufacture attempted it in America," noted one treatise. "We presume it was the high price of labour and fuel which led to his abandoning the attempt." In 1863, Coignet and Albright consumed thousands of tons of bone ash from the skeletons of hundreds of thousands of cows and hundreds of tons of guano, phosphates burned using 5 million pounds of coal and dissolved using 5 million pounds of sulfuric acid to produce over a half-million pounds of phosphorus.[32] And all of it was destined for match factories, where the struggles over phosphorus's materiality raged at their most intense, entangling the politics of age, class, and gender.

Ironically, it was how these struggles became entangled with the politics of European states that produced the best evidence of how Americans would have transformed this European phosphorus into lucifer matches. Generations of German, English, and French workers struggling to stay healthy in match factories, or to stay out of them altogether, had helped to launch government investigations by resisting and by raising awareness of the disfiguring jaw disease. But they needed powerful, activist, and reformist government allies to move their hidden struggles into official records, conditions that appear not to have existed in the United States. The practices and politics of making matches were, it was true, deeply embedded in the particular conditions of labor, environment, and place. Match making in Lyon did not look exactly the same as in Paris. The London mixture of small workshops with large factories made for a different geography of labor than that in Manchester. Nor were conditions identical in cities and factories across the United States. But some practices and processes were close to universal. Everywhere children constituted the great majority of the workers. Everywhere the chemistry of production dictated that the work had to follow certain paths. And everywhere the inescapable ecology of phosphorus formed the terrain of struggle.

A FEW dozen miles from Manchester, at six o'clock in the morning, fifteen-year-old Thomas Fisher approached a large building on the outskirts of Liverpool, overlooking the sea. He'd been making this same trek to N. Martindale's match manufactory for the last four years, and few knew it better than he did, now that for much of his twelve-hour workdays Fisher ran messages to its various sections. The nearly 300 people — mostly children, mostly "poor Irish" — who worked at peak production were, according to one observer, "very ragged and dirty, very few with shoes or socks." Fisher himself was missing a finger and probably arrived barefoot like the rest. Thankfully, because of the weather and the season, the windows running along both sides of the long building were open today, allowing the sea breeze to blow through, helping to at least somewhat clear the air of the overpowering sulfur and phosphorus vapors. In winter, when production was highest and the workdays longest, the cold and storms meant even window ventilation wasn't possible, and the steam-powered fans were never enough.

Martindale's was one of the larger match producers in England, but it was far from the largest. It employed no more than half the number of people working at Dixon's massive complex outside Manchester, through which the world's most spectacular march of glowing children passed every morning and night. Nor did Martindale's appear to be notable for any significant innovations in technologies of production or management. It was, in a word, typical. Because of its size, its location in the city, its labor force, and the chemistry of its match heads, Martindale's was also likely representative of many American establishments. But because of what Thomas Fisher and the other boys and girls working in this particular factory would tell John Edward White — the author and lead investigator of the report titled "The Lucifer Match Manufacture" for the 1862 Children's Employment Commission — during his visit to Martindale's that day, no other lucifer workplace entered the written archive with such comparably rich descriptions of the materiality of production. No other set of interviews gave a better sense of precisely how people made, contested, and navigated the spaces of match manufacture, of what the work actually felt like.

Lucifer Matches and the Violence of Phosphorus

Entering through the door in the middle of the factory, which consisted mostly of one very long, one-story room, Thomas Fisher passed the boys helping the two dippers, each man dipping the frames of sulfured splints in a hot phosphorus paste that the boys slathered on a dipping stone by the door. Nearby, under steam-powered fans, drying racks held the completed frames of dipped matches, from which, John White wrote, the luminescent "vapour may be seen going up." Although the "composition"—the phosphorus and chlorate of potash dipping paste—was "mixed and prepared in a separate building by two men alone," Fisher would still have been greeted by boys engaged all day in stirring batches beside the dippers to keep the mixture from either hardening or separating. Fisher's jobs at the factory varied from day to day, but if he was running messages this morning, he might have gone to one end of the building, where "a great number" of boys were filling the frames with already-sulfured splints, to deliver quotas from the manager. Or he may have walked straight across the floor from the dippers to collect and tally piecework tokens from the children cutting dried matches and putting them into boxes lined with strips of sandpaper. At the opposite end from the frame fillers, and in a large basement room below, women and girls were making those boxes, and the manager may have tasked Fisher with carrying finished ones to the cutters. Fisher may have run messages to the man and five boys who were dipping bundles of splints in sulfur and rolling the bundles out in an "adjoining room, cool and with a current of air through it." [33] But if he wasn't running messages, Fisher almost certainly went straight to the engine room, separated from the rest, where the splints were made, the first stage in the process of manufacturing lucifer matches. As he moved about the factory, Fisher slipped between the interstices of a disaggregated chemistry of flame. The sandpaper on the boxes for friction, the phosphorus and chlorate of potash to spark, the sulfur to catch, and the wood to burn were all essential for a lucifer flame to be born. Just as essential was that the components of this reaction be kept separate and stable until a consumer wanted to strike a match. But this was never entirely possible. Disaggregating and reaggregating flames was dangerous work, as Fisher well knew, and it began with trees and power saws.

Match factories consumed astonishing quantities of wood. At Dixon's works in Newton Heath, which produced between 6 million and 9 million matches a day, the three-acre timber yard was piled high with huge trunks of American pine worth £10,000. And though matches were incredibly cheap, the wood was of the highest quality. For the machinery to work, the wood had to be the best pine, perfectly clear-grained, as any knots could jam up the rapidly whirring blades. In Martindale's engine and machine room, an adult engineer and his assistants tended the engine while boys "wait upon the men who work the splint-cutting machines to pick up the splints as they fall." Thomas Fisher's current job was to clean "the engine on Friday and Saturday. It is stopped then. Never cleaned it when it was going. The engineer and [Fisher] do it." He and the engineer cleaned during breakfast at eight or dinner at noon, as the "engine always stops for those times." He used to clean part of the machine it powered. That was how he lost his finger, and only "12 months after he came. Got it in the slide when cleaning it. It is the slide that makes the splints. Used to clean them every day." But as Fisher learned, although his finger could never be replaced, he could be. "Since he has lost his finger another boy does it," probably ten-year-old John Fannan, who now worked the middle of the three match machines.[34]

These match machines were, to contemporary observers, wonders of the age. "Seven thousand two hundred Congreve splints for four-pence three-farthings," rhapsodized one writer for *Household Words*. "This shows how we cut wood in the nineteenth century." As with Martindale's factory in Liverpool, the London factory visited by the *Household Words* writer had three such machines operating about ten hours a day, producing together around 500 splints a second, 3.6 million an hour, and more than 10 billion in a year. Keeping these machines running and from suffering damage meant boys like Thomas Fisher and John Fannan had to risk damage to themselves. Fannan cleaned "the machine Saturday, and every night. Never does it while it is going. If he did he might cut his hands off." Such injuries were common. Boys regularly lost fingertips and even caught their heads on the circular saws. While the men expertly fed blocks into the cutting machines, Fannan, alongside a few other boys, scrambled to pull "down the matches with his hand as they are cut" without losing fingers to the blades, and crawled along the shop floor collecting the splints that fell. He and the other boys then carried their splints into an adjoining room where they straightened and tied them into half-gross bundles. As Fannan described the whirlwind work of bundling, the "lads do that with their hands, and make a bundle as quick as the machine can make three cuts," that is, bundling and tying seventy-two splints in under two seconds. Once the splints were bundled, Fannan piled them up on drying racks near a hot stove.[35]

The next stage in the production process was sulfuring, work that required navigating a more chemical form of violence, different from the mechanical violence of splinting. As John Fannan collected splints from the machine and bundled them in the drying room, he encountered the "lads that work with the man what dips the sulphur," who were bringing "the bundles to him out of the store room, and when they are dipped roll them out." The work of sulfuring and dusting coated the people who did it inside and out with stinking, burning sulfur. "The scorching wood is pungent to the eyes," Commissioner John White wrote; "the sulphur causes irritation of the throat and coughing." The work of dusting, or rolling out, the bundles filled the air with yellow dust that "covers the face and clothes, and is of course taken in largely at the nose and mouth, causing cough, choking, &c."[36]

Now came the stage of frame filling, the work that employed "by far the larger proportion of all children engaged in match manufactories." Taking the finished bundles of sulfured splints to workbenches at one end of the Liverpool factory, dozens of girls and boys separated them into loose sticks and set about placing them into "frames," or "clamps." White described the frames in common use at the time as "formed by successive layers of from 30 to 50 thin strips of wood, each strip notched on one surface with about 50 cross grooves." Frame filling was done entirely by hand: a "strip is placed between two supports, grooves upwards, and a match placed in each groove. A second strip is placed over the first and the process repeated till the frame is full, thus holding (50 × 30) 1,500 matches or upwards."[37]

Most of the people working in match factories filled frames, but it was probably the least documented process of production. Because the job was repetitive, observers usually mentioned filling only when imagining machines that might replace fillers. Because fillers were all children, few considered their work a craft. Fillers knew better. Women's advocate Virginia Penny learned that at one American match factory she visited, "Girls are paid for filling the frames in which they are to be dipped, sixty-two cents 100 frames, each frame containing 1,500 double, or 3,000 single matches." At another factory, she "saw small girls and boys putting matches in the frames to be dipped. They are paid sixty cents 100 frames, containing 1,500 double matches. They can seldom fill more than 85 frames a day." Stop and think about what this meant. These mostly German and Irish girls and boys worked thirteen-and-a-half-hour days in American match factories and were paid by the piece. Assuming a girl worked constantly without stopping even for meals, to fill 85 frames with 127,500 splints in that time meant she placed splints securely in their grooves at a minimum speed of 2.6 per second. And these were only the average girls and boys. "A brisk hand," Penny was informed at that same factory, "can earn from $5 to $7 a week," and a "lady told me she knows a girl that earns $6 a week" for filling frames. Again, even if we make the most conservative assumptions, the slowest a girl or

Filling and dipping a frame. Children filled these frames layer by layer at incredible speeds, slotting splint after splint into grooves cut in the wooden strips strung and stacked across the frame. Meanwhile, other children assisted the dippers (usually adult men) as they dipped the splint ends of completed frames in "compo"—a paste of gum, chlorate of potash, and phosphorus—and then flipped the frames to dip the other ends of the splints. *Left*: Filling a frame by hand. Engraving in Louis Figuier, *Les merveilles de l'industrie* (1860), 573. *Right*: Dipping a frame. Engraving in Louis Figuier, *Les merveilles de l'industrie* (1860), 575.

boy could work to earn $7 in a seven-day week without a single break would be the astonishing speed of 5.1 splints a second.[38] Considering the limits of exhaustion and hunger, the actual speeds at which children filled frames were almost certainly higher, perhaps considerably so. But before these children could make their fingers blur across the frames, leaving wooden strips and splints perfectly in place, they would have been as slow and clumsy as any seven-year-old. For employers to exploit these children to their fullest potential in time, children growing up in conditions of extreme poverty had to retrain their nerves and muscles willingly, by force or by desperation, to become superhuman fillers.

In Martindale's, whenever ten-year-old Patrick Lovan or seven-year-old Richard Jones or any of the dozens of other boys finished filling a frame, they would "bring the clams up to the dipping slab and put them on the ground." There, they received proof of their work as a "man gives each a ticket," which they would later exchange for wages, around one shilling a week. Most of the fillers brought these wages home to their mothers, who might "give" their children a small share. Richard Jones's "Mother brought him" to Martindale's and "gives him 1d. The 'missis' pays him." Arriving as early as four in the morning and working as late as nine at night in the winter, the boys worked a minimum of twelve and a maximum of seventeen hours each day, with an hour and a half for meals, filling their clamps at blinding speeds until the bell rang loudly for breakfast and dinner. Patrick Lovan testified that he "never has tea till he goes home. Does not even in winter. Is very hungry then." Indeed, food and hunger were ever-present forces shaping how the children worked, struggled, and allied with one another. Lovan explained that he did not have "anything in his pocket to eat, because he eats all his dinner at dinner time. Some of the lads get bread and their tea fetched." His story was not only one of poverty, but one of betrayal. Patrick's mother used to send tea to her son, "but the lad who lived in the house and had to bring it used to eat it, and 'Mother wouldn't send it no more.'" Antebellum courts allowed minors, unlike adults, to break labor agreements at will, but the parents, poorhouses, orphanages, or other adults who contracted out the children they claimed, or from whom they demanded money in exchange for food and shelter, deeply circumscribed children's freedom to find new work. And so while the children starved and stole food, sometimes hijacking another's lifeline to resources outside the factory, most still had to persist in their alarmingly fast work.[39]

Both parents and employers could rely on the overlapping structures and desperations of class, age, gender, and hunger to force child workers to find their own nourishment, even if they had to steal or share among themselves. The "overlooker" at a Lambeth factory claimed that "boys are not beaten here," but perhaps that was because hunger exacted its own, sufficient discipline. The witness claimed that "the children often steal bacon, cheese, 'and all manner of things.' Has known frequently boys eat all they bring for the day early in the morning, and then have nothing until 9 P.M. Then they become so hungry that they will steal anything they can get." Stealing food to survive the workday, these children were engaged in a complicated moral economy of cooperation and competition, taking sometimes from one another, other times from their employers or merchants, and sharing with some but perhaps not with others. Employers, meanwhile, fought back against such practices of taking; one manufacturer claimed the boys who worked for him were "generally of the lowest class and characters: they are searched each time they leave; was formerly very much robbed, till searching was adopted."[40]

William Flenn, age thirteen, had been working at Martindale's for six months alongside one of the two dippers when he spoke with John White. "This boy," White noted, "has splashes of composition on his forehead." Here was the source of the glow, the stink, and for far too many match workers, the indescribably painful and disfiguring violence of "the jaw disease," a disease most workers simply and revealingly called "the compo." Flenn was so covered with this luminous, blue-colored combustible paste of gum, phosphorus, and chlorate of potash — so covered with "compo" — that even "his hands are blue, especially under the nails." As the fillers like Patrick Lovan left their clamp frames on the ground by the dippers, Flenn picked them up, "shoving the 'clams' on the stone for the dipper." Pick up, shove on, smooth out the compo with a knife, pick up, shove on, smooth out: this was Flenn's principal task all day. The dipper, always an older man, took the frames that Flenn shoved on the stone and carefully dipped both ends of the splints in the compo the boy slathered on a heated slab. Experienced dippers could dip more than two frames a minute, or 360,000 matches an hour.

A "disease," in one sense, was just what people called violence that they couldn't understand, where the effect was obvious but the cause was mysterious. Calling such violence "the compo" or "the jaw disease," then, was not so different from calling it magic; but one name located the danger in matter and place, was a warning to others, while the other name located it only in the body, in its effect, and did nothing to identify the cause of suffering. "The compo" wasn't a direct accusation of the owners of match manufactories; it wasn't quite a calling out of the Martindales and the Dixons for mysteriously (and invisibly) wielding the industrial magic of phosphorus to steal life and riches for themselves while leaving pain and suffering and death in the bodies of match workers, but it came closer than "jaw disease." We would do well to follow the workers' lead, to let the structures of violence allowing some to knowingly and deliberately thrive at the expense of others to come into focus. The precise mechanisms or agents of the violence may not have been clear — indeed, they may have been sufficiently opaque that those who organized and owned the means of production could claim irresponsibility—but that didn't mean it wasn't violence, and it didn't mean no one was responsible.

The more proximate the work was to the composition, the more dangerous it was. "While the dipper is dipping," William Flenn explained, "the 'compo' flies off sometimes and gets on their flesh, and burns them. (Shows a festery burn on wrist.) It burns holes in his waistcoat sometimes. (I saw them.)" The work continually covered Flenn's hands in hot compo, "but he washes them every 5 or 10 minutes in the can. The water is changed about every half day," but not so often that the water was ever any color but toxic blue. And anyway, the regular washing couldn't get rid of all the compo on his hands and forehead. Cleaning his clothes was even more difficult. "His waistcoat shines in the dark," White noted, although Flenn claimed it did "not smell except it catches fire, then it smells very nasty. Catches itself if he sits near the fire." Not only, then, could Flenn find no relief from his stink and his glow; he had to do so cold, unable to risk getting warm.

William Flenn's dipper employed another boy as the "taker out," who carried the completed frames to the drying racks. In many establishments, like at Dixon's in Newton Heath, the racks were situated in a separate, intensely heated room, and the taker out would have to continually pass in and out of hot clouds of phosphorus vapor as he left the dips to dry.[41]

Along with dipping and working in the drying room, mixing compo exposed those who did it to particularly concentrated quantities of absorbable phosphorus, accumulations of toxins that workers tried to navigate by distributing exposure over time and bodies in rotating shifts. Twelve-year-old John McKay sometimes took an hour out of his day of filling frames at Martindale's to stir a batch of compo, each batch enough for four or five dippings. When it was his turn to stir, McKay almost always worked alongside another boy named Patrick. At any given hour, four boys would be busy stirring, but no one was assigned to the work full time. Years of children collectively refusing to sit for prolonged periods over what they understood to be the most poisonous material in the factory had likely convinced employers to instead attempt to spread the pain around. And the work of mixing the compo was "attended with danger not only from the fumes given out, but from the risk of explosion unless carefully conducted." Sitting "up on the bench beside where master dips," McKay made sure to keep his nose away from the pot of compo as he stirred, watching as it "smokes up." Once he started stirring, he couldn't stop until it was done. This was one reason he tried to avoid stirring, "because it makes him so sore in his arms. Makes them tired up at his shoulders." Nor did it help that his hands and cuffs quickly were covered in "the glue and the phoss from stirring the batch," which stuck even though he "always washes it off after he has done stirring the basin." As employers reassigned boys like John McKay to stir batches throughout the day, spreading the violence of phosphorus in its hottest, most toxic form widely and thinly, the boys continued to avoid the work, forcing employers to seek fresh bodies to serve as temporary toxic sinks. Such collective action meant that McKay, though at the factory for four months, and though that factory required dozens if not hundreds of batches to be prepared every day, "has not stirred more than eight times altogether."[42]

Cutters were also heavily exposed to the phosphorus vapor. Directly across from the dippers and drying racks, John Fen spent his day at Martindale's fetching frames from the racks and carrying them to his work station. There, he pulled the splints out of the frame, ambidextrously made loose bundles, grabbed a bunch, quickly knocked the ends even, and then used his foot and a hinge knife to cut the double-ended splints into twice the number of matches. Last, he put those matches into boxes, 100 a box, and started the process again, all "with a rapidity almost unexampled; for in this way, two hundred thousand matches are cut, and two thousand boxes filled in a day, by one boy." That was on a good day. In hot weather, cutters could not "cut them so fast for fear of fire, and they have to stop to put them out," which they often did by hand. John Fen told John White how, after he brought a frame to his bench and grabbed a handful of matches, sometimes "they catch fire when he cuts them. Puts them into a sand pot beside him to put them out. Often burns his hands then. 'Look here at my hands with burns' (shows his palms covered with scaly scars, one showing a little blood). 'It smarts me a bit then.'" He had only being working for four months, and was so young he didn't know his own age. Fen's brother claimed he was only seven years old. Still, he had already learned, through trial, pain, and instruction, much of the embodied knowledge necessary for cutting. Fen knew when the splints were "dry enough by feeling at them with his hands. If they are not dry the stuff comes off. Goes and washes his hands then for fear the composition should come afire on his hands. It does catch alight sometimes, but he cannot see it much, it only smarts him. 'It smarts the other boys too o'their hands; you will see them a swinging them like this' (shows)."[43]

Most of the other girls and boys his age filled frames, and John Fen seemed envious that they were less fully trapped by the structures of work and violence than he was. "A bell rings for the clamp fillers to leave off," he noted, "but the boxers stay later till they have boxed all the matches that are dry enough." Even lunch was a comparative luxury. "The clamp fillers go to dinner at twelve when the bell rings," Fen observed, but not so the boxers like him. They "always stay and mind their work at dinner time. If he were to go out in the yard 'it would be stole,'" as boxers were paid for the number of boxes of matches they handed to employers, not for their time, and not directly for their work. Fen had to be vigilant lest another boxer steal and get paid for the products of his labor. Because frames were so much larger and heavier than boxes, this was a much bigger problem for boxers than fillers. Fen did earn more than double a filler for similar hours, being paid by the piece and making 2s 6d a week, but with his parents taking all his wages, he "never has more than 1d himself," the same as any filler. John Fen's parents thus daily sacrificed their son for twice the shillings collected by the parents of fillers, but he saw not a penny more for it. And Fen, although only seven years old, seemed entirely aware of the injustice.[44]

At Martindale's, box making was solely the work of women and girls, and the way the managers organized the division of labor also produced segregated spaces in the factory. At one end of the long floor, boys filled frames, while at the opposite end, women and girls made boxes. Even more conspicuously gendered was the "large flat-roofed basement room." The dim basement had formerly been used for match making; some smell and shine almost certainly lingered, no matter how they had cleaned it. With allies in town and in government, Martindale's workers had recently pressed the factory owners to "improve" the buildings with better drainage and ventilation; the basement match workers had seized the opportunity to escape to the newer, airier spaces. In their absence, the basement was "now filled by women making match boxes, a much more suitable work for the place."[45]

The spatial politics was complicated. The boys and men who had previously made matches in that basement had managed to exploit their relatively valuable position in the production process and use the increasingly popular knowledge of phosphorus toxicity—knowledge that thousands of match workers had made visible through their maimed bodies and the stories they told to doctors and officials over decades—to place public and state pressure on their employers to change their conditions of life and labor for the better and move into the relatively healthier spaces of the "improved" main floor. But they had also mobilized the power of their gender. The women who now moved downstairs into a poorly ventilated room that had absorbed years of phosphorus vapor would have surely preferred a healthier space to make boxes, too. But it is notable that of the two spaces for box making, adult women had secured the basement, the space least exposed to new fumes, while all the girls, along with a few women, made boxes in the same airspace as the dippers and cutters. Unable to wield the power of age, the girls did manage to manipulate gendered expectations into securing separate sources of water to wash in — water blessedly free from the accumulated phosphorus that made the boys' wash cans into toxic basins. As thirteen-year-old Isabella Sumner told White, she "washes in the boilers. The water is clean (Bridget Maclachlan says the other girls wash in the same way)." Significantly, Sumner was not paid directly by the manager. Rather, she, and probably the other girls, "works for Bridget Maclachlan, who gives her 3s. a week."[46]

The Liverpool factory's owners welcomed the nested hierarchies and power relations of wage labor, contracting, apprenticeships, and what might best be described as the industrial family directly into the spaces of the factory proper, but most match manufacturers, even behemoths like Dixon's, relied on outworkers to assemble most or all of their matchboxes. The same was true in the United States. The work of assembling boxes at P. T. Ives's match factory in Wallingford, Connecticut, was put out to the village: "'The match boxes are manufactured mostly outside the factory, by boys and girls of the village, and at an apparently very low figure. The materials are all prepared at the shop; and the box-makers receive only seventy cents per thousand for the smallest boxes, and one dollar and twenty-five cents for the larger ones.'"[47]

The same politics that pressed women into sewing men's shirts deep into the night were navigated by other working-class families by taking in matchbox making. Here, too, to meet the demands of landlords, patriarchs, and match manufacturers, the work had to extend into darkness, and that meant either a dangerous reliance on brighter camphene lamps or straining eyes by the nonexplosive but dimmer light of stearine and tallow-dip candles. It also meant relying on some of the very lucifer matches for which they were making boxes.

Nowhere were the contours of nineteenth-century industrial childhood clearer. As a woman led writers for Charles Dickens's *All the Year Round* "up a dark passage and a darker stair" to her domestic box-making workshop, they passed "huge blocks of wood on the first landing, of the size and shape of those strewn about shipwrights' shops and dockyards, and now walk into an atmosphere redolent of deal-shavings, sulphur, and dye." The thirty or so laborers in the two garrets were children, many between ages three and seven, younger even than the youngest match frame fillers. "'My master and me ain't got no family of our own,'" the woman explained, "'so we call these girls and boys our children, and though they've to work hard, they're well off compared to hundreds of others.'" This was more than simply a factory on a smaller scale. The possessive language—"our children"—and the implication of charity in the midst of an industrial workshop could only have so seamlessly masked the exploitation of child labor inside the social relations of households and marriage.[48]

The industrial chain of outworked boxes and the social relations shaping it could extend even further from the factory. "Besides making up boxes on the premises," her husband also sent "out the creased slips of wood and the paper labels to women and children who work at home; he acts in short, as middle man between the dealer and the labourers." Aided by the privileges of age, marriage, gender, and property, this man had climbed near the top of a pyramid of outwork, trading the products of that work with those who commanded the far taller pyramids of match factories, "earning" himself a comfortable living of about 15 shillings a day. At the bottom of the outworking pyramids, this twice-sweated work formed a major sector of employment for women and children among the most precarious reaches of the free urban poor. In the United States they were usually Irish and German immigrant families. They were also often the same families sending children to the match factories.

And in the tenement domestic workshops of outworking box makers, though far from any phosphorus vapor, the chemistry of flame still shaped the work and rubbed fingers bloody. To make the completed boxes, the women and children had to assemble five separate articles pre-made at the factory or by a "middle man." Using paste they provided themselves, they put together by hand two "slender shavings of wood, one each for its inner and outer part; one label of coloured paper for the half containing the lucifers; one printed label bearing the dealer's names for the outer box; [and] a square piece of sand-paper to strike the matches on the bottom." Most of the work was tedious but bearable. The friction necessary to ignite the matches, however, meant that not even the box makers working miles from the factories could fully escape the violence of industrial phosphorus. "The manipulation of this sand-paper is the most painful part of the work," the children informed the magazine writers, as the "rough surface cuts the children's fingers, and leaves them raw and bleeding, much as if the cuticle were rubbed off with a file; for each bit of sand-paper is smoothed and patted down by hand, and many hours of this work produce their inevitable effect." Some of these work-bloodied children were astonishingly young. At one house, the "youngest, who 'will be three the 7th of next month,' is an active member of the staff, and has worked regularly for more than a year," meaning the child had been at most twenty-three months old when she or he began this work.[49]

Lucifer Matches and the Violence of Phosphorus

Testimony suggested that box making involved thousands of "families" stealing the work and wages of their own and their neighbors' children, of children from the workhouses where cities forced the poor and "vagrant" to work in exchange for a roof, and of inmates in juvenile "reformatories" like New York's House of Refuge, feeding but not paying these "domestic" workforces. One witness had "heard that they get them from the workhouses. Believes, as they only receive 3d. a-gross, that the neighbours' and other children are only recompensed by having their food, such as it is. A child must be a good worker to earn 3d. a-day, making a gross; not 1 in 50 could earn this much." Other employers pressed their advantages even further, outright stealing, or "keeping," the wages of each "fresh set" of children before discharging them without pay. As one manufacturer testified in defense of the late hours he kept "his" children working, he had "known the case of a person, a maker, who would take several girls 'on liking,' keep them a month, and then make an excuse that they would not suit him, and discharge them, without paying any wages." [50]

The production of lucifers exploited a family crisis driven from below with assistance from above, but it did not destroy the power and labor relations of the family. Rather, it respatialized them. Some children escaped the often literal toxicity of household match workshops to work in larger factories, but they still brought wages home to their mothers, who were usually responsible for organizing their children's labor, whether at home or hired out. These children gained some real measure of independence, able to individually and collectively shape the match industry by selling labor to particular factories they knew or had heard to be healthier, even when they didn't pay better — pay they would keep only a fraction of anyway. The very real changes in windows, fans, architecture, and new spatial organizations of work were concrete expressions of the power and resistance of lucifer-making workers, including children. [51]

Because the match industry was comparatively older in Germany, one witness claimed experience had bred disdain, and there was "a great dislike in Germany generally amongst the workpeople to working in a match manufactory, and many of the manufacturers whom he knows complain that in consequence they have great difficulty in getting enough hands." Here was, John White implied, a vision of England's future, and so he reported, not disapprovingly, that many German employers solved this labor problem by conscripting "prisoners for the work. The manufacturer contracts for their labour because he can get it cheaper. Ten years ago the labour of a full grown man could be got for about 2d. a day," or one-fifteenth the pay for, in

Liverpool, the seven-year-old boxer John Fen. Arthur Albright, the owner of probably the largest phosphorus manufacturing works in the world at the time, testified in 1862 that based on his many connections in Germany, now that free workers refused to work with phosphorus, "the labour of criminal prisoners was farmed out by the government, and in Hesse Darmstadt also I know of a case where prisoners, several hundreds I believe, were so employed."[52]

In England and the United States, meanwhile, where state officials were less interventionist, the spatial politics of lucifers had reshaped the compo itself, replacing some of the toxic phosphorus with the nontoxic, but differently dangerous, chlorate of potash. "The act of mixing the composition itself is dangerous, especially in England," noted one observer, "where chlorate of potash is used, which is very explosive. In Germany they use no chlorate of potash, but much more phosphorus." In 1862, a worker-now-owner in Birmingham claimed that it was the German-style "silent matches that are the worst; would sooner work for a week at noisy matches than for an hour with the silent. The common blue are silent and nearly all phosphorus." In Germany, free workers had largely escaped the ecology of production, but factory owners made sure that unfree workers, although they had less to fear in terms of fast violence, absorbed far more slow poison. In England and the United States, workers had succeeded in forcing employers to eliminate some of the slow violence of phosphorus, but only by increasing the fast violence of explosive potash. Voicing objections similar to those they made for potash, many British and American manufacturers claimed that the entirely nontoxic, but considerably more expensive, red phosphorus—made by superheating common phosphorus and which reformers alarmed by the epidemic of the jaw disease would push, in vain, for manufacturers to adopt from the 1840s until the end of the century—was explosive and so, for the safety of their workers and their property, wouldn't use it, especially with chlorate of potash.[53]

The chemical makeup of compo was an expression of factory politics circumscribed by the conservation of phosphoric violence. Where capitalists most dominated production, as in Germany, a pure phosphorus compo saved employers money and was the least likely to start property-damaging fires, but it produced the most toxic working bodies and also exposed consumers to higher risks of poisoning. In Britain and the United States, the mix of phosphorus and chlorate of potash balanced and diffused the risks of poisoning and combustion. But the introduction of red phosphorus made almost everyone unhappy: it cost more, accelerated burns and fires, and then didn't even light as easily for consumers. Absent larger changes in the practices and politics of production, eliminating poison was neither cheaper nor, in an absolute sense, safer, and workers, consumers, and capitalists all had reasons to resist. The chemistry of matches meant that capitalists and workers and parents were always engaged in a highly uneven, but still real, three-way tug-of-war between explosions and toxicity and profit, redistributing and navigating risks that extended beyond the factory all the way to consumers.

L UCIFER matches may have been small, cheap things, but they remade consumers' worlds, and people knew it. In 1850, the first item to be featured in *Household Words*'s series on cheapness was the lucifer match. "Some twenty years ago," the article opened, "the process of obtaining fire, in every house in England, with few exceptions, was as rude, as laborious, and as uncertain, as the effort of the Indian to produce a flame by the friction of two dry sticks." Before matches, to get a light in a tenement or cottage at night required planning, skill, and time. When, for instance, a crying baby awoke, her mother, who "was soon on her feet," had to have already made and stored tinder in a box, known how to find the box in the dark, been able to wield flint and steel well enough to light the tinder without losing too much skin, and then transferred the tinder flame to a piece of kindling or a candle. "Click, click, click; not a spark tells upon the sullen blackness. More rapidly does the flint ply the sympathetic steel. The room is bright with the radiant shower. But the child, familiar enough with the operation, is impatient at its tediousness, and shouts till the mother is frantic. At length one lucky spark does its office — the tinder is alight." If lighting tinder was tedious, making it was worse. "The domestic manufacture of the tinder was a serious affair," the magazine reminded its readers, when at "due seasons, and very often if the premises were damp, a stifling smell rose from the kitchen," where women and children spent the day drying out wood shavings and scraps of fabric, while the "best linen rag was periodically burnt, and its ashes deposited in the tinderman's box, pressed down with a close fitting lid upon which the flint and steel reposed." With phosphorus and chlorate of potash, however, "a blessing was bestowed upon society that can scarcely be measured by those who have had no former knowledge of the miseries and privations of the tinder-box. The Penny Box of Lucifers," each holding 100 incredibly cheap and easy-to-light matches, was, the writers proclaimed, "a real triumph of Science, and an advance in Civilisation."[54]

Matches truly did make lighting easier for people, especially in cities, where maintaining continual fires was both expensive and dangerous. By making it possible to cheaply create and extinguish flames, matches also helped cash-strapped working families better conserve precious illuminants; no matter how cheap lamp fuels were becoming, they never approached the inexpensiveness of lucifers. Camphene in particular multiplied in an unstable symbiosis with lucifers. When heated, such as by a burning lamp, camphene evaporated even faster, making it incredibly dangerous to walk around with a lighted camphene lamp if any other open flames were nearby. But matches made it easy to quickly put out and relight lamps and fires — so easy that many impatient or time-pressed camphene users tried to extinguish, refill, and relight their lamps before the burning fluid had cooled sufficiently to stop throwing off flammable gas, and a number of explosions happened during these transitions. Cheap lights coupled with cheaper fires meant increasing numbers of households could afford, and could be forced into, the kind of night labor trapping ever more outworkers, and they made it feasible for manufacturers to displace the costs of lighting, and the risks to property, into the homes of the working poor.[55]

As thousands of children made hundreds of millions of matches a day in factories, other children of the working poor purchased and peddled those matches from either manufacturers or shopkeepers, who generally tried to get such flammable wares out onto the street as fast as possible. In 1853, a writer for the *New York Daily Times* set out with some friends to see how the poor young boys of the city, "the news-boys, dock-loafers, and whole class of vagrant children amuse themselves," and so they visited the National Theatre on Chatham Street. There the reporters found the "pit was crowded full of the worst specimens of that great class — the New-York ragged boys" who, away from adult authority, spent their time "peddling matches . . . all the cold day." Another *Times* writer told the story of Maggie, an orphan girl he met selling matches on the street. She told him she had made 2 shillings and 6 cents on her rounds, working from seven o'clock to half-past four. She began by going down "to the Battery, Sir. I buys 'em in Reade-steet; . . . I gives a cent and a half a box, Sir, and sell 'em for three cents. Sometimes I don't make three cents a day; and sometimes two shillings." "They buys 'em best in the stores," Maggie said, "and sometimes the gentlemen asks me questions, and they don't believe I'm an orphan and working for my aunt, and then they comes here and finds out all."[56]

The economy of lucifer peddling was organized around both extremes of working age. Many peddlers were old women and men selling matches to survive and contribute to extended households, people like the "feeble

old man" interviewed by London's Henry Mayhew: "I've been selling lucifers about five years, for I was worn out with hard work and rheumatics
when I was 65 or 66. I go regular rounds, about 2 miles in a day, or 2½, or if
it's fine 3 miles or more from where I live, and the same distance back, for I
can sometimes walk middling if I can do nothing else. I carry my boxes tied
up in a handkerchief, and hold 2 or 3 in my hand." Peddlers tended to divide
sections of the city between them, and they protected their markets. On his
routes, the man sold matches to "gentlemen," "mistresses and maids," and
"needle-workers" who "see me from their windows, and come down to the
door." Another match peddler worked London "in ten rounds, or districts,
but six is better, for you can then go the same round the same day next week,
and so get known." He estimated there were no more than 200 "real" street
sellers in the city, all old women and men, "and depending a good deal upon
them [matches], for they're an easy carriage for an infirm body, and as ready
a sale as most things."[57]

Elderly and child peddlers, soothed babies and mothers, aunts and
mothers collecting wages, "families" of outworkers, and child match makers
glowing in the dark: the making and moving and burning of lucifers stitched
together working families with a global web of bones and shit. Maggie,
whether she knew it or not, was connecting Afro-Argentine slaughterers,
West Indian guano miners, French and English mouth pipetters, and poisoned migrant children with an outworking seamstress, who released some
of the potential energy pumped through this web every time she struck a
lucifer to light her camphene lamp. As the woman created the night workspace through which she tried to scrape by, she thus bound herself and her
family into dependencies not only on exploitative clothiers and the enslaving geography of turpentine, but on the inescapable ecology of phosphorus.

And lucifers were everywhere. Loose matches ignited in the coat
pockets of the rich and of the poor. Babies ate them and died of phosphorus
poisoning. People stepped on them and set their feet on fire, and children
"played with matches" pretty much anywhere they wanted. "From the extreme cheapness and the extreme convenience of lucifers, they swarm, like
the frogs in Egypt," observers complained, "in every chamber and, what is
worse, in every kitchen. They intrude into your house, and into your bed-
room, and upon your bed and under it, and into ovens, and into kneading-
troughs. They fall into coffee and into soup." Those who smoked, an increasingly popular practice, "carry them loose, in their waistcoat-pockets, in their
trouser-pockets, in their coat-pockets." Lucifers were strewn everywhere and
underfoot, "in passages, on staircases, in outhouses, and stables, amongst

straw, sawdust, shavings, leaves." And they knew no boundaries of class. "In any third class railway carriage," observers wrote, "you have only to ask your neighbour for an allumette to have half a dozen placed at your disposal. The lucifer is a sort of common property to which every one present has a claim." So commonplace were these matches that as early as 1835, Locofocos, an early American brand of phosphorus matches, became the name of a political party, when a group of New York Jacksonians, hearing that Tammany men were going to shut off gaslight service to the hall where they were meeting, used the matches to light candles and continue their deliberations.[58]

Fires, unsurprisingly, were rampant, whether arson or accident. In France, before lucifer matches were introduced in 1838, French statisticians estimated there had been 2,262 destructive fires a year. But only six years later, that number had doubled, rising even more alarmingly to 10,753 by 1854, a nearly 500 percent increase in less than two decades. Residents in American cities and towns desperately sought to confront this rapidly multiplying urban ecology of fire. In 1834, New Bedford tried to ban carrying lit matches in the street. In 1838, Baltimore considered outlawing the sale of matches by unlicensed vendors. In 1840, Hartford passed an ordinance requiring all Locofoco matches be kept in a box of iron or other nonflammable material under penalty of $10. Most railroad companies strictly prohibited transporting matches. New Yorkers pressured legislators to force landlords to build iron staircases outside tenements so tenants had a chance to escape these ever-more-common fires, while insurers considered providing policyholders with safes for keeping lucifers, as many destructive fires were started by children or by mice nibbling loose matches.[59]

Such a pervasive, hang-thread ecology of instant fire surely made the mid-nineteenth-century city a more dangerous place, the lives lost and burns suffered impossible to calculate. But this ecology also placed considerable power in the hands of the smallest and the most oppressed. And if read for carefully, one can spot traces of a hidden transcript of flame, with lucifer matches newly popular weapons of the weak. Ubiquitous, easily concealed, and capable of doing violence distant in time and space, lucifers left fires that were impossible to ignore, while the keepers of official transcripts struggled to identify who had set them, and why.

In cities across the United States, children were actively, accidentally, and disruptively multiplying a new anarchy of flame. "Playing with matches," children burned tenement buildings and houses, lit lucifers in basements, and set fire to beds. Apprenticed woodworkers burned carpenter workshops and woodsheds, furniture makers, and organ factories. Children burned one

Lucifer Matches and the Violence of Phosphorus

another, their parents, and sometimes themselves. In the interstices of their workdays, whether for idle amusement or purposeful arson, with ever-ready lucifers, girls and boys hired as servants set fire to boardinghouses, started fires that blew up gunpowder warehouses, burned book and stationery stores, and set clothiers, milliners, and broomcorn works ablaze. They also repeatedly started fires in stables, some of which spread devastatingly through large swaths of cities. Although every single report of fire attributed to children described them as "playing," such apparent claims of careless-ness or accident need to be read against fire insurance policies, which would pay out for an accident but not arson, and not if the proprietors themselves were accused of carelessness. In the nineteenth century, "children playing with matches" were considered acts of nature or God. They were firestarters for which policy owners could collect. Many of the children playing with matches may have been convenient, believable fictions, invented to collect insurance or even to disguise child murders as accidents. But for such claims to be believable, "children playing with matches" must have reflected some-thing real and widespread enough. In New York alone, between 1855 and 1860, newspapers reported at least eight separate fires started by children in stables, killing at least eight horses and causing over $15,000 in damage to property, much of it not insured.[60]

Did the children really not realize that lighting matches in a stable could easily start an uncontrollable fire? None of the stable fires started during winter, so they probably hadn't been lit for warmth. Were the children per-haps trying to light a lantern to see something in a dark corner? Were they at work or just trying to find a place to play? If they were stable hands, might they have hated their employers or even the horses? Did these unnamed chil-dren even exist at all? These questions are as unanswerable today as they were in 1859, but such evidence of uncertainty is itself a kind of evidence. Be-hind the endlessly repeated story of "children playing with matches" lurked terror and possibility.

Matches might also be deployed to contest the charged housing prac-tices of New York City, entangling phosphorus in the politics of service, white supremacy, and property. In April 1855, a free black woman named Harriet Johnson Douglass sought lodging in a building on Pearl Street. After Douglass had unloaded her belongings, the white landlady intervened, took Douglass's things, and refused her lodging. On what grounds was not clear, but the refusal seemed to have led to an argument and a threat. The landlady later testified that Douglass replied coolly, "'Never mind—I'll leave you all without a home before long.'" According to witnesses, Douglass then "left the house," but when she returned to get her things, she "was seen again to go into the room, and after remaining there some five or ten minutes, she was seen hurrying out of the room and going down stairs, when, at the same instant, smoke was discovered coming from the room she left, and instantly the place was found on fire. There were two boxes of matches on the mantelpiece when Harriet went into the room, and after the fire only one box remained." Douglass protested her innocence, stating before the court, "I am 34 years of age, born in New-York; live at No. 150 Anthony-street; I go out to service; ... in the sight of God I am not guilty." Unmoved, the magistrate "committed her to the Tombs for trial, in default of $500 bail."[61] Guilty or not, the case revealed a fascinating field of plausibility, of a world where matches were at once carelessly unguarded but nervously watched and readily blamed, tools for convenience and democratized weapons for revenge.

Lucifers even appear to have become weapons of antislavery. "We lately mentioned that a twelve-pound cannon-ball had been found here in a bale of cotton," the *Providence Journal* complained in 1860, noting the common practice of cheating cotton manufacturers who purchased their bales by weight, "and we then took occasion to remark that the substitution of iron for sand as an article to increase the weight of the bale showed a slight moral improvement in the dishonest packers." But a new practice had the mill owners worse than annoyed. They were terrified. Writing with alarm to alert the public, an owner claimed "something worse even than sand has been found in a bale which recently arrived. That is lucifer matches. They were in a pine box which was partially broken, so that they could not fail to ignite in passing through the picker. Had they not been accidentally discovered, they might have caused the destruction of one of the most valuable mills in this State." And it was not the only case. A few weeks later, the *Providence Journal* received a copy of a letter that a cotton manufacturer in Coventry, Rhode Island, had written to a cotton dealer, "inclosing a sample of friction matches found in a bale of cotton purchased of him. The situation of the matches was such as to leave no doubt that they were put there when the cotton was packed."[62]

Were these direct actions by abolitionists to destroy cotton mills and terrorize manufacturers who were reaping profits from the products of slavery? It seems unlikely that this was an accident of ubiquity, especially after the second incident, and it's hard to imagine that anyone would try to circumvent transport bans on matches by smuggling spontaneously combusting phosphorus inside cotton bales. The way cotton was ginned, baled, and transported made it unlikely that the matches found in the middle of bales had originated anywhere other than a packing room or gin house. The *Provi-*

Lucifer Matches and the Violence of Phosphorus

dence Journal was probably correct. And if this was true, it almost certainly meant that an enslaved person or persons were responsible. "Packing is very hard, oppressive work," the escaped slave John Brown recalled. "The dust and fibres fly about in thick clouds, and get into the chest, checking respiration, and injuring the lungs very seriously. It is a common thing for the slaves to sicken off with chest diseases, acquired in the packing-room or jin-house, and to hear them wheezing and coughing like broken-winded horses, as they crawl about to the work that is killing them." In Brown's account of the work and violence of packing, as with guano miners and match workers, "we see the outlines of the gradual process," the historian Walter Johnson wrote, "by which human life was turned into cotton: the torturous conversion of labor to capital, and of living people to corpses." It seemed some, perhaps many, of these men and women refused to be ground into cotton quietly, into the stuff of textiles and wicks. "The frequency with which these destructives," these lucifers, "are found in cotton is really alarming," the *Chicago Press and Tribune* worried, "and must greatly increase the risks of underwriters."[63] In which gin house the matches were packed in the bale and who packed them and why were — and this was partly the point — questions nearly impossible to answer in the geographically and historically mystifying processes of cotton commodification. But these were very likely untraceable acts of revenge by enslaved people on the very global chains of cotton and profit that ensnared and fed upon their life and labor.

Enslaved people, and those who feared their freedom, were certainly aware of matches as weapons of revolt in 1860. In the Virginia trial of an enslaved man named Jerry, accounts emerged of slave arson and white fears of lucifer-lit fires. Jerry was tried for plotting a conspiracy for rebellion in Clarke County, Virginia. According to the witness, Jerry spoke excitedly about the radical abolitionist John Brown (not the escaped slave), claiming he and four of his sons had wanted to be a part of the Harpers Ferry raid but had not known when to go. But now that patrols were strictly limiting the movement of blacks, "there had been some burnings since the patrol commenced, and," Jerry supposedly threatened, "'we will keep burning until they are stopped.'" A week or so later, the witness returned and claimed Jerry "told him that there had been more burnings since he last saw them, stating that the patrol had not been out that week, and that he and others had made a plot the night before to burn the house of Daniel H. Sowers in the dark of the moon. At that juncture Mr. Alfred Castleman appeared in sight, passing along the road, and Jerry commenced abusing him most violently to the witness, stating that he intended to burn him out himself, that he had been to Berryville the Sunday before to get matches, but could not get any." Slaveholders tried to strictly limit access to matches. Determined slaves still found ways to procure them. Solomon Northup poached the means of flame and light from his enslavers when he "managed to obtain a few matches and a piece of candle, unperceived, from the kitchen, during a temporary absence of Aunt Phebe." Other enslaved people took lucifers from their work in mining, and still others just bought them, their affordability extending even to the enslaved. Jerry knew he could afford the matches and had likely purchased or swiped some before, but this time white sellers may have known what he wanted them for and so tried to prevent an enslaved man from arming himself with instant fire. White Virginians could not, however, completely snuff out this slow-motion, fire-aided rebellion.[64]

The worlds that people, all people, made or could make with lucifers reinvented material and imagined landscapes of power. The fires were real and impossible to deny, but they burned in impenetrable smoke clouds of uncertainty — of who, when, and why. It was a dialectic that determined the terrain of action for countless actors in the nineteenth-century country and city.

L IKE pushing a sack of mercury uphill to increase its potential energy, the labor- and energy-intensive project of rerouting the ecology of phosphorus from organic to inorganic relations was an inherently unstable and leaky process. Unmade from bone and guano and remade into flames-in-waiting on truly colossal scales, those structures of energy-elevated phosphorus continually bled, and slid back in new now toxic forms, into the bodies and living landscapes conscripted into this industrial chemistry. Each stage and site of this organic and inorganic life cycle was shaped by workers struggling with those who sought to claim the products of their labor, struggles over wages, hours, enslavement, and the limits of parental mastery of children.

But these inescapable ecologies were also deeply shaped by struggles between workers and the "objects" of their labor, means of production that fought back, poisoned, and spontaneously ignited. Cattle refused to give up their bones without attempts to gore and escape their captors and killers. The smoke and stench of roasting bones saturated the lungs, clothes, and atmospheres of workers, workplaces, and neighborhoods. The spaces of phosphorus manufacture became sacrifice zones where the clouds of toxic phosphorus, coal pollution, and sulfur spewed forth in the production process so poisoned landscapes that not a single plant could grow.[65] The phosphorus that children mixed, dipped matches into, and breathed and absorbed all

Lucifer Matches and the Violence of Phosphorus

day as they made matches stuck to their clothes and to their bodies, making them sick, stink, and glow in the long term and burning and splitting their hands in the short, the phosphorus tips continually igniting in the friction of production. Even the final production of flame repeatedly slipped from people's control, as matches circulated so extensively that children, even the babies of the poor, could access and light them, whether purposefully or not. In the bustling wooden and straw worlds of nineteenth-century cities, any flame could transform, either according to or with perfect indifference to its creator's intent, into an inferno threatening lives and buildings.

The natures of producing and reproducing lucifers, then, meant that for everyone bound up in this ecology of light, the potential energies of phosphorus were at once inescapable and continually escaping control, were both the means and the prize of an uneven struggle in an unstable geography of risk, with sometimes surprising alliances and winners. And it was, alone with the lard lights and pigpen archipelago of the Ohio Valley, an ecology of violence and struggle and power that survived the cataclysmic reorganization of the means of light that took place during the American Civil War.

Chapter Six

ROCK OIL, CIVIL WAR, AND INDUSTRIAL SLAVERY INTERRUPTED

The myths of light we still retell today were written by the American Civil War. It might have happened differently; indeed, it was already beginning to happen differently. But so closely would the war follow on the heels of the discovery of petroleum in western Pennsylvania, and so thoroughly would kerosene and coal gas conquer urban lamps by the cessation of hostilities, that the ascent of mineral, fossil-fuel lights would seem natural, inevitable, and an obvious sign of progress. That's how it was understood then, and that's how it's been understood since. By the end of the Civil War, the geography of the means of light in the United States would be transformed so dramatically as to be almost unrecognizable. The whale fishery would be crippled; camphene would be nowhere and never again to be found; and instead of looking to sail and slavery for the means of light, Americans would turn almost entirely to the Midwest, with its industrial, free-labor bituminous coal mines, free-labor oilfields, and free-labor steam-powered lard factories. Only the global ecology of phosphorus—a geography in which all but the final processes of match-dipping were located far outside the borders of the United States—would remain relatively undisturbed. When observers looked back from the end of the century, the Civil War seemed an indisputable triumph of a free-labor, mineral-powered industrial capitalism over an organic, slave-tainted past. But this extraordinary revolution should have come as a shock, and it should continue to surprise us today that un-free labor and technological "progress" became, even if only temporarily and only seemingly, uncoupled.

It did appear obvious to nearly all contemporary observers that some shift in the mid-nineteenth century from the east coast to the Appalachians would occur in the production of the means of light, and that production would take on a more mineral flavor. But that this shift would happen so quickly, that it would center in western Pennsylvania instead of, or in addition to, western Virginia, or that western Virginia would join the Union as the separate state of West Virginia in 1863, was neither obvious nor predictable, not even after the war had begun. Nor was it clear in 1861 that slave-based turpentine would suddenly disappear from lamps while Ohio and Illinois hogs would supply an *increasing* share of Americans' light through candles and lard oil.

Subsequent generations mostly forgot it, but a carbon-powered revolution in light had steadily taken shape in the Ohio River Valley over the 1850s—a revolution that had begun with salt. For over half a century, enslaved workers in the Kanawha River Valley of present-day southern West Virginia had drilled wells to pump up underground reservoirs of salt water, mining local deposits of coal to boil out the salt. Without this crucial and scarce mineral, necessary both for animals to thrive and for preserving their dead flesh, there would have been no revolution in lard lights around Cincinnati. But the story of Kanawha salt and light did not end there.

Kanawha miners and manufacturers soon realized that the coal they had been mining to boil salty water was perfect for gasworks. For decades they desperately tried to connect their coal with lamps in American cities. First they tried attracting New York capital and persuading the eastern planters who dominated the Virginia legislature to invest in rail, canal, and river-improvement projects so that Kanawha cannel coal could reach gasworks in the East and in the Mississippi and Ohio Valleys. When such projects failed or proceeded too slowly, mine owners developed a new industry transforming their coals into something they called coal oil. Once fractionally distilled, this more concentrated and more easily shipped coal oil yielded a substance that manufacturers called kerosene and which they marketed as an illuminant to compete with camphene. But on the eve of war would come the greatest change, as salt well drillers hit upon huge repositories of petroleum that local coal-oil makers knew just what to do with. These well-drillers struck petroleum in western Pennsylvania, and they struck it in western Virginia. Borrowing the name used for the coal-oil illuminant, petroleum processers flooded the American lamp market with "kerosene" just as the cotton states began to secede.[1]

In 1859, Titusville, Pennsylvania, was little more than a sleepy agricultural town with the occasional visitor interested in the strange oily substance that sometimes leaked into the streams running through the area. But that year, Edwin Drake, who had been slowly drilling a hole into the ground in the fading hope that he might find the source of that oil, stumbled into fame. Having lost the faith of his employers, the townspeople, and probably even himself, Drake was desperate and in debt, and when oil started to seep out of the wellhead on August 28, 1859, it was the hired well borer, "Uncle Billy" Smith, who realized what they had done. The rush to Titusville and the region around what was soon renamed Oil Creek was extraordinary. In a few years, Oil Creek went from a quiet countryside lacking a rail connection to the most important site in the world in the production of the means of light. With wooden oil derricks covering the landscape as far as the eye could see, oil spilling out of the ground and into rivers, and fortunes made and lost overnight, "Petrolia" became what one historian called a "sacrificial landscape" in the face of capitalist excess. At the center of the new landscape was a city unlike any the world had seen, or likely ever wanted to see. Oil City, one visitor remarked, was "so impregnated with oil in all its forms and odors that it seems almost impossible to exist there to one uninitiated. In wet weather the rain mixing with the oil oozing from half a million barrels of Petroleum exported from the town forms a mud that destroys the clothes and all things with which it comes in contact."[2]

While dramatic and obviously significant, the petroleum boom of Oil Creek was not particularly unlikely or surprising. Nothing in history is inevitable, but given the events and developments of the preceding decades both in America and in Europe, a coal and oil rush in western Pennsylvania taking place at some time in the mid-nineteenth century was probably more likely than not. But the same was true of western Virginia and, in the late 1850s, appeared poised to happen. Indeed, in the region surrounding Burning Springs, Virginia, along the Little Kanawha River, deep in the heart of coal and slave country, such an oil rush began only a few months later.

The question we have really got to answer, then, is not so much why Titusville was special but what kind of industrial future did the Civil War foreclose in Burning Springs. Both revolutions were already on track on the eve of the Civil War. But only one succeeded. To really understand the past, we have to explain the defeats along with the victories. We have to take seriously the nearly successful emergence of a massive industrial engine of coal, oil, and slavery centered in the Kanawha Valley and bound by rail and steam to the Atlantic coast and the Ohio River—a new industrial landscape of coal mines, refineries, and oil wells reorienting the economic, political, and energy geography of the United States and transforming the industrial possibilities of the Confederacy. We need to be clear about the full consequences of the timing of the Civil War, of West Virginia joining the Union, and of the tremendous struggles that destroyed slavery on what may have been the eve of its industrial revolution.

Would the future of industrial capitalism in the United States, and even the world, be powered and illuminated by organic or by mineral sources or by both? Even more importantly, would this revolution in fuel and light be based on slavery, free labor, or a combination of the two? In many ways, the Civil War decided these questions before they were even asked, but that should not mean we should ignore them now. If each of these were viable futures, it says something profound about the history of capitalism, about the forces and logics governing its motion. In short, it may be just as important to determine what almost happened in mid-nineteenth-century American fossil-fuel energy landscapes as what actually did. First, we must demonstrate the existence, endurance, and viability of the least known and most thoroughly defeated of these almost futures: that of industrial slavery in the Ohio River Valley.

I N April 1819, an advance guard of industrial slavery disappeared into the Blue Ridge Mountains of Virginia. Eager to expand and secure his mineral empire, Harry Heth, the leading figure in the Richmond coalfields, had ordered one of his overseers to march a coffle of enslaved pit hands from his Richmond coal mines to his new saltworks in the Kanawha Valley in present-day West Virginia. Heth planned to force some of the men he enslaved to mine some of the hundreds of thousands of tons of coal consumed each year by the furnaces of the Kanawha salt boilers. The enslaved colliers had other ideas. As Heth's overseer marched the coffle progressively deeper into the loosely policed and lightly settled forests of the Blue Ridge Mountains, the men were able to unravel the power relations holding them in slavery. First, "Billey and the 2 Johns" ran away and made it more than sixty miles before David Street, the overseer, managed to find them. But Street soon lost them again and then disappeared himself. Neither slaves nor driver were ever heard from again. Though thousands of enslaved people were repeatedly forced to complete the passage between these two centers of Virginian industrial slavery, at least some were able to interrupt their journeys, escaping into the unindustrialized expanses separating the state's two major coalfields.[3]

Over forty years later, a newly freed young man named Booker picked up the journey where the fugitives had left off. He and his family were headed to the new state of West Virginia to reunite with loved ones and to work in the salt industry. The two journeys were separated by half a century but in many ways were remarkably similar. The Civil War was over, millions of enslaved women and men had broken their chains, the world-historical revolution of the Confederacy had been defeated, and the state of Virginia had been cleaved in two. Meanwhile, Booker T. Washington, as he would soon rename himself, retraced overland paths worn by thousands of enslaved workers driven from eastern and central Virginian plantations to western Virginian saltworks.

Washington's first taste of freedom was to be heavily seasoned with salt. "In some way, during the war, by running away and following the Federal soldiers," Washington recalled, his stepfather had "found his way into the new state of West Virginia." Whether as a black soldier fighting to overthrow slavery or as "contraband" deserting a plantation and denying his labor to enslaver and Confederacy, Washington's stepfather had traveled from Franklin County west over the Blue Ridge Mountains. And as "soon as freedom was declared, he sent for my mother to come to the Kanawha Valley, in West Virginia," where he "had already secured a job at a salt-furnace, and . . . a little cabin for us to live in."[4] This was likely no accident. The planters of Franklin County—near Roanoke—had, for generations, forged unusually close ties with Kanawha saltmakers by hiring out the young men they owned to work in the coal mines feeding the salt furnaces. The arrangement enriched both planters and saltmakers. And this geography of labor not only determined the spatial formation of communities of the enslaved but shaped where and how they would carry their communities out of slavery.

The apparent lack of change in the South over these forty years is a story that has been told before. Cotton before, cotton after. Tobacco in the seventeenth century, tobacco today. The South was stagnant, and the political economies of slave societies could not tolerate industry or capitalism. The South was held back by its culture, by the absence of a Yankee work ethic and celebration of progress. Western Virginia had begun the century with salt worked by black men, and generations later, West Virginia would enter the Union with salt worked by black men. These before-and-after stories also happen to be entirely wrong.

By the time that Harry Heth's overseer and enslaved miners disappeared along their westward journey, the Kanawha River salt industry was fast becoming one of the most critical sites in the geography of U.S. industry, and it was doing so through slave labor. It was, however, still a new outpost of industrial slavery, and so western industrialists collaborated with eastern planters to form a new, mutually beneficial geography of enslaved labor. Large planters in eastern Virginia placed agents on the Kanawha to arrange hiring out the people they enslaved to western saltmakers. The contracts were usually for one-year terms (paid upon completion) starting on New Year's Day and ending on Christmas, when the enslaved would typically return east to their plantation communities. Meanwhile, salt companies routinely sent representatives east to search for enslaved people planters were willing to lease. With salt companies leasing enslaved men for common labor at rates trending from a low of $100 in the 1830s to a high of $200 in the 1850s, eastern planters realized substantial profits by hiring out some of the men and boys they enslaved. In 1840, the state sent a University of Virginia professor to survey the mineral industries of the Kanawha River. He reported that ninety furnaces along the river annually produced about 3 million bushels of salt and consumed 5 million bushels of coal, mined on-site by 995 miners and workmen. Even before the coal-oil boom of the late 1850s, the Kanawha Valley was, according to the professor's description, one of the most important and productive coalfields in the world, where "more than twice the coal is consumed every year than is furnished by all the coal mines of eastern Virginia put together."[5]

As was the case with turpentine, the increased distance, risks, and profits of the coal-fired salt industry translated into higher lease rates. In the 1830s, Virginia courts heard evidence planters were leasing enslaved men out to the Kanawha Salines at rates 25 to 30 percent higher than in eastern counties. In 1838, a letter from an eastern planter to a western saltmaker claimed that enslaved men who could be hired out locally for $90 could be hired out to Kanawha for $150. This is not to suggest that eastern planters were ripping off western saltmakers. Relocated to a tributary of the Ohio River, enslaved Virginians could, and often did, escape from the Kanawha Salines to the free state of Ohio, both over land and on the steamships carrying salt and coal to western markets. With death and terror lurking in the coal dungeons on one side and the beacon of the nearby Ohio River shining on the other, many leased people attempted to escape the coalfields. In 1844, when an enslaved man named Gatewood escaped from Lewis Ruffner's coal mine, Ruffner posted an advertisement describing Gatewood as "25 or 26 years

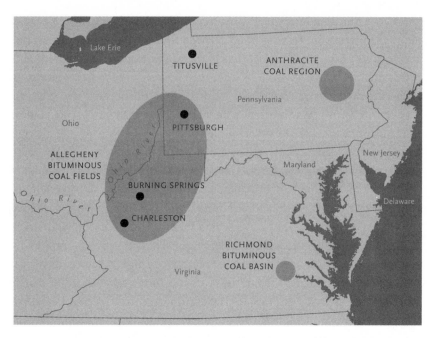

U.S. coal and oil regions east of the Mississippi in 1861.

old, about 5 feet 7 inches high, tolerably black, speaks gruff when spoken too," warning that "there is reason to suppose that he is lurking about in the neighborhood, but may if not soon taken up, make for Ohio."[6] It was this geography of freedom and slavery that inflated the hiring price for enslaved workers in western Virginia, a state of affairs that some Kanawha petitioners to the Virginia General Assembly blamed on Ohio abolitionists. The inflated lease rates were market indexes of the fears of enslavers, of enslaved people's own will and power to be free, and of the political networks of enslaved, free, freed, and fugitive people who continually chipped away at slavery in the Kanawha Valley.

Coal operators fought back against this geography of freedom, many by literally imprisoning the workers they enslaved. The enslaved who were hired to work in cannel coal mines alongside free white miners were, according to one historian, "maintained in slave quarters when they were not on the job. In Mason County, which faced the Ohio River and was within sight of free soil, coal operators were forced to take extreme measures to prevent their slave workers from escaping. Thus, R. C. M. Lovell confined his hired hands behind a stockade after work." The practice of confining off-duty enslaved workers was so commonplace that its absence had to be explained. When

Rock Oil, Civil War, and Industrial Slavery

asked to testify for a case investigating a salt furnace explosion, the company owner had to clarify that at his saltworks, "the slaves stayed in a cabin 100 feet from the engine when not on duty, but that the company did not confine them there and permitted the slaves to run at-large."[7] The intensity of surveillance and captivity likely varied from mine to mine, and with distance from free soil. Yet there remains little doubt that to combat the alternative geographies of freedom formed by enslaved and free abolitionists working along the Ohio River, the owners of Kanawha salt furnaces and coal mines tried to make their operations even more prison-like.

As the resort to guards and captive slave quarters demonstrated, industrial enslavers were not all-powerful. They recognized the limits of their power over the people they claimed as property, and so in addition to coercive carceral practices, they also implemented an incentive system to pay enslaved men directly for "overwork," which was often done on Sunday. As one former coal bank manager recalled, "The coal diggers generally dug their coal for Sunday's run on Saturday; but it was paid for extra. It was generally hauled to the furnace on Sunday." Paying for overwork was common practice across all forms of industrial slavery in the United States. Some analysts have misinterpreted this as evidence slavery was giving way to wage labor. Giving some enslaved people some money for some of their work may have appeared closer to free labor, but it did not change the fundamental fact that the enslaved remained chattel, commodities to be sold. Nor did it spell the gradual disappearance of slavery. In the factories and cities where enslaved people could earn some wages, slavery remained strong, and the small amounts of money they could hope to earn may actually have served as a safety valve for whites concerned with real revolution.[8] Payments for overwork constituted a new tool in the coercive arsenal of enslavers — perhaps one wrangled from them by the insistence and resistance of enslaved people, but still a part of the power structures that enslavers maintained to keep the enslaved in place and working for a property-owning white class.

In the Kanawha Valley, salt and coal operators were laying the foundations for a new kind of industrial racial slavery. By separating owners from their chattel, the system of slave hiring thinned the direct knowledge and power of enslavers over the people they owned while empowering industrial managers to treat leased people as pure forced labor rather than living property. Unable or unwilling to know whether salt manufacturers were employing the leased people underground or at the furnace, eastern planters were ill-equipped to insist on the kind of measures and rules that would protect the lives of their property. Kanawha manufacturers, meanwhile, protected the company-owned laborers from working in the mines, or at least in the most dangerous parts, a dynamic that led leased people underground into dangerous coal mines while skilled, company-owned men were kept above ground, and free white workers were able to further distance themselves from risk.[9]

Given such a contested, expensive geography of enslaved labor at the far edge of slavery's dominion, Kanawha saltmakers, many of whom were originally from Ohio, might very well have attempted to establish an industrial free labor regime at Virginia's western border. But they did not. Even at the highest rates, leasing an enslaved person rarely cost saltmakers more than 75 cents a day (after factoring in food, shelter, and clothing), while wages for free white workers were never less than $1.00 and often rose as high as $2.50. Whenever given the chance, Kanawha operators replaced free workers with enslaved. Because payment came at the end of the contract, the people saltmakers leased were basically capital loans, an arrangement greatly to the saltmakers' advantage and that further eroded the bargaining position of free laborers.[10]

Saltmakers tasked the majority of the enslaved people they leased with mining coal. They weren't supposed to. Many eastern planters stipulated in their leasing contracts that their slaves should not work in the mines, but violation of these safeguards was the norm. Not only was leasing enslaved workers cheaper than hiring free ones, but if killed in a mining accident, a leased collier cost company owners only the amount of work done up to that point. Despite the yearlong contracts, operators typically paid the owners of enslaved people who ran away or died only for the weeks they had worked. Given the danger of the mine workscapes, it was no surprise that operators like Lewis Ruffner tried to smuggle enslaved people across the protective paper boundaries of contracts to work underground. One "woman sued Lewis Ruffner for damages incurred when her slave, Ben, was killed in a roof fall in Ruffner's mine. In her $800 damage suit, the plaintiff contended that Ruffner had agreed not to employ the slave in his coal mines."[11]

For their part, slave owners lessened their own financial risk through life insurance policies, just as they had in the coalfields around Richmond. Although older and more securely established, the deeper, even more dangerous coal mines skirting the Atlantic worked in the favor of Kanawha mine operators. As correspondence among Baltimore Life Insurance Company agents demonstrated, insurers who were concerned about the spate of injuries and deaths in eastern mines were more than happy to recommend policies for Kanawha coal diggers. Because in "the Black Heath pitts near Richmond," one agent wrote, "a number of accidents have occurred from gas," he no longer, "at present," advised issuing life "insurance on hands in these pits at any premium," while in contrast, "in Kanawha in digging . . . coal there is no gas. There is I am informed no pitting there. This is the place where the hands of Mr. Doswell where application for policies is now before you has hired his hands." The agent concluded his report to the secretary of Baltimore Life with a thorough endorsement of insuring enslaved colliers working in Kanawha mines, writing, "I have thought for a long time that coal pitts are more healthy places for negroes than factories or R. Roads," and he suggested that the company "should not charge more than ¼ per cent Extra premium on coal pit hands. Insurance on negroes can only be made profitable by insuring a large number."[12]

The Kanawha mines were safer, but not safe. Lewis Ruffner, in whose coal mine Booker T. Washington would later work for a time as a freed young man, owned and leased at least forty-eight enslaved people at his saltworks in 1850. And Ruffner would have assigned most of the leased people to work in his coal mines. Their experiences likely mirrored Washington's, as the mines themselves apparently changed little until decades after the war. They may not have been as gassy as the Chesterfield pits, but like all coal mines, they were still dangerous, sometimes terrifying environments. As Washington later wrote of Ruffner's saltworks, "Work in the coal-mine I always dreaded. One reason for this was that any one who worked in a coal-mine was always unclean, at least while at work, and it was a very hard job to get one's skin clean after the day's work was over." Kanawha coal mining was a dirty, daily underground migration, "fully a mile from the opening of the coal-mine to the face of the coal," Washington recalled, "and all, of course, was in the blackest darkness. I do not believe that one ever experiences anywhere else such darkness as he does in a coal-mine." To make matters worse and to "add to the horror of being lost, sometimes my light would go out, and then, if I did not happen to have a match, I would wander about in the darkness until by chance I found some one to give me a light."[13]

At the mines of the Winifrede Mining and Manufacturing Company, which were managed by local operators for New York investors, overseers compelled enslaved miners to pry coal from the seams using iron hand picks mounted with short steel bits. To keep track and charge of the men, operators instituted a task system to discipline the labor, assigning each enslaved miner a number. After hacking and loosening enough coal, each Winifrede miner would shovel and load it into a car, and then "he attached a tin car check bearing this number to the car so the foreman could determine each slave's daily production. The cars had wooden wheels which carried them

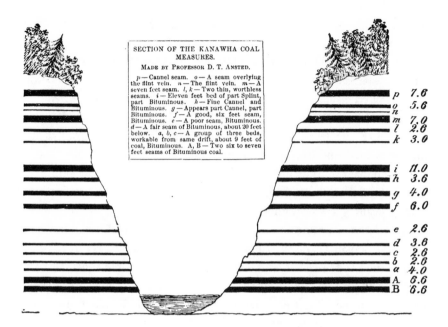

SECTION OF THE KANAWHA COAL MEASURES.

MADE BY PROFESSOR D. T. ANSTED.

p — Cannel seam. *o* — A seam overlying the flint vein. *n* — The flint vein. *m* — A seven feet seam. *l, k* — Two thin, worthless seams. *i* — Eleven feet bed of part Splint, part Bituminous. *h* — Fine Cannel and Bituminous. *g* — Appears part Cannel, part Bituminous. *f* — A good, six feet seam, Bituminous. *e* — A poor seam, Bituminous. *d* — A fair seam of Bituminous, about 20 feet below. *a, b, c* — A group of three beds, workable from same drift, about 9 feet of coal, Bituminous. A, B — Two six to seven feet seams of Bituminous coal.

p	*7.6*
o	*5.6*
n	
m	*7.0*
l	*2.6*
k	*3.0*
i	*11.0*
h	*3.6*
g	*4.0*
f	*6.0*
e	*2.6*
d	*3.6*
c	*2.6*
b	*2.6*
a	*4.0*
A	*6.6*
B	*6.6*

Crosscut of the Kanawha River coalfields. Not only was there vastly more coal in the western part of the state than in the east, but the exposed coal layers of the Kanawha Valley meant that, unlike near Richmond, sinking shafts was usually unnecessary. In effect, the river and glaciers had sunk shafts, allowing colliers to simply cut drifts and chambers horizontally into exposed seams. Illustration by D. T. Ansted, in Andrew Roy, *The Coal Mines* (Cleveland, 1876), 318.

over the wooden tracks to and from the coal bank outside. Slaves who completed their daily tasks were given supper; those who did not received a flogging instead." The Winifrede company used mules in its mines to haul the cars back to the mouth. Many other companies forced the enslaved under their command to "get on their hands and knees, place their heads against the small coal cars and push them in and out of the banks."[14]

Unlike in the deeper, gassier Richmond mines, which men and machines had painstakingly excavated, the Kanawha River itself, with help from the advance and retreat of massive glaciers over hundreds of thousands of years, had done most of the difficult vertical work of exposing the coal faces in western Virginia. And there was a truly extraordinary amount of coal made readily accessible by the freely provided work of river, ice, and erosion. "At the falls of Kanawha," one Kanawha newspaper wrote, "we are informed, that upon actual examination of the several coal seams in the mountain, lying one above another at different intervals, the aggregate thickness of the whole is one hundred and twenty feet of pure coal. Among these is the vein of cannel coal now extensively mined and manufactured into oil." In a global market for coal, the local geology and topography of the Kanawha Valley acted as a natural subsidy to colliers. Instead of spending months boring, blasting, and pumping their way down through rock, sand, and water just to reach the coal, Kanawha coal diggers had "only" to cut horizontally into the sides of the mountains through the relatively softer coal.[15]

Compared with the infrastructure projects in northern states like Pennsylvania and New York, Virginia's remained relatively modest and were designed more to serve the interests of eastern planters than the salt and coal manufacturers west of the Alleghenies. Virginia's constitutions, despite several conventions and protests from the state's western citizens, based representation on ownership of land and slaves rather than on the number of white male voters. This meant that eastern planters were consistently overrepresented in Virginia's general assembly, and their interests were allowed to dominate the allocation of state resources. The largest improvement project of the period, the James River and Kanawha Canal, became an almost independent agency of its own, but it never realized its name. After plans for connecting the two rivers by water were abandoned, funds for improving the Kanawha River (part of the project's mandate) remained practically nonexistent. Only 3 percent of the $5.16 million of state funds expended by the James River and Kanawha Canal made it to the Kanawha region.

But even before the discovery that Kanawha cannel coal could be transformed into illuminating oil, there was enormous interest in the mineral as a source of gaslight. Coal gas could be manufactured from any kind of bituminous coal; but none was better than cannel, not even the Richmond bituminous, and until the 1850s, the only source of cannel in Atlantic markets came, with tariffs, from British mines. The discovery of cannel coal along the Kanawha River in the 1850s, the largest known source of cannel in the United States, if not the world, promised to change the political economy of gaslight, if only it could reach the major markets to connect mines with gasworks. And even if a major railroad connecting the western Virginian city of Charleston to the eastern city of Richmond were out of the question, improvement of the Kanawha River would have greatly increased the competitiveness of Kanawha cannel in the Ohio and Mississippi Valley markets dominated by Pittsburgh bituminous.[16]

The Coal River and Kanawha Mining and Manufacturing Company, established in 1851 by New York City investors, dreamed of riches realized through a new national geography of gaslight, fueled by the slave-mined cannel coal of the Kanawha Valley. At a meeting of the board in New York City in March 1854, one company officer excitedly reported on "the progress of the efforts to raise working capital, & stated that the Manhattan Gas Co have proposed to contract with us for Ten thousand tons of Cannel Coal delivered in this City." This was what they had been waiting for, and the board quickly voted to appoint a committee "to negotiate with the Gas Company and other purchasers of cannel coal for terms upon which they would contract for the purchase of a certain quantity of coal including in each negotiation the price that would be paid; the time when required to be delivered and the advances which would be made for the outlay that would be necessary in the transportation."[17]

One week later, the board reconvened to hear the report of the committee, which had met with the officers of the two Manhattan gaslight companies and the one in Brooklyn to discuss contracts for Kanawha cannel coal. They found that "the present consumption of Cannel Coal by those companies amounts, in the aggregate to about 60,000 tons, that the increase of business leads to a rapid augmentation and will probably within another year reach to that of 90 or 100,000 tons." Up to 1854, the report noted, these gasworks had purchased all of their cannel coal from "European Mines at a cost delivered here, of $12 per ton, upon contracts made some time since, and which contracts are now expiring; that a renewal of them, even to the extent of five thousand tons, cannot now be made upon the same terms, and that any considerable future supply, even at a higher price, is very problematical." With the European contracts ending and the price of Scottish cannel rising, an excellent opportunity was opening up for western Virginian cannel companies. New York gas companies were, the committee claimed, "anxious to secure supplies at home, if it be possible, and would readily make contracts now for prompt delivery, at the rate of 14 or even 15 dolls per ton, to the extent of from 10,000 to 30,000 tons" on the condition that "the Coal River Mines were in a workable condition and thus prepared to commence the delivery of Coal in this City, . . . and anticipate the payments upon evidence of the Coal being in barges and on its way for delivery." The gas companies were familiar with Coal River cannel and considered it of excellent quality, but they remained uncertain of the current size and dependability of the supply. According to the report, the gas company officers expressed "great anxiety . . . for prompt and energetic action on the part of the Coal River Co," but after assurance by the mining agents, the gas manufacturers pledged "that all that could be delivered would be taken by them at satisfactory and liberal prices and for cash." Here was a proposed vision of industrial circuits running straight from the dark, slave-worked coal faces of the Kanawha Valley through New York gasworks to the gas fixtures multiplying in the factories, offices, department stores, clothiers, and bourgeois homes of the largest, most important city in the capitalist United States. The only problem was that it was a vision based on fictional mining. To make fiction into fact, and thereby realize the kind of corporate partnership between mine and gasworks upon which fortunes could be made, the investigatory committee thus "urge[d] the adoption of the most prompt and energetic measures to insure the speedy working of the Company's mines to the production of at least 100,000 tons per annum."[18]

Unfortunately for the Coal River and Kanawha Mining and Manufacturing Company, production was not the only issue. There was also, as the correspondence made clear, the problem of transportation, and these New York contracts were put on hold pending landscape engineering projects to improve riverbanks, riverbeds, and river flow and the completion of eastern rail connections in the Kanawha Valley. The cannel coal companies did manage to secure a foothold in Ohio and Mississippi markets. The Coal River and Kanawha Mining and Manufacturing Company spent $10,000 during the winter of 1855 and 1856 on dams, locks, and other improvements to the Coal River, a tributary that emptied into the Kanawha at St. Albans. But with minimal improvements to the Kanawha itself, and the promised eastern rail and canal connections never seeming to materialize, the major buyers on the Atlantic seaboard began to look elsewhere for cannel. Meanwhile, many gas companies, all of which charged consumers for the volume of gas consumed, found they could make larger profits by substituting ordinary bituminous for cannel coal. The cheaper coals produced a similar volume of gas but far less hydrogen, and so to get the same amount of light, consumers had to burn much more gas. This became a major point of contention after the Civil War, when activist municipal reformers sought to challenge the practices, profits, and power of gaslight monopolies.[19]

Unsurprisingly, cannel coal companies and their vocal supporters in the regional newspaper the *Kanawha Valley Star* were increasingly frustrated with the Virginia General Assembly during the 1850s. "Here, along the very banks of the Kanawha, Elk and Coal rivers," one editorial proclaimed, "lie deep and inexhaustible veins of bituminous and cannel coal unsurpassed for variety [or] richness by the coal mines of Pennsylvania or of England." Why, then, did so much of these mineral lands remain unworked? Why did "not cannel coal of Kanawha fill the coal yards of Alexandria and Richmond and [be] shipped thence to New York, and become the successful competitor of the Lehigh and Cumberland coal, thus enriching Virginia and making our

Atlantic coast the coal yards of the Union?" Even faced with the disadvantages of navigation on an unimproved Kanawha River, the editorial continued, Kanawha coals were filling the downriver markets of Cincinnati and Louisville. So why not the rest of the nation? "Why is all this? The answer is easily given. It is because Virginia is so foolish, so very suicidal in her policy of Internal Improvements, that she will not prosecute to a speedy construction the great line of railway between Covington and the Ohio, to which she now stands pledged and which Nature seems on every hand to have designated as the great artery of commerce between the Seaboard and the Valley of the Mississippi." Nature, the Kanawha industry boosters argued, intended western Virginia to be the nation's primary source of coal, light, and industrial power, and only unnatural Pennsylvanian interference to subsidize its coals and unconscionable Virginian legislative inaction had prevented this inevitable outcome.[20]

N OTWITHSTANDING these considerable obstacles, a revolution in the means of light still took hold in the Kanawha Valley. These obstacles might even have helped. Faced with unreliable routes to larger markets, Kanawha coal producers did what corn growers had done for generations through hogs and whiskey. They transformed cannel coal into a more concentrated and more easily shipped form called coal oil. In August 1855, the Coal River and Kanawha Mining and Manufacturing Company, having failed to secure contracts with the New York gas companies, explored just such an oily path to market and lamp. At a meeting of the board, the company agreed to pay "to George W. Gussman one hundred dollars to defray his expenses to the Mines of this Company, for the purpose of making a preliminary report upon the expense of erecting works for the manufacture of Carbonic Hydrogen Oils from Cannel Coal." In 1855, coal oil was still relatively unknown, but interest was rising quickly in "the manufacture of Carbonic Hydrogen Oils destined to supercede Spirits of Turpentine, Fluid & Camphene, all substances dangerous to be used of which the consumption is so great in the United States." What the company wanted to know was how much oil their mines contained. "Mr Gussman having tried our Cannel Coal reports that he distilled at the rate of 75 gallons of liquid matter per ton which on purification & analysis produced" twenty-five gallons of nonexplosive oil for lamps, eight gallons of lubricating oil, and a variety of other substances. It was exactly what they'd hoped to learn: there were liquid fortunes congealed in the cannel coal buried in their mines. In July 1857, a nearby company was already "making *two hundred gallons of oil per day in a crude*

Rock Oil, Civil War, and Industrial Slavery

state, from cannel coal," and with its new retorts in place, it would soon be making 1,000 gallons daily, anticipating a forty-retort, 3,000-gallon-per-day capacity the next spring. And the market price for this crude oil was 60 cents per gallon in New York.[21]

According to the *Kanawha Valley Star,* "The manner of extracting the crude oil from cannel coal is very simple." First, workmen broke up the coal into small pieces, "not larger in size than a hen's egg," and then placed the pieces in cylindrical retorts, much as would be done in a gasworks. Once the coal was in the retorts, the first stage of distillation took place. Heating the coal retorts to temperatures below 800 degrees, half as hot as they would be in a gasworks, workmen oversaw the fractional distillation of the coal into volatile oils and gas (the lower temperatures compared with a gasworks were designed to produce relatively more oils and less gas). After the retorts were heated externally on the tops and sides, "a stream of steam, *intensely hot,* is thrown *into* each retort," the resulting heat and pressure further separating the coal into gas, liquid, and solid states. Pipes leading from the bottom of each retort allowed the liquids and gases to escape from the retorts, after which they were passed through water-chilled pipes, like the worm of a turpentine still, and separated into various components.[22]

Because company owners were not optimistic about transportation along the Kanawha, considering Virginia's consistent failure to deliver on the promised improvements, many intended "clarifying and refining the crude oil at Maysville, Ky.," where the final fractional distillation of the coal oil into kerosene would take place. "The uncertainty of navigation in the Kanawha river prevents them," the *Kanawha Valley Star* noted, "from erecting a refining establishment in this county." But so great were the anticipated profits from coal oil that many other companies responded to the same problem by investing substantial sums of their own money (much of it New York capital) to improve local river navigation. The Coal River and Kanawha Mining and Manufacturing Company recorded in the spring of 1856 that it had "expended Ten Thousand Dollars during the past winter towards the erection of a Lock & Dam (No 5) on Coal River which will complete the navigation from the Company's property to tide water & enable them to ship their products from their own Lands to any market." And markets were thirsty for coal oil. "The discovery of coal-oil and the invention of coal-oil lamps supply wants which have long been felt in the community," proclaimed one editorial. Across the country, "an excellent light, in a cheap, safe, clean and convenient form, has long been a desideratum," and coal oil was precisely that light, for "coal-oil furnishes the substance, and coal-oil lamps the means of obtaining a light as bright as gas, cheaper, in the long run, than any other light, perfectly safe, as clean as camphene, and as convenient as candles."[23]

It was one thing to have coal oil. It was quite another when no lamps existed to transform it into light. Coal oil had been known of for some time, and attempts to manufacture it at scale had even been attempted in the area from at least 1854 at the Breckenridge Company of Cloverport, Kentucky. Hopes were high, for the "discovery of coal-oil was supposed to be one of the greatest of the age — and so it truly was; but an insuperable difficulty was experienced in burning it." And so the Breckenridge "company had expended $300,000 in its manufacture; but it turned out to be a dead weight on their hands. Under these circumstances, application was made by the Breckenridge Company to the firm of Dietz & Co., of New York, who were largely in the lamp business, . . . and, in the course of two years and a half, made his invention of the Dietz burner, which is as important a contrivance in the history of lights as the Howe needle in that of machine-sewing."[24]

Along with the similar Knapp and Drake lamps, the Dietz burners, first produced in 1856 by the same company that had made its fortunes manufacturing and selling Doric camphene lamps, had a revolutionary effect on the coal-oil industry. As one newspaper noted in 1859,

> The manufacture of coal oil, since the invention of the burner, has become quite extensive. The Breckenridge Company, of Cloverport, Kentucky, have $300,000 invested; the Union Company, at Union, Kentucky, $200,000; the Quincy Company, at Pittsburg, $100,000; the Albert Company, in New Brunswick, $300,000; and it is now extensively manufactured in England. As for lamps, Dietz & Co., of New York, have nearly $100,000 invested in the business. All these, however, seem to be but the commencement of a mighty business in this line of industry, which is destined to assume an importance to be counted by millions instead of thousands of dollars.

Another article estimated that there was already, in 1858, a demand for at least 20 million gallons of refined coal-oil products for lubrication and light (half of which was for household use), requiring 165 refineries.[25]

The biggest question in the last few years of the 1850s, then, was not if coal oil would be a boom industry, but where that industry would be. Would it be dominated by Boston and New York firms distilling oil from Nova Scotian and British cannels, or would it be centered in the Ohio Valley cannel coalfields around western Pennsylvania, western Virginia, and Kentucky? In

Rock Oil, Civil War, and Industrial Slavery

1858, the older, better-capitalized firms in New York and Boston still had a lead over Ohio Valley manufactories, but they were losing ground. By February 1860, there were over thirty coal-oil refineries in the United States transforming 75,000 gallons of crude coal oil into 22,750 gallons of refined "kerosene" oil each day. This was the product of 60,000 bushels of cannel coal. By the end of 1860, manufacturers were daily transforming nearly 100,000 gallons of crude into 30,000 gallons of refined coal oil, with an annual value of $5 million, and retail prices had dropped from $1.25 in 1858 to 75 cents and sometimes as low as 35 cents a gallon in 1860, cheaper than any illuminant except camphene. Meanwhile, manufacturers were mass-producing coal-oil lamps through a system of interchangeable parts, and an estimated 1.8 million lamps had been sold to consumers at prices between $3.50 and $8.00 a dozen, while an additional 1.8 million had been purchased by hopeful dealers. It was an industry that seemed poised to only continue growing.[26]

In February 1860, the *Kanawha Valley Star* reported the "number of workmen employed in the several coal oil works in this country will reach 2,000; that of the miners engaged in mining cannel, 700 or more. — Besides this, there are a large force of men employed in making lamps, burners, wicks, chemicals, etc." By April 1861, the paper claimed there were five cannel coal-oil factories in the county, producing 5,000 gallons of crude oil daily.[27]

Commentators anticipated that the future geography of crude and refined coal oil was going to revolve around cannel coal. Just as gasworks couldn't manufacture gas from anthracite and so needed bituminous coal, coal-oil manufacturers couldn't turn most bituminous into oil and so needed cannel coal. Finding and mapping cannel coal was, therefore, of immediate concern. As one surveyor noted, coal "oil is not made from bituminous coal, strictly so called, but from the *cannel* or candle coal," and while deposits of cannel were scattered all across the Ohio Valley, "the best that I know of is found on the waters of the Big Kanawha river. . . . The veins of it are about six feet in thickness, and can be worked to almost any extent." There was no doubt that manufacturing coal oil from cannel was tremendously profitable "from the fact that it is made in Boston and New York out of cannel coal imported from England." Kanawha cannel, it was hoped, would undercut the eastern refineries, "because here the coal is not to be moved, the manufacture being carried on right at the coal bank, and freight being charged only on the oil." Indeed, by 1859, the future seemed clear and bright for the western Virginian industry: "New York capital is looking up these Kanawha coal banks, and in a few years there will grow up an interest in that region in the coal oil manufacture, which will astonish the country. . . . Its cheapness and reputation for light is now established. . . . This coal field is to become the great center of production and population for the Union."[28]

Yet even with the solution of concentrating the coal into oil at the mines and Kanawha companies' private investment in river improvements, transportation remained a problem, and western coal-oil manufacturers grew increasingly insistent that Virginia needed a central, east-west railroad trunk line if the state wished to realize the potential of this new industry. In an 1857 article titled "Internal Improvements," the editors of the *Kanawha Valley Star* tried to sell this rail line through a kind of geological providence. Noting that cannel coal seams constituted a five-foot-thick layer 150 miles long and 50 miles wide, the editors calculated that there were nearly 1 trillion bushels of cannel coal buried in the Kanawha hills, worth, at the average market price of 25 cents per bushel, over $200 billion. "Let us, then, by the magic light (not of Aladin's lamp) but by the real light of retorts and alembicks and chemical analysis," they exhorted, "turn these 'black masses' on which our mountains rest into oil, of which the world has great need to lubricate its rusty joints, and illuminate its dark alleys." Conservatively estimating a yield of two gallons per bushel of cannel, the editors claimed the crude oil contained in the cannel coalfield of western Virginia would be "oil enough, we should think, to grease and light the globe; and certainly much more than ever swam in all the whales since Jonah's time. The value of 1,672,704,000,000 gallons of oil at the present market price of sixty cents per gallon, will amount to $1,006,622,400,000."[29]

If the money alone were not sufficient to persuade, then surely, Kanawha industrialists believed, the impact that coal oil and its refining could have on the political economy of light in the U.S. should finally convince the Virginia General Assembly to leap forward with western railroad projects. "Why cannot our State make her rail road connexion with the Central road, and let the trade in this material go to Richmond to be refined and sold, instead of sending it to Cincinnati and New York?" one editorial asked. Not only was coal oil, the writer claimed, a limitless industry, "impossible to be overdone, for the demand must always equal the supply," but it provided a perfect opportunity to humiliate and impoverish the hated New England whale fishery. If the Kanawha Valley were brought into a southern rail system, the writer asked, would it not be cause for celebration among all slaveholders? "Will not the whole South, our own dear people, rejoice; for your correspondent looks to see the time when the whale shall have rest from the persecution of the Yankee, and the grass grow (if the soil will permit it) in the streets of

Rock Oil, Civil War, and Industrial Slavery

that abolition hole, the town of New Bedford, whose people expend all their wordy sympathies for the far away negro, and use the poor sailor, their own kindred flesh and blood, worse than dogs."[30]

As coal-oil boosters were happy to point out, this new mode of producing light was superior not only for the underground oceans of oil it promised to unlock but for its rate of exploitation, for the amount of surplus value and oil that could be produced through each worker. Comparing the operations of the whale fishery with those of the coal-oil works in Breckinridge, Kentucky, one article calculated that "the present product of the Breckinridge works, is, with 30 men . . . 675,000 [gallons] annually. The same number of men and the same amount of capital as the whale fishery requires, employed in the production of oil from the Breckinridge coal, would produce in twelve months the enormous amount of $275,000,000, instead of $10,500,000 as" in the fishery.[31]

Cannel coal and coal-oil companies sometimes began operations by recruiting free laborers, but, like saltmakers, they almost always replaced the workforce with enslaved workers once they were established. The Kanawha Cannel Coal Mining and Oil Manufacturing Company, for instance, replaced free coal diggers with enslaved miners, cutting the cost of digging coal from 2 cents per bushel to 1½ cents. Nor were the enslaved restricted to the mines. In 1860, at least eleven enslaved people were working at the oilworks of the Great Kanawha Coal and Oil Company. Some residents seemed worried that the outside capital pouring into the coal-oil industry from Ohio, New York, Boston, and Philadelphia would do little to benefit local property holders. In response, the Kanawha Valley Star reminded its readers that the social relations in the valley ensured that Kanawha whites would be able to share in any profits, for "we have in our oil works, owned by citizens of free States, the hired slaves of our own citizens; thus using our labor and scattering their wages amongst the slaveholders in our very midst."[32]

In 1858, J. G. Dumas, a chemist and engineer from Charleston, Virginia, tried to improve upon the distillation process by almost completely eliminating the need for direct human work in an oilworks. In labor-poor western Virginia, manufacturers hoped that Dumas's system, along with that of slave leasing, would undercut and circumvent the power of free workers by alienating them from the production process entirely. The Kanawha Valley Star reported that Dumas "prepared the Plan for the Oil Works at Peytona, and we understand it has met their approval." The plan began by using rail and gravity to replace muscle power, "receiving the coal at the mines, in cars, from which it is emptied directly into a powerful crusher; the crusher is so elevated, that a car stands under it and receives the crushed coal, carrying it thence to the front of the retorts." Yet even more important than the gravity machine, "the method of carrying on the operation after it is once placed in the Retort, is much simpler than the old method, and is," the article emphasized, "entirely self-working—that is, there is no hand-labor employed; the oil is carried from and into the different vessels required, entirely by machinery."[33]

In the antebellum Kanawha Valley, an "entirely self-working" oilworks was more than an example of technological progress. As industrialists along the Kanawha repeatedly seized chances to replace living labor with capital, and free labor with enslaved, a new set of social relations, complete with its own internal contradictions and momentum, began to emerge in the coalfields of western Virginia. This was critically important. It meant that technological improvements, increases in efficiency, and the accumulation of capital were asymmetrically concentrated in the oilworks, displacing and subordinating human labor to mechanized production processes, while the work processes of the mines changed almost not at all. With oilworks able to produce more oil with fewer men, more enslaved miners were needed to extract the coal, to accumulate the products of nature. As operators forced the underground and aboveground workscapes of coal oil to diverge—as managers arranged to produce one through the dangerous handwork of cheap, disposable, enslaved men while making sure that the other was more and more made through "self-working" machines of steam, iron, and coal—they sharpened and determined a set of divisions across capital, nature, and labor that increasingly came to define the production of light from fossilized carbon energy.

W HEN petroleum burst onto the scene a few years later, it spilled over into channels, categories, and conflicts in Kanawha that the makers of coal oil had already established. The story of how Edwin Drake's discovery of oil in Titusville, Pennsylvania, launched the world's first oil rush to the region that became known as Oil Creek is well known, told and retold in both academic and popular accounts.[34] But this is the story of a different oil boom. Shorter lived and starting a few months after Titusville's, the petroleum boom in the Kanawha Valley was yet another example that history might have happened differently, that the industrial politics of light on the eve of the Civil War were moving in multiple directions without showing any clear signs of settling in any one place or in any one form. Emerging at a moment of growing sectional crisis, the impact that the production of petroleum and its distilled product, also called kerosene, would have on the future of slavery, free labor, and industrial capitalism in the United States was an open question. Natural history may have determined where oil could be found, but human history determined where, when, and how it would actually be discovered.

As an author for the *Kanawha Valley Star* noted in 1861, "That there is oil in this Valley, and in great quantities, has not been a secret to our citizens, for as far back as the discovery of coal, the petroleum has been gathered from the salt-wells and cisterns for the use of the coal-miners, when at work, in the coal banks, for illuminating purposes." Indeed, petroleum had seeped into the cultural landscape of places and names in western Virginia long before Titusville drew the world's attention. According to the article, "Oil may be seen at low stages of the river, in the summer season, for miles above and below this place, oozing from the banks of the river, and floating down the stream. Any of our boys, in the habit of bathing in Kanawha, will testify to the fact, as the oil is a source of annoyance to them. Again, the Great Kanawha river has been long nick-named 'Greasy River,' from the fact of the oil upon its surface." Mostly, though, this petroleum had been considered waste, a nuisance that salt drillers had to overcome. "A few miles below Charleston, near the mouth of Davis' creek, on Mr. Shelton's farm," the article began, "an old salt well was abandoned years ago, before our salt makers learned the art of tubing wells to keep out fresh water, oil and other matter deleterious to the manufacture of salt, because of the large quantity of oil that made its appearance on the salt water." Oil had long been known to well-borers as a source of frustration, and in "nearly every well they bored, they found, more or less, oil. In some instances, we are told, the oil came from the wells in large streams, and flowed into the river, where it could be seen for miles floating upon its surface."[35]

It was the combination of the knowledge and practices of the salt and coal-oil industries that propelled petroleum from a cultural landscape into an industrial and economic one, from waste into lamps. The coal-oil industry had not only demonstrated the value of rock oil but raised the capital, infrastructure, markets, and technologies necessary for distilling, distributing, and burning petroleum-based illuminants. The very word "kerosene" had originally referred to a brand of refined coal oil manufactured in New York as early as 1856. Most producers believed that petroleum was just naturally occurring crude coal oil, so similar were the fractional distillation processes and so similar were the products that could be distilled out.[36]

Of even more immediate importance, however, was the accumulated knowledge, equipment, and skill of the famed salt well drillers of the Kanawha Salines. Just as in Titusville, oil along the Kanawha was born of salt. Edwin Drake, hanging around Titusville and suspecting that petroleum might be reached by drilling, traveled 100 miles south to Tarentum, Pennsylvania, where he recruited William A. Smith, a local salt well borer. Working alone and far from other expert well-drillers, Drake and Smith took two years to drill a mere seventy feet. Kanawha drillers took only six to eight months to bore a 1,000-foot well. It is not entirely clear whether these Kanawha well-drillers were free or enslaved. But they were in very high demand. Well-drilling specialists like Jabez Spinks had "their own workmen," but they may have been free men Spinks had hired for wages or enslaved men he'd leased or both. If well-drilling was a free-labor trade, it was one in which, as with slave life insurance in coal mines, white workers leveraged their whiteness to skim off some extra value for themselves in an economy founded on the exploitation of enslaved industrial laborers. It is clear, however, that once wells were sunk, manufacturers tasked enslaved workers to run the steam engines pumping brine and to tend to the wells and pipes leading to the furnaces. When the western Virginian petroleum boom began in Burning Springs, oil well operators could draw on deep local knowledge among free and enslaved saltworkers to drill and maintain these never-stable wells.[37]

Drake finally struck oil in Titusville in August 1859. Only a few months later, in November, the first oil well in Kanawha was in operation. And this was no backwater, minor industry. By the spring of 1861, before the Civil War interrupted production in western Virginia, the Kanawha oilfields were daily producing 800 barrels of oil, compared with 1,300 in Pennsylvania.[38] Oil Creek had a lead, but hardly an insurmountable one. At the time, the Little Kanawha River and Oil Creek were simply two centers in a greater Ohio Valley oildom.

Oil was first struck in the Kanawha Valley in November 1859, but the real rush began in the spring of 1860 when oil was discovered on the farm of John V. Rathbone. Interest was redoubled in December 1860 when J. C. Rathbone bored a well on the farm that pumped 8,000 to 10,000 gallons of oil each day. The Rathbone farms "soon became a city of huts. Nothing could be seen but great piles of barrels, derricks, scaffolds, and cisterns; nothing heard but the puff of the steam-engine, and the click, click, of the drill!" The oilfields of the Kanawha Valley, centered around the farms of the Rathbone family, were most often known as Burning Springs, the name of a town famous for salt wells that caught on fire from the ignition of escaping pockets of natural gas. They were also called, optimistically, Eternal Center.[39]

Burning Springs lay up the Little Kanawha River, which joined the Ohio at the suddenly important town of Parkersburg, Virginia. By March 1861, "the Parkersburg (Va.) Gazette, noticing the oil discoveries, says that two barrel factories are being built in that town, capable of turning out 400 barrels per day, and that at Burning Springs a factory is being erected to manufacture 1,000 barrels per day." That same month, the Virginia General Assembly passed bills incorporating the Coal and Oil Company of Braxton County, the Burning Springs and Oil Line Railroad Company, and the Laurel Valley Oil and Coal Company in Mason County. Reporters traveling to Burning Springs described a rapidly improvised patchwork of transportation linkages desperately attempting to get the liquid fossils to market. Taking a boat downriver to Parkersburg in March 1861, correspondents for the *Cleveland Plain Dealer* found that the town's "levee was covered with barrels of oil awaiting shipment." The barrels had been floated downriver "in flat boats from burning Springs and Rathbone wells, a distance of 35 miles." Steamers then carried the barrels of oil accumulating on the docks of Parkersburg "to Pittsburgh, Baltimore, Cincinnati and other places," to be distilled into kerosene. The gathering point for all the oil of Burning Springs, Parkersburg became a city of oil and for oil. "There is a large influx of strangers at Parkersburg from all parts of the country," one visitor described, noting that the new armies of "lawyers about the place night and day, do little else than make out leases. Provisions and hardware houses feel little or none of the present panic. The steamboats and the railroads are having heavy receipts from the oil interests." But where there was movement and money and work, there was also politics. In March 1861, the Little Kanawha River was "literally covered with flatboats and the boatmen are now on a strike. They ask two dollars a barrel for taking the grease to Parkersburg. The producers are only willing to give a dollar and fifty cents."[40]

In both Oil Creek and Burning Springs, the free coopers, rivermen, and teamsters who controlled the crucial bottlenecks of transportation could wield surprising power. In Pithole City, an oil town along Oil Creek that rose from almost nothing to a frontier city of 15,000 inhabitants in less than a year, at least 3,000 were teamsters. Teamsters charged $3.00 a barrel to carry, while the rivermen of Oil Creek charged rates between 15 cents and $1.00, and "these skilled, hard-drinking pilots averaged between $100 and $200 for the trip down Oil Creek." Scheduled two or three times a week while the rivers remained unfrozen, freshets were crazy times. Men released dammed waters to flush downstream 10,000 to 20,000 barrels per freshet, barrels handled by up to 800 boats that had been towed upstream by horses and mules driven to death by the thousands. Timing and navigating freshets were enormously difficult; the crowded river usually saw competitive pilots smash or ground at least some boats during each one, and hundreds of gallons of oil were always lost.[41]

With thousands of teamsters, rivermen, and coopers working with and against one another to claim a sizeable share of the value of the oil they moved, producers and refiners flexed their considerable muscle to try to eliminate these workers. From 1860 to 1864 the total cost of getting a barrel of oil from Oil Creek to seaboard market ranged from $8.00 to $15.00, with over 50 percent of those costs accruing from transport. This was much more than a problem of efficiency. This was politics. The legions of clerks, capitalists, and merchants from which the "respectable" oil pioneers emerged—the refiners and producers—resented the power of the teamsters and despised them as a class. In Pennsylvania, capitalists eventually defeated the teamsters through the construction and vigilant defense of pipelines. If the capitalists who controlled the Burning Springs region had been given time to consolidate their power, there, too, teamsters may have been replaced with pipes. But they also might have been replaced by enslaved porters and rivermen like those who managed most of the commerce in the lower Mississippi.[42]

Following their arrival in the new Virginian oil depot of Parkersburg, the Cleveland journalists then traveled twenty-two miles southeast on the Baltimore and Ohio Railroad to "Patroleum station," the oil-laden flatboats riding freshets past them in the opposite direction. The remainder of their journey would be far more difficult. "The distance from Patroleum station to 'Eternal Centre,'" the Cleveland reporters wrote, "is eighteen miles over one of the worst roads that can be imagined. The hills are terrific, being from 400 to 1000 feet high and very steep. The party were advised that their best course was to proceed on foot. They did so, and after a toilsome day's journey night overtook them within four miles of the 'Centre.'"

As the reporters neared Eternal Center, they encountered scores of other pilgrims destined for the oilfields. Some decided to share the work, company, and shelter of travel with the correspondents, forming a temporary migrant community, and "by that time the party had increased to seventeen, a number having been overtaken on the road, who were also on foot. They all slept that night in a small log cabin." In that cabin, and the dozens of others like it that sheltered and structured the journeys of hundreds of people flocking to Burning Springs, travelers circulated important news of the oilfields along with cautionary tales of the limits and solutions to the practices of transporting and containing oil. That night, the correspondents

Rock Oil, Civil War, and Industrial Slavery

"were informed that the Camden well, situated on the Rathbone farm, had 'blown out,' filling two flat-boats and a canoe with oil, besides all the barrels that were on hand, and that about sixty barrels ran into the Kanawa before the sudden flow could be checked."

The roads to Burning Springs were packed; people jostled together in the mud and up mountains on hoof, wagon, and foot. "The excitement is tremendous and the rush of people almost incredible," wrote the Cleveland travelers, who were "met on the hills in dozzens with knapsacks on their backs, hurrying frenziedly to the oil regions. Boring is going on along the Kanawa from Rathbone's to Elizabethtown, — 8 miles. . . . Mr. Barron says that out of 78 passengers who came up the river from Parkersburgh when he did 70 were oil men." The sudden influx of people into rural hinterlands transformed the Rathbone farm into a hastily constructed oil camp, into an oil landscape. By March 1861, articles claimed the total "number engaged in the production of oil from Parkersburg to Burning Springs Run is not less than 4,000." The Cleveland expedition reported that "the rush of people to this section is great and the accommodations very limited. The Cleveland party had to content themselves with a miserable hut, where they took their meals, having brought some provisions." Of the fifty wells bored on this one farm, fourteen were pumping oil, "the least of which yields forty barrels per day. . . . The rates for which land is leased are $1,000 bonus per acre, $1,000 when oil is reached, and one-third of the oil in iron bound barrels." And most of the oil was "found at from 125 to 225 feet, for which distance the cost of boring is about $2 per foot." A mountain forest had been transformed into an industrial thicket of wooden derricks: "Over five hundred derricks up here within sight, and more going up daily. The ground is leased nearly to the top of the mountain. . . . The whole space where the oil has been found is not half a mile square."[43]

One of the most famous wells in Burning Springs was called the Lewellyn Well. Over 100 feet deep and 4 inches wide, this well actually had to be restrained lest it produce too much oil, faster than workmen could cooper it in barrels. "Before this plug was put down the flow of oil was enormous," learned one visitor, who was told by the workmen that the original, unplugged "well threw a stream of oil to the hight of from 10 to 20 feet. Men who came to the well to plug it up rode through oil around the well which was *up to the horses knees!*" And even later, after the plug had slowed the flow of oil, "they were filling seven barrels at a time. Seven barrels would be filled in four minutes. It is impossible for them to get barrels enough to hold the oil. It is estimated that if the plug was withdrawn altogether and the well allowed uninterrupted play it would throw *two thousand barrels of clear oil in twenty four hours!*"[44]

Nor was the Lewellyn well an anomaly. "Mr. Braden's well is one of the curiosities of the age," one Cincinnati correspondent wrote, noting that Braden's well "flows regularly at intervals of fifteen minutes flowing out some fifteen or twenty barrels at a flow." When oil was first struck there, the imperfectly predictable rhythms and patterns of work of the oil itself forced the workmen to scramble in an attempt to catch up and match their own work to that of the oil. "The workmen were putting in the top of the well to conduct the oil into the vat," the article recounted, but before they could get the top "secured it commenced blowing and forced the oil some forty feet high, the tube, scattering in every direction, and in trying to choke it down it whistled louder than a locomotive, scaring the whole neighborhood." First came the sound and the fury. Then came a pillar of fire, the shrieking, igniting geyser of oil "burning the hands of the workmen and scorching the hair and whiskers of others." But the men eventually brought the well under sufficient control. By the time the Cincinnati reporter visited, "there were four hundred and fifty barrels filled at this well on Friday, up to half-past three o'clock."[45]

By 1861, Burning Springs had become a second star in the constellation of Ohio Valley oil. Even with the political and economic crisis that would erupt during that year, and with pro-Confederate western Virginian guerillas terrorizing the countryside, Burning Springs operators produced 4 million gallons of oil. "But these hopes were of short duration." When Union armies claimed tentative control of the region, local Confederate rebels destroyed the oilfields rather than let them produce for their Union enemies. While some producers desperately sought to hang on, the guerillas never let up, and the Union armies never fully secured control of the oilfields, even after West Virginia was officially admitted to the Union in 1863. Over the course of 1863, raiders destroyed 150,000 barrels of oil in Burning Springs.[46]

I N hindsight, the western Virginia coal-oil and petroleum industries seemed doomed by the destruction of the Civil War. Cut off from the Confederate armies of eastern Virginia, how could this outpost of industrial slavery possibly survive? But in the southern and western counties of what became West Virginia, in the Kanawha Valley, the future remained an open question for at least a few months.

In March 1861, just as the Burning Springs oilfields seemed to be coming into their own, the industrialists of the Kanawha Valley debated whether Vir-

Rock Oil, Civil War, and Industrial Slavery

ginia should join the already-seceded cotton states of Mississippi, Alabama, Louisiana, Texas, South Carolina, Florida, and Georgia in the formation of the Confederate States of America. "All coal shipped from Western Virginia into the Southern Confederacy has to pay a tariff of 24 per cent," the *Kanawha Valley Star* complained.[47] And while petroleum may have provided a setback to cannel coal miners, cannel coal was not finished, and the prospects of war and tariffs might have revived it. But whether or not coal oil experienced a resurgence, raw coal was still in enormous demand both in and outside gasworks, not to mention the considerable stimulus secession might provide to Burning Springs petroleum producers as the Mississippi Valley cut itself off from Pennsylvania oil.

"Ought we to go North or go South?" This was the question. Considering the new tariff, the editors of the *Kanawha Valley Star* believed the answer was clear. "If Virginia secedes and joins the Southern Confederacy," they suggested, "coal can be shipped from the Kanawha Valley into the Southern Confederacy *free of duty*. . . . Just think of it! Kanawha coal shipped to New Orleans, free of duty, and Pittsburgh required to pay 24 per cent. duty!" They argued that secession might be exactly what Virginia needed to solve all of the spatial and economic problems of its industrial slavery, asking rhetorically,

> Is it not clearly our interest for Virginia to be a member of the Southern Confederacy? Would not the coal lands of Western Virginia be wonderfully enhanced in value were Virginia to join the Southern Confederacy? The tariff shows it plainly. Millions upon millions of dollars of capital would speedily be invested in Trans-Alleghany Virginia were Virginia to join the Confederate States. Our coal lands would become immensely valuable. Capital from Pittsburgh, and other places, would seek investment in our coal property — and coal operations on a grand scale would speedily be commenced in this portion of the State.[48]

What would it have meant if this vision had been realized? What if the Kanawha Valley had become the coal mines and oilfields of the Confederacy? As the history of underdevelopment in western Virginia demonstrated, the voices of western industrial slavery remained dwarfed by the interests of eastern planters. But they were not silenced, and the war shifted the balance of power. Although initially justified as an anti-industrial, antimodern defense of planter society, the Confederate project was quickly seized by those who envisioned a centralized state pursuing a fully modern industrialized economy of factories, railroads, and heavy industries wholly dependent upon racial slavery. The propaganda of states' rights ideology notwithstanding, the racial and class chauvinism of Confederate officials led to the formation of what one historian argued was the most centralized and powerful state in North America until the Second World War. The Confederate state owned industries, instituted massive and draconian conscription policies, and employed 70,000 civilians as bureaucrats, tax collectors, and conscription agents, and the "police power of the Confederate state was sometimes staggering."[49]

There is a long tradition of viewing the Old South as a stagnant agrarian society and the Confederacy as a futile attempt to extend the antebellum moment. But the reality was far worse. Had the Confederacy prevailed, the contributions of slave-worked industries like the enormous Tredegar Iron Works and the rapidly built-up coal mines around Richmond would have been impossible to deny. Industrial slaveholders, long marginalized by planter society, would almost certainly have joined the ranks of planters, politicians, and generals as Confederate heroes, the institutional momentum of the Confederate state further entrenching and celebrating a system of industrial slavery. Add the practically limitless coal and oil of the Kanawha Valley into the mix, and the implications of a successful Confederacy, of an industrializing military state founded on white supremacy and racial chattel slavery, become even more terrifying. This was one future that the world's greatest slave rebellion defeated. As W. E. B. Du Bois first argued in 1935, and white historians have finally fully conceded, by deserting their enslavers by the hundreds of thousands, by slowing and stopping work by the millions, and by joining Union forces to directly fight against Confederate armies, enslaved men and women forced the upcountry southern whites and the Union armies fighting to defeat the Confederacy militarily into helping them wage a war to destroy slavery. And in their world-historical success, they nipped an emerging slave-powered white-supremacist industrial war machine in the bud.[50]

T HE Civil War also helped to violently change the possibilities of producing the means of light in the United States. The story has usually gone like this: First there was fire. Then, well, nothing terribly interesting happened for a few hundred millennia. But then, in the Age of Enlightenment, scientists and American whalers changed the world with whale oil, and just in the nick of time, kerosene saved the whales. This story is false on a number of counts.

"Grand Ball Given by the Whales in Honor of the Discovery of the Oil Wells in Pennsylvania." Despite the fact that sperm whale oil had, except in the lamps of the very rich, been replaced years before by other illuminants, illustrations and stories directly tying the decline of whaling to the rise of petroleum were, and remain, common. If the petroleum industry did deal a final death blow to a whale fishery already weakened by difficulties with supply, camphene, and labor unrest, it did so by outcompeting whale oils not as illuminants but as lubricants. Kerosene's real competitor, in both material and market terms, was camphene, an industry conveniently crippled at the same time oilfields in Pennsylvania and West Virginia were discovered. Engraving in *Vanity Fair*, April 20, 1861, 186.

When Drake's well famously struck oil in August 1859 and the petroleum rush began in Pennsylvania, the whale fishery had already fundamentally transformed from its peak in the 1830s and 1840s, when it sat at the center of a thriving web of light. By the time oil started flowing up from American soil, the transformation of whales into lubricants was practically the only process keeping voyages buoyant. The Civil War caused even this lifeline to fray to nearly nothing. From 1861 to 1865, more than fifty whaleships were captured and burned by Confederate privateers. Many others were commandeered by Union forces for the war, and twenty-four whaleships were deliberately sunk in Charleston Harbor by the Union to strangle the South Carolinian port. Meanwhile, in 1862, the U.S. Lighthouse Board decided to replace sperm oil with lard oil, snapping the final surviving thread of whale light. And petroleum continued to flow. No, erupt is probably a better word.

A spectacularly successful whaling voyage might return with close to 3,000 barrels of oil, collected over a period of no less than two years. A single Pennsylvania well yielded the same amount in one day. In its most productive year, the whale fishery brought a little over 13 million gallons of whale oil to American shores; the petroleum industry surpassed that figure in only two years. And it was only the beginning. Petroleum-based kerosene flooded the market for lamp oil. Prices plummeted, and kerosene exploded onto the scene as an illuminant eclipsing anything achieved by the fishery.[51]

But before kerosene could dominate American lamps, something was going to have to be done about camphene. Here, too, the Civil War was the real agent of change, making a bloody politics of light appear in hindsight as technological inevitability. With the outbreak of war following the attack on Fort Sumter, turpentine producers found themselves virtually cut off from major markets by the Union blockade, and camphene practically disappeared from urban markets overnight. Turpentine, the key ingredient in camphene, was devastated by the Civil War, with prices skyrocketing from 35 cents before the war to $3.80 a gallon by 1864. In 1860, consumers burned around 15 million gallons of camphene, at 47 cents a gallon, in lamps ranging widely in size and portability. By 1862, camphene was almost nowhere to be found. The price of kerosene, on the other hand, was only 36 cents a gallon in 1862, and even after taxes and wartime inflation, it rose only to 75 cents in 1864.[52] Kerosene and coal oil may eventually have come to replace camphene anyway, but without the destruction of the turpentine industry during the Civil War, the process would likely have been different. There is little reason to believe that kerosene, by itself, would have completely displaced camphene.

What did destroy the foundation of slavery and turpentine camps was the Civil War. At the camps he oversaw along the Fish River in Alabama, Benjamin Grist pushed the enslaved men under his command to keep dipping, chipping, and distilling turpentine for the rest of the summer and fall of 1861, but he was worried. "I am in trubel for I never saw such times in my life," Benjamin wrote to his cousin James R. Grist, who was still attempting to manage his turpentine empire from Wilmington, North Carolina. Benjamin complained that the war was crushing turpentine makers and, perhaps more ominously, was giving new hopes to the enslaved men he had carried south with him from Wilmington. "I have had more trubel with the negros this yare thank I ever had," Benjamin wrote, claiming they were "sckary to death," but that James need not "be afrade of the yankey giting any of our negros. I will keap them out of the way. I have my plan lade & I will stay them." One month later, Benjamin once again assured his cousin, "I shall look out for the negros & tacke care of them if the yankees lands hear." That fall, Benjamin Grist and his overseers made a tremendous effort to find, capture, and drive in the enslaved woodsmen who had run away, giving him so much "trubel," and on October 9, 1861, he wrote "all of the negros in at both plases except Jesper & I cannot hear a word of him." Like so many other turpentine operators during the war, Benjamin Grist was terrified of both the men he enslaved and the Union armies, fears that the woodsmen likely encouraged. But while the turpentine industry crumbled, these black men remained trapped in industrial slavery, and by 1862, Benjamin Grist had leased sixty-nine of the eighty-eight men enslaved in his Fish River turpentine camps to the nearby Shelby Iron Works for yearlong contracts for a total of $6,234.75, or about $125 each.[53] What happened to the remaining nineteen men is unclear.

As the price of turpentine plummeted and turpentine producers desperately tried to move enslaved woodsmen out of the camps and into other wartime industries before they could run away, the Union armies steadily cut their way through the piney woods. During Sherman's march from Savannah, Georgia, to Goldsboro, North Carolina, the abandoned turpentine camps became flammable, weaponized landscapes. While guides — many likely fugitive turpentine workers who had escaped to the swamps while their enslavers fled the forests — led Union armies across plank roads and through turpentine paths in North Carolina and Alabama, the advancing and retreating forces burned thousands of barrels of resin.[54] On March 7, 1865, Sherman's troops halted at Station 103 on the Wilmington and Raleigh Railroad, after passing "2,000 barrels of rosin on fire — a magnificent sight," likely an inferno left by retreating Confederates. From the swampy terrain at the Brunswick River Ferry, the commanding Union officer reported that "there can be little doubt the rebels are evacuating. They have made immense fires, the smoke of which you must have seen, indicating that they are destroying turpentine, &c." A member of the 51st New York Volunteers recalled encamping near New Bern, North Carolina, in an abandoned turpentine camp, where "the trees became saturated with pitch, and were in a highly inflammable condition. Some luckless night, on mischief bent, set fire to some of these trees and soon the entire forest was a mass of crackling flames, that not only illuminated our camp during the night but rendered it both night and day very unpleasant and difficult as a breathing place, on account of the dense smoke.... In a short time we were all as black as the darkest darkey in North Carolina."[55] Barrels of turpentine sat idle at blockaded ports, Union and Confederate armies torched the flammable remnants of the turpentine camps, and enslaved people in North Carolina and Alabama dealt the final blow to the geography of piney light by emancipating themselves by the thousands.

The Civil War was also a transformative period for gasworkers. Gasworks laborers understood that the fear they inspired in xenophobic and middle-class American consumers was nothing compared with their fears of a dark city. While striking railroad mechanics struggled unsuccessfully for days during February 1853 to make any progress with the owners of the Baltimore and Ohio Railroad, the "employees at the gas works struck to day for 15 per cent. advance, which was immediately accorded to them." In November 1862, the Irish laborers employed at the two major gas manufacturers in New York City struck for higher wages. According to the *New York Times*, the gas companies "readily complied with their demand, and it was supposed that the difficulties were all amicably arranged, and the workmen—some four hundred in number—would all return to their duty." Again, it appeared that gasworkers, positioned at the heart of an industrial system upon which so much money and fear were heaped, wielded unusual power over their wages. But apparently something went wrong, and the Manhattan Gaslight Company discharged all of its Irish workers and replaced them with Germans. Perhaps the Irish workers had overreached, or perhaps the gas companies saw a chance to break some of the power of labor by playing off ethnic divisions. Instead of returning to work, the Irish workmen at the other gasworks formally organized and began a militant strike, barring others from entry and attacking the German men trying to work at the Manhattan gasworks. The police were called, the strikers retreated, and for the time it seemed that the gas companies had won. Following the war, in July 1868, 600 firemen at Philadelphia's gasworks struck for a 25 percent increase in their wages, plunging the city into darkness. For three days city officials resisted, and while the fears of dark anarchy remained unrealized, the workers remained organized and undivided. The officials, perhaps responding to the extraordinary public pressure, finally relented and gave in to the workers' demands.[56]

The pressure came because as gaslights continued to expand, more and more urban Americans of means came to rely on them, came to act for them. In New Orleans, antebellum officials had let poorer parts of the city drown in order to save gaslight for the rich. During the Civil War, Union occupiers were willing to keep people enslaved to achieve the same. In the 1850s, the number of Philadelphians who burned gas as private customers more than quadrupled. In 1860, those 41,200 customers burned gas in nearly half a million fixtures. Gas streetlamps, too, continued to multiply, and mostly to illuminate the night neighborhoods of the middle and upper classes. In 1850, there were 1,576 gas streetlights in Philadelphia. Ten years later, 5,604 burned in the city, and by 1890, gas streetlights had nearly quintupled again to 26,043. These streetlights, entangled in spatial relations of class, even more starkly demarcated race. In 1873, gaslights only illuminated the streets of the affluent northwest quadrant of Washington, D.C., and its commercial and political center. By 1891, Washington's 5,607 gas streetlights reached most of the city. Along major streets, officials were going so far as to replace gas burners with electric lights. But even then, the streets of the poor, black neighborhoods in the southwest of the city were still "lit" by only 434 dim oil lamps.[57]

Alongside the adaptation and expansion of coal gas, it would be tempting to see in the Civil War's devastation of camphene and whale oil a turning point in the shift from organic to mineral energy regimes. Yet the deathscapes of Cincinnati and Chicago had never been busier, or deadlier for hogs. Southern markets may have been cut off, but the enormous demand of the Union armies for pork, soap, and candles propelled millions of hogs to death over the course of the Civil War. According to the Army Regulations of the United States, Union soldiers were supposed to receive as part of their rations "1½ pounds of tallow, or 1¼ pounds of adamantine [stearine] or 1 pound of sperm candles." Writing to his brother from camp in Tennessee, Union soldier William Allen Clark realized, "I never told you the list of Rations. It is ¾ of a pound of Crackers or 1¼ pound of Bread perday, 1 pound of pork or 1¼ lb. of Beef, 1 lb. of Beans or peas to eight men per day, ⅛ of a pound of Sugar, the same of Coffee perday, a pound of soap and the same of Candles to eight men." Correspondence with family and friends was a central part of camp life for soldiers, and candles helped to structure the practice in time. Each evening in camp, with the "letters duly read," one observer wrote, "there succeeds a busy season of writing replies. Candles are lighted, fires kindled, and the camp soon presents the leading features of a writing school." Between dusk and the order for darkness, soldiers maintained paper ties to home and created new relationships among one another through the light of candles and campfires, until "at nine o'clock 'taps' are beaten, when all lights are extinguished, and the soldier wraps himself up in his blanket for the night." One soldier wrote to his sister of girls and marriage, which were "common talk in the tents after the candles are lit until bedtime." Illuminated through millions of government-issued candles, the regulated times and spaces of wartime letter writing helped to knit together a national community and a shared experience of the war.[58]

In and around the army camps, processed goods like candles that were at once difficult to produce and such important parts of the rhythms of everyday life became valuable beyond their illuminating power. They became credit and currency. According to one article addressing complaints among volunteer troops about rations, regular troops "draw *all* the rations to which they are entitled," after which "it will always be found there is a small surplus of many articles, such as salt, vinegar, soap, candles, and sometimes beef, pork, or hard bread." One soldier's Civil War diary recorded that he and a friend "take pork, soap & candles to a stingy grocer in Alexandria & trade them for potatos, onions & molasses." Officers were required to provide for themselves and their horses, but the typical practice was for officers to purchase supplies "from the regimental Commissary Sergeant, out of the rations already drawn for the men, but not distributed." One article complained that this practice led to the soldiers "virtually selling to the officers their surplus provisions before they knew whether there would be enough for themselves," with the result that "I have known the men to be stinted in sugar and candles for weeks, because they had not drawn their full rations, and the officers had first taken all that they wanted." Another published soldier's account noted that "our Sergeant bought a good ham this morning with the soap and candles that he had not drawn."[59]

By the end of the war, the enormously productive and enormously profitable pig by-product industry of candles constituted the last surviving organic light, led by such corporate juggernauts as Armour & Company and Procter & Gamble. In 1860, slaughterers and packers killed and disassembled 434,499 hogs in Cincinnati. In 1863, they unmade 608,457. Chicago's rise as the world's largest center of animal death was even more spectacular. In 1860, Chicagoans packed 151,339 hogs, far behind Cincinnatians. By 1863, they would slaughter 970,264. Riding this surge of hog death, candlemakers also turned the Civil War to their advantage. From 1861 to 1862, the number of candles exported from Cincinnati nearly doubled to 245,997 boxes, then rose to 263,912 in 1863. In 1862, while debating a new tax bill before Congress, a representative from Cincinnati claimed the number of candles exported was "an amount very far below—I cannot say how far—the amount manufactured."[60] From commercial reports, hog lights, as both candles and lard oil, appear to have been rescued by the Civil War. Hit hard by the entrance in the late 1850s of coal oil, kerosene, and paraffin candles made from oil by-products, hog lights experienced a noticeable upswing in exports in 1861. Though we may want to draw clear lines between organic and mineral, past and future, the history of light proves otherwise.

In the 1860s, it wasn't clear if lucifer matches were organic at all, as the

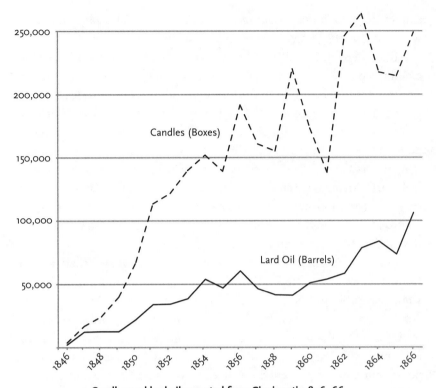

Candles and lard oil exported from Cincinnati, 1846–66.
Figures from Cincinnati Chamber of Commerce, *Annual Report* (1866).

nature of guano was still widely debated. And beyond questions of their nature, matches continued to escape authorities' knowledge and control during the war. Even when and where there was no battle, matches could kill
soldiers. In July 1861, some Rhode Island men were performing a drill when
the limber-chest of one of the cannons, where the ammunition was stored,
exploded, killing two soldiers and wounding several others. "The supposi-
tion is," the *New York Times* reported, "that the explosion was caused by friction matches, but whether placed in the chest by an enemy, or dropped in
through carelessness, cannot be decided." Faced with such threats, officers
tried (and failed) to carefully regulate who had access to matches. When
Union troops boarded a ship, officers would search their knapsacks for lucifers and, if any were found, confiscate or dispose of them. Authorities also
worried about what these elusive strike-anywhere flames could mean for the
war's other front, that being fought by the enslaved. One writer who feared
making emancipation a goal of the war was particularly concerned about
what he called the "John Brown method," which he understood to mean to

simply "collect a sufficient number of blacks, free or slave; [and] arm them with pikes, scythes, knives, pitchforks and lucifer matches."[61]

But where state and military officials worried about the lucifers they couldn't control, many capitalists saw opportunity. With war halting imports of European lucifers and the match stamp tax of 1862 doubling the price of matches, those with the capital to purchase their own stamps cornered the market. Prices for phosphorus more than doubled from 80 cents a pound before the war to $2.00, while sulfur rose a similar amount, from 2½ to 6½ cents a pound. The companies, like Swift & Courtney of Wilmington, Delaware, that amassed enough capital to survive and thrive amidst the price surges and newly protected markets began to quickly consolidate monopoly control. By 1881, this process of consolidation had culminated in the formation of the giant match monopoly of the Diamond Match Company.[62]

Finally, kerosene. As the Civil War crippled the whale fishery and the turpentine camps, multiplied the tensions and production of monopoly gasworks, gave a boon to hog lights, and helped American match manufacturers consolidate their own monopolies of light, so important and so productive were the oilfields of northwestern Pennsylvania, while guerilla warfare decimated the western Virginian oilworks, that many in the Union interpreted petroleum as divine proof of the justice of their cause. The very moment cotton disappeared, petroleum flowed in to make up for some of the loss. The *Philadelphia Public Ledger* published an article in 1862 exaggeratingly titled, "Petroleum Oil as Valuable as Cotton," touting an English market circular predicting "that if the rocks and wells of Pennsylvania, Canada and other districts continue their exudation at the present rate of supply, the value of the trade in this oil may even equal American cotton."[63]

Distilling kerosene also amplified the tensions of an incredibly flammable landscape built to refine and store camphene. In just the eleven months from May 1861 to April 1862, there were at least sixteen separate fires and explosions in at least fourteen different kerosene oil refineries (most of which occurred in the several-mile radius encompassing New York City, Brooklyn, Williamsburg, and Jersey City), killing dozens of people and destroying tens of thousands of dollars' worth of kerosene and petroleum. Docks, storage depots, and even groceries became routine settings for kerosene explosions that could be truly horrific. Describing a 3,000-barrel fire in Pittsburgh at the depot of the Pennsylvania Railroad, one article claimed the fire was so intense, "All the water of the Niagara turned upon it would have been without effect."[64]

Kerosene was supposed to be the savior of all those risking their lives around camphene. Not only was it cheaper even than camphene, but advertisers claimed it was perfectly safe. By 1860, the light of the future was said to have arrived as if it were Nature's gift to the United States, petroleum "flowing in some localities literally like rivers, and prepared directly in nature's own great distillery," where "Nature distils free of charge."[65]

But for consumers, the more things changed, the more they stayed the same. In the 1850s, tens of thousands of outworking women had sewn men's clothing around the clock by the dangerous light of camphene for starvation wages. Neither Civil War, kerosene, nor sewing machines would change this during the 1860s. Instead, even more women worked even longer hours for even lower wages, sewing clothing and knapsacks for the men of the Union armies with lights just as dangerous as, if not more dangerous than, camphene. Every new kerosene lamp explosion seemed to shock reporters, but the claims that the new oil was a safe replacement for camphene proved to be utterly false.

The public outrage following the spate of explosions that began in 1861 was directed, first, at grocers accused of being kerosene "adulterers" and, later (and more accurately), at the manufacturers who cut corners and padded profits by failing to fully distill out the more volatile materials known as benzole and naphtha. Claims continued to be made that "pure" kerosene was perfectly safe, that all would be well if regulations for "flash point" testing were implemented. Yet by the 1870s, the problem had only become worse. The United Fire Underwriters of America estimated "the total number of lamp explosions each year in this country at ten thousand," following an earlier estimation by the Underwriters' Association of the Northwest that "between five and six thousand people annually go hence via the kerosene route."[66]

For those who survived their kerosene lamps and sewing machines "ticking all day long and far into the night," women in the 1860s could only hope to make 30 cents for fourteen hours of work. Most made much less. But some women began organizing to resist their exploitation and to demand that employers increase wages. Much of the campaign involved testifying and winning allies in newspapers. One skilled woman working by sewing machine from seven in the morning to nine at night made four pairs of cotton drawers in a day, each pair requiring 1,800 stitches, and received 16¾ cents for the day's work. Another New York woman sewing flannel pants for the army complained to the Working Women's Protective Union that she could no longer endure the strain, while newspapers reported that a further "inquiry revealed the fact that the wealthy firm who employed her, paid five and a half cents per pair for these drawers, of which she could make two pair per day, remarking, 'If I get to bed about daylight and sleep two or three hours, I feel satisfied.'" Like thousands of other women, one older New Yorker struggled with the "coarse flannel army shirt, large size, made by hand sewing." A younger woman "might make two or perhaps three in twelve hours," reporters noted, but this "old lady occupied, with another woman, a damp, dark basement, where she strained her eyes in the day time, and sewed by the light of her neighbor's lamp during the evening." Others protested the starvation wages they were paid to furnish the army knapsacks equipping the Union soldiers, as only "three of these knapsacks can be finished in one day by an ordinary good seamstress, working from 6 o'clock in the morning and quitting about 11 P. M." In Philadelphia, the quartermaster general reported that from 8,000 to 10,000 "work people" were employed "in the manufacture of clothing and equipage" for the military. But that was probably an under-count. Many of those "work people" likely sweated the sewing to outworking women who were never recorded in the quartermaster's figures.[67]

The spread of kerosene intensified the democratization of men's exploitation of women's work and time that camphene had made possible. The Civil War enriched male clothiers and clothed male soldiers through the exploitation of women on the "home front," while the temporal violence of capital and gender was disguised as self-inflicted violence in self-illuminated night spaces. It was a process of mystification that unloaded the costs and risks of time expansion onto working women while displacing that expansion away from shops and into homes. This kerosene-lit democratization of male supremacy also masked class violence, as the social relations of conscription, contracting, and war forced both free working-class women *and* men to bear more of the violence and risk of provisioning and fighting the Civil War than anyone other than the enslaved, while contractors grew rich.[68]

Forced to survive in this new but, since camphene, all too familiar ecology of light, people developed new strategies for knowing and living with explosive lamps. Catherine Beecher and Harriet Beecher Stowe urged women to carefully prepare and test all the kerosene they brought into their homes before pouring it into a lamp. "Good kerosene oil should be purified from all that portion which boils or evaporates at a low temperature," they wrote, "for it is the production of this vapor, and its mixture with atmospheric air, that gives rise to those terrible explosions which sometimes occur when a light is brought near a can of poor oil." To perform this all-important test of kerosene, they advised women to "pour a little into an iron spoon, and heat it over a lamp until it is moderately warm to the touch. If the oil produces vapor which can be set on fire by means of a flame held a short distance above the surface of the liquid, it is bad." Good kerosene, on the other hand, "should be clear in color and free from all matters which can gum up the wick. . . . It should also be perfectly safe. It ought to be kept in a cool, dark place, and carefully excluded from the air." Keeping lamps clean from grease buildup was just as critical. "The inside of lamps and oil-cans should be cleansed with" a mixture of one tablespoon of soda dissolved into a quart of water. "Take the lamp to pieces and clean it as often as necessary. Wipe the chimney at least once a day, and wash it whenever mere wiping fails to cleanse it. Some persons, owing to the dirty state of their chimneys, lose half the light which is produced. Keep dry fingers in trimming lamps. Renew the wicks before they get too short. They should never be allowed to burn shorter than an inch and a half."[69]

Living with kerosene lamps meant more light, but it also meant the endless, drearily monotonous work of cleaning and meticulous chemical analysis. With the increased risks of explosions and fires from the new lamps, properly "cleaning and lighting the lamps was skilled and painstaking work, labor that the mistress of a household usually reserved for herself even when help was available." When middle-class Lydia Maria Child compiled a summary of her activities for 1864, lamps featured prominently: "Cooked 360 dinners. Cooked 362 breakfasts. Swept and dusted sitting room & kitchen 350 times. Filled lamps 362 times. Swept and dusted chamber & stairs 40 times." Feeding and caring for lamps was demanding, never-ending work, but lamp cleaning was about more than just getting the most amount of light from each ounce of kerosene. It was an attempt to save life. The sooty grime that kerosene flames deposited on lamp glass and layered across every nearby surface was also highly combustible and increased the risk of fires and explosions. The women who had the time and resources to clean lamps and perform routine flash point tests on all their kerosene (or have others do so for them) reshaped the ecologies of domestic spaces produced through the metabolisms and movements of people and flames.[70]

The middle-class families who could make the time and space to properly test their kerosene and clean their lamps thereby gained an ecological advantage over working-class households. Urban workingwomen lived in quarters that "were often cramped—filled, not only with people, but also perhaps with the tools and materials of outworkers." Crowded into smaller, dirtier, leakier homes—the "oily soot of cheap coal stoves and charcoal burners collected on floors and walls, their fumes lingering in the air," while mixing with the combusted remains of wicks, kerosene, and candles—

workingwomen still had to spend extra time scrounging, peddling, and out-working in order to make rent and meet the bare needs for life. Regularly cleaning lamps under such conditions was next to impossible. What middle-class women loathingly called "spring cleaning" was in large part another strategy to negotiate living with open-flame lights. Along with the soot from dark, closed wood- and coal-burning stoves, lamps and candles deposited layers of grime over every household surface, a process that accelerated in winter when shorter days and closed windows meant more combustion and even less circulation.[71]

The spaces and times produced through camphene and kerosene lamps left a sea of human wreckage in their wakes. Lamp explosions killed and dis-figured thousands of women who desperately sought to make ends meet around dangerous lights, but even those who escaped the direct violence of a lamp explosion did not escape the temporal politics of sewing unscathed. The strain of sewing hour after hour, day after day, year after year in tene-ments that were dark by day and dimly lit by carefully economized candles and lamps by night reshaped the eyes, minds, and skeletons of tens of thou-sands of outworking women. So common and so severe was the work of sewing that the American reformer Virginia Penny could spot a seamstress simply by her posture: "'by the neck suddenly bending forward, and the arms being, even in walking, considerably bent forward, or folded more or less upward from the elbows.'" For nineteenth-century workingwomen, sewing constituted, in the words of the historian Christine Stansell, a "biological ex-perience of class." A doctor in 1860 estimated that every year, on top of the fatalities from burns and explosions, 1,000 women died from "causes related to sewing in the outside system. Malnutrition, fatigue, cold and bad ventila-tion in the tenements bred pneumonia and consumption, the major killers of nineteenth-century cities." More work simply translated into more illness: "A newspaper investigator in 1853 heard that the hardest-working women could squeeze as much as double the average earnings out of piece rates, but the extra money usually went to medicines."[72]

BUT for the secession of slaveholders, the massive revolt of the enslaved, and the corresponding assertion of demands and power by northern white workers, the landscapes of labor, light, and industry may have looked radically different in 1865. In 1861, industrial slavery was coming into its own, and camphene, coal gas, and kerosene might easily have become a triumvirate of slavery, industry, and light had Virginia invested heavily in Kanawha oil and coal a few years earlier, or if Virginia had remained one state, pulling its western regions into the Confederacy. If the Civil War had been delayed, the Ohio Valley might have combined the labor of both free and enslaved workers into a new trans-Allegheny, Ohio-to-Mississippi industrial corridor. If the South had won, the postbellum Confederacy wouldn't have been the antebellum South any more than was the case for the North. A victorious and industrialized Confederacy, even more swaggeringly confident than it ultimately was in defeat, having demonstrated the powerful ability to wage industrialized warfare with black slaves and white soldiers, would almost certainly have sought to violently extend this revolutionary war machine to the Caribbean, the American West, and possibly Mexico and South America, as southern boosters had long envisioned. This was a process made partly possible by the history of light and unmade by the revolution of hundreds of thousands of enslaved men and women who found or forced allies in the armies, governments, and industries of the North.

Industrial slavery was interrupted, but not by the mythological march of progress or technology. The coincidences of emancipation, warfare, and Oil Creek's serendipitous ascent on the eve of secession made it possible to misunderstand the relationship between petroleum and providence. Later, it allowed people to convince themselves that American lucifers had always been on the road to transcendence from toil, a road that had merely stopped at the waystation of kerosene on its way to electricity. But the first century of the industrialization of light was at least as much a story of increasing human and animal unfreedom as it was an escape from labor. Electricity didn't fundamentally change this. But its boosters and consumers did make it ever harder to glimpse the dark history of artificial light.

EPILOGUE

This is how Americans blinded themselves with light, and why it's still so difficult to see.

On December 11, 1882, a Boston crowd gathered on a cold winter evening to witness an unusual convergence of debuts. At quarter past eight, the doors to the new Bijou Theatre were thrown open for the first performance in Boston of Gilbert and Sullivan's comic opera *Iolanthe*, a satire about magical fairies at comedic odds with Parliament. Those fortunate enough to have secured tickets became the first audience to grace the Bijou's auditorium. Yet what was truly novel, and what no one had ever before seen in the United States, was a theater illuminated entirely by electrical lighting. Thomas Edison himself, the so-called Wizard of Menlo Park, was in attendance, and he "personally superintended the electrical apparatus last evening, remaining in the engine-room during the entire performance, and not looking once into the theatre."[1] It was a public gamble, an electrical performance that Edison dearly hoped would dazzle.

At the time of the Bijou's opening, incandescent electric bulbs accounted for a vanishingly small fraction of lighting in America. Without question, gas, kerosene, candles, and matches remained securely entrenched as the principal means of light in urban environments. Electricity had been used to illuminate factories, ships, and the mansions of the nation's elite, but there existed no power grids, with the exception of the small and unprofitable Pearl Street station near Wall Street. To promote popular confidence in electricity's still-uncertain future, many boosters, including Edison himself, staged public demonstrations juxtaposing the material with the amazing. *Iolanthe*, one of Gilbert and Sullivan's most fantastical compositions, was a deliberate choice for the theatrical debut of electric lighting.

This electrical performance extended well beyond the stage, as its directors recruited local newspaper writers into the cast as well. For part of the grand opening of the electric theater, Spencer Borden, the director of the New England department of the Edison Company for Isolated Lighting, invited "several newspaper men" to a demonstration of how all 644 lights operated, from the lamps in the lobby and auditorium to the onstage arch comprised of stacked arcs of colored lamps used to produce different light blends. "The exhibition yesterday showed how perfectly each series and all the series were under control," wrote one *Boston Herald* reporter, eager to play his part. Awed by "how the lights of each could be turned down, or put out altogether, or turned on in a flash into a state of brilliant incandescence," the newspaperman proclaimed that Edison had rendered "electricity a thoroughly pliable and practicable thing." It was progress so momentous the reporter was struck with prophecy, suddenly able to foretell that now, after the Bijou, there was "no doubt" electricity was "destined to be the illuminant of the future."[2] Such performances promised that, like the magic of *Iolanthe*, electric lights would grant their users a heightened, almost supernatural degree of control over their surroundings. But all was not as it seemed. This performance of control and mastery relied on keeping its surprisingly deep and decidedly unmastered backstage hidden from the audience. Otherwise the illusion, and Edison's gamble, would fail, revealing electricity to be a light of work and pain and human politics like any other.

Way, way backstage, in the mountains of the American West, a different gamble was about to end mortally for Jerry Toomey in 1887. He was part of a crew working a copper vein 258 feet below and 500 feet in from the main shaft at the Parrot Mine in Butte, Montana. "It was a treacherous wall," his fellow miner John Sullivan would later explain before an inquest, "and we were warned to keep timbered up" to prevent the stope's ceiling from caving in. The warning had come from the supervisor on duty, Mr. Tibby, who had already told the men "to be careful" and had "cussed" them the day before for "being so careless." Nonetheless, they "had gone a little beyond their work and were ahead of timber about 10 ft." This placed them in "the worst place in [the] mine," where a miner should "take every precaution." Despite the risk, Tibby would testify, "they take they chances rather than go to the trouble of putting in sufficient stulls."[3]

Tibby's warnings went unheeded that day, not because his crew was particularly cavalier, but because of a cascade of simple calculations. Tibby and the other shift bosses, following orders from the mine superintendent, would have made absolutely clear to their crews that full wages meant meeting a daily production quota of copper ore. Timbering up meant time not getting ore. Like colliers, copper miners categorized everything not directly involving the extraction of ore as "dead work." And in the late-nineteenth-century copper mining industry, where the economics of capital-intensive entry and falling prices were driving a spiraling system of overproduction, the pressure to shortcut dead work could be overwhelming. With families depending on full wages and with few alternatives for employment, miners would have to "take they chances."[4]

So, like countless others in Butte's underground forced to navigate between staying alive and making a living, Jerry Toomey went for the ore before timbering. At 1:45 P.M., Sullivan watched as "the cave came from the hanging wall," immediately crushing Toomey to death. "Don't know how much fell on him," Sullivan testified in conclusion. "It was waste and not ore." As such, it had no value, and who bothered measuring trash?[5] Here was the paradox of dead work laid bare: human life in tension with money. Sullivan's words revealed how the relations of dead work and commodity production transformed some rock into waste, not even worth securing, while making some into ore, worth risking life itself.

In the 1880s, Boston's Bijou Theatre and Butte's Parrot Mine were materially connected by a commodity chain of copper ore, copper wires, and electric machines, but being connected did not mean they were the same or that all those the chain entangled shared a common politics. Because of the electric lights that linked them, copper mines and electric spectacles were spaces growing further apart—as much as, and perhaps more than, they were growing together.

To understand this spatial contradiction is to understand a hidden politics of class at the heart of the Gilded Age. It is also to understand when and how Americans began to blind themselves to the history of light you've just read. It was a blindness that let white middle- and upper-class Americans feel safely disconnected from past and present systems of oppression they wanted to ignore and to forget, systems that had made and continued to make their lives and lights possible but that they desperately didn't want to see. Pushers like Edison and the Copper Kings of Butte thought there were fortunes to be made in selling these anxious Americans electric narcotics. They were right. The lights were certainly useful. But what really got people hooked beyond all pragmatic reason was the myth, the feeling of power, the promise of a better future and an escape from past and present. And to grow rich from the manufacture of the copper means of electric visions of a safer, easier, more harmonious, and more predictable future, the capitalists organizing its production relegated men like Jerry Toomey to lives of greater danger, drudgery, conflict, and uncertainty. Though electricity's boosters pretended otherwise, for American lucifers it was a familiar story.

E DISON's gamble seemed to be paying off. "The Bijou Theatre was opened last evening amid a blaze of glory," the *Boston Evening Transcript* reported. "The Edison incandescent lights worked to universal admiration, and it was shown on more than one occasion how beautifully manageable and tractable this mode of lighting can be made." Such a reaction must have been a relief to Edison and those under his employ. "Well, this theatre business is a new one to us," he remarked to another reporter, adding that he had been "somewhat anxious as to this experiment." Fortunately in his view, "everything worked without a hitch, and the public seemed satisfied."[6]

Edison's assessment of a satisfied public was something of an understatement. One account noted that "the lighting and ventilation of the new house were the most perfect we have yet seen." These were no minor concerns for the audience at the time. All around Washington Street, which served as the core of Boston's theater district, were stages and auditoriums illuminated by the yellowy flames of gas lamps. Not only did they flicker or, as footlights, produce hellish patterns of light and shadow onstage, they raised temperatures to more than 100 degrees, consumed oxygen, and frequently left patrons with intense headaches. For individuals who had set type or otherwise worked by the "torture of the brain-frying" heat and eyestrain of gaslight in Boston's newly electrified printing houses and newspapers, such material differences were even more obvious.[7] Electric lights impressed as much for what they did not do as for what they did.

But the electrification of the Bijou was still something of an illusion. Even as headlines proclaimed that the Bijou was the first theater in the United States to be illuminated entirely by Edison's miraculous bulbs, gaslights continued to be used for rehearsals and stage building throughout the day. The electric dynamo only ran during evening performances, and according to one historian of the company, Boston Edison "routinely reimbursed the theater for the cost of the *gas* it used during daylight hours!"[8] In other words, the spectacle of an electrical theater was made directly possible by the gaslight that electricity was supposedly rendering a dead technology.

Electrical lighting systems were, it was true, involved in producing visible spaces, bodies, and practices. But they were also actively constructing *in*visibilities. In the Bijou, Edison's incandescent lighting system helped to further hide the gritty and mundane backstage and stage crew from the audience by operating "automatically."[9] Front stage, now better-illuminated than was ever possible through combustion-based illuminants, became an increasingly magical space, at once more divorced from the audience's normal reality and more believable.

These staged dreams of an automatic present and future, of a stable and predictable world, were inseparable from their broader social and political context. The labor movements of the 1870s to 1890s were deeply surprising and unsettling for Americans who wanted to believe they were white and lived in a classless society. The strikes, destruction of property, unionizing, and formation of national organizations like the Knights of Labor—which included not only European-born workingmen but black workers and women of all colors—frightened white, Anglo, middle-class Protestants even more than the new massive corporations like the railroads, perceived by many as threats to the republic and self-government.[10]

Edison's marvels, in a kind of inversion of the theatrical suspension of disbelief, asked audiences to look behind the curtain—but not too far—and *believe* in the magical solutions offered by his technology.[11] The largely middle-class audiences of Edison's electric performances eagerly incorporated electric lighting into technocratic visions of a future free from what they saw as the twin evils of an un-American working-class politics and a wild economy mismanaged by speculators and monopolists. But these spectacles represented only one-half of the processes helping to lay the material and ideological foundations of an electric future. How copper became a commodity—became the means of electric light—was a story all its own.[12]

T HE risks and dangers of copper competition tumbled down social relations and down mine shafts to determine the pace, violence, location, and intensity of work. While investors tried to recover their multimillion-dollar gambles in lode mining by paying miners to produce as much copper as quickly and cheaply as possible, miners tried to get paid and survive their underground work without being held responsible for freely maintaining the structural integrity of the mines. This struggle did fuel production of incredible quantities of copper ore, but the contradictions between producing copper *ore* and copper *mines*, between commodities and dead work, also made the capital-intensive mining districts of the U.S. West some of the deadliest spaces in the world.[13]

As the dead work of sinking shafts and cutting drifts progressed, extraction, the central focus of the labor process, became a continual struggle. Each morning, the mine superintendent delivered quotas to the foremen, who passed them along to the shift bosses. Responsible for hiring their own crews, these men spent twelve to fourteen hours underground each day, sometimes in temperatures over 100 degrees, directing and supervising the miners and muckers. Shift bosses usually divided their crews into pairs, each responsible for working one of the rock faces. With only the flame of a candle to light the workspace, one man swung a sledgehammer at the tiny target of a long steel drill while the other held it with his bare hands, often using the reflection of his thumbnail to better mark the drill head for the man with the hammer. He skillfully twisted and rocked the bit with every blow, pulling his thumb back at the last second. Trust between the two laborers was essential. Blow by twist by blow, these teams would spend their days drilling six- to eight-foot-deep holes, one to two inches wide, into the granite rock face. The men would pack the holes full of dynamite, carefully cut and lay the fuses, retreat to a safe distance, and ignite the charges. Blasting was done in the afternoon right before the men ended their shift. Following the explosion, men called muckers approached the workface and carefully checked that all the dynamite had detonated. The lowest in the copper mining labor hierarchy, muckers were tasked with picking and shoveling the shattered ore into carts, which mules and men hauled to the shaft station.[14]

Shift bosses were caught between the downward economic pressure of the quotas handed out by management and the upward social pressure of the miners' demands to protect the crew from harm. Shift bosses usually resolved these contradictory pressures by delegating responsibility for safety to the miners themselves, who then often passed risk down to muckers. Thus risk trickled down through social relations and mine shafts, accumulating in the stopes and shafts and bodies of miners, while value in the form of ore flowed back up through the same social relations of production. At the bottom of this cascade of risk, muckers were caught between the material dangers of mining and the full weight of the economic power structure bearing down on the mines.[15]

All of this work—dead and paid—required lights. Lots of them. And in much the same way that electric boosters like Edison relied on the unacknowledged illumination of gaslights to sell incandescent spectacles, the magnates of copper—in electricity's deep backstage—wrested their fortunes from the humble underground light of tens of millions of candles. Until the 1890s, these candles provided the only means by which men could navigate the mines. For an eight- or ten-hour shift, companies usually issued miners three candles per man at the shaft collar with other supplies. These candles had to be cheap enough to be purchased and provisioned en masse, hard enough not to break easily, and able to withstand a range of temperatures from below freezing to 140 degrees without cracking or bending. In the nineteenth century, this meant the candles pioneered in the hog-stearine archipelago of the Ohio Valley and reproduced across the industrial West.

These were not Martha Ballard's candles. Shaped and standardized over the course of the second half of the nineteenth century by exchanges between miners and manufacturers, miners' candles evolved into molded cylinders ¾ of an inch wide and between 6 and 9½ inches tall, while steel and wire candlesticks consisting of a ¾-inch candle thimble (holder), with a handle, hook, and spike of varying length, became the norm. The specially designed hooks and spikes allowed miners to hang their candlesticks from their caps and drive them into the timbers near their work area. In 1898, even after the introduction of at least 1,000 incandescent electric lights at the Anaconda Mine and more at the Mountain Con, the Anaconda Mining Company consumed between 2 and 2.5 million candles, at a total cost of $41,761.49, spread across 783,435 "days worked" underground. Given that miners worked seven days a week with no holidays, this meant that Anaconda company miners consumed 6,000 candles every day, the equivalent of the street-lighting needs for a city.[16]

As the drive to produce copper and suppress labor costs came in conflict with miners' struggles to stay alive and a culture of independence and pride in work, western miners sought to push back against what they saw as eastern and English-style capitalism. In 1887, the organized workers of Butte demonstrated their power by securing a closed shop, with thirty-four different unions under the Silver Bow Trades and Labor Assembly representing nearly all 6,000 workers in the copper camp. The wage relation in Butte was directly shaped by these politics; most miners were hired on contracts that would pay them for the ore they produced, but always on top of the union-negotiated $3.50 daily wage guaranteed for all underground workers. These hard-fought wages, nearly twice what they were in the East, meant that miners who survived their work could do relatively well in the West, despite the higher costs of living.[17]

Experienced and well-connected miners sought out not only good pay and reliable work but "work in the cooler mines or in the cooler shafts of the same mine." This meant that the most powerful groups of miners — in Butte these were unequivocally the Irish — secured for themselves what they believed to be an ecological and biological advantage over younger, unskilled, and non-Irish immigrant workers. Those outside the Irish and Cornish labor aristocracy were not so fortunate, even when protected by unions. Irish shift bosses often paired immigrants and unskilled miners with more experienced Irish miners, who made the new recruits, isolated from a community of work and knowledge through language barriers, handle the most dangerous labor. New immigrant miners were usually assigned as muckers, handed a shovel, and given little training or guidance before being dispatched to their level and stope.[18]

Working in cooler, drier mines might have seemed a desirable arrangement for many miners, but it exposed them to hidden hazards. The dry silica dust created by blasting and drilling slowly but steadily shredded the men's lung tissue, causing silicosis over the long term and rendering them susceptible to tuberculosis and pneumonia in the short run. Paradoxically, such respiratory diseases were concentrated among the Irish. They may have captured the most privileged jobs and most desired workspaces; but their knowledge was imperfect, and their very power and success were killing them. For years, Irish unions and fraternal orders had fought for Irish-only hiring practices at the Mountain Con Mine. High paying and seemingly comfortable though it was, steady work in the dry air of the Mountain Con also exposed the "privileged" miners to the most silica-laden underground atmosphere in Butte.[19]

But miners' politics were never just about class and safety. Butte was not only a closed shop but a white shop, where union charters explicitly restricted access to "white men" only. The labor unity so celebrated in Butte, "the Gibraltar of Unionism," was inseparable from racist politics. The Irish and Cornish who ruled Butte's working class were often at odds but nonetheless united under a politics of white supremacy to keep out Chinese laborers and other workers they claimed were not white.[20]

The internal dynamics that led to overproduction and overaccumulation continued to plague the copper industry, but the rise of electrification increased demand sufficiently to rescue it from the worst of these crises.[21] As a result of this relationship between copper and electric light, and copper's exclusion from the global specie crises, Butte was better able to weather the Panic of 1893 that so crippled the silver industry. Without this stability, Butte might never have become the center for the Western Federation of Miners or the strategic coordinating center for labor actions across the far more radical reaches of the hard-rock mining region, such as the silver camps of Coeur d'Alene, Idaho, which saw continual eruptions of violence during this period.

That the labor wars of the 1890s, spurred by the collapse of silver, were often just as much about hospitals as about wages, was another indication of how important dead work and the politics of life underground were in shaping the labor movement in the West. The politics of dead work were even at the heart of the constitution of the Western Federation of Miners, which first convened on May 15, 1893, in Butte, a relatively safe haven from the concerted attacks mine owners were waging on unions and wages across the silver and lead camps of Idaho, Colorado, Montana, and Nevada. Of the ten objects federation organizers declared in their constitution, three addressed concerns about workplace safety.[22]

WHILE Western miners demanded recognition and justice and proclaimed industrial solidarity, Americans elsewhere celebrated a city-sized monument of denial. The 1893 World's Columbian Exposition in Chicago awed visitors with idealized spaces rendered perfect by science and technology. These visions of future worlds of order were political statements in visible contrast with surrounding industrial and urban and racialized struggles. Fair exhibit designers deliberately placed electrification "at the apex of an evolutionary framework," mobilizing electric light into a discourse of progress versus primitiveness, and a hierarchy of racial and cultural superiority, to make the catastrophic violence and terrorism of the ongoing projects of Jim Crow, Indian dispossession, and European imperialism disappear into the language of "civilization."[23] Such electric dreams also helped white Americans to achieve the remarkable feat of convincing themselves that the legacy of slavery and the continuing upheavals of black freedom struggles and white terrorism could somehow be safely and permanently quarantined in the past; the future, after all, was already here, and the Wizard of Menlo Park had promised Americans absolution from work, workers, and history, from camphene and coal gas and all things that burned. But try as they might, fair organizers could never completely disguise the struggles they so artfully sought to deny. Even in the "Dream City," labor and laborers and struggle were inescapable.

The Columbian Exposition paid Westinghouse Electric to install 90,000 incandescent lamps at $5.25 each for a total of nearly $475,000, while General Electric furnished 5,000 arc lights to illuminate the fair. This was a lot of light. Just three years earlier, there were only 68,000 arcs and 900,000 incandescents in the entire United States, making the temporary fairgrounds—that "city" of many names, the White City, Magic City, Dream City, Enchanted City—home to more lighting than any real city in the country. With this much artificial light, the fair could be kept open profitably after dark, allowing the lights to impress even larger audiences. According to one archaeologist of the fair, during "the 6 months that it was open, May 1 to October 30, 27,529,400 tickets were sold, representing an estimated 12–16 million unique visitors who spent days, weeks, and for some, months, taking in the sights of the White City."[24]

The construction of the White City, so named in contrast to the "Black

City" of Chicago, mobilized a transient labor army. "At its height," one historian noted, "the Exposition was the largest employer in the Chicago area, and, after the railroads, one of the largest employers in the country. Employing as many as sixteen thousand workers at any one time and approximately twenty-five thousand over the course of the construction and running of the Exposition, the Chicago World's Fair became a magnet of the unemployed." In less than two years, workers raised "some four-hundred buildings (one of which was larger than the Great Pyramid at Giza)," while laying hundreds of thousands of brick pavestones, "seventy miles of sewer plumbing, and 415 miles of electrical wiring."[25]

Work as such was not invisible in the White City, but absent was any recognition of the problems facing rural or industrial workers. Fair organizers rendered workers particularly invisible in the Mines and Mining Building, located immediately across from the Electricity Building. The wires and finished copper products on display offered subtle acknowledgment that copper mines produced material used in electric technologies, but the building was arranged as a museum more of natural history than of human endeavor, let alone of the labor conflicts erupting in mining districts across the country. Michigan's and Arizona's exhibits at least featured a *model* of a mine; Montana's display failed to include any evidence that mines and the men who worked in them existed at all. As the Western Federation of Miners met in Butte to protest conditions, violence, and wages, visitors to the Montana Mining Exhibit would have been greeted with pretty rocks and famous theater stars. "The best exhibits are from the great mining camp of the state, i.e., Butte City," one observer wrote, praising the "large quantities of sulphide copper ores, and the metallic copper made from them," on display. "The most prominent feature of the exhibit, however," he noted, was "a solid silver, life-size statue of the celebrated actress, Ada Rehan, standing on a globe which in turn rests on a base of solid gold. The whole work represents several hundred thousand dollars worth of precious metals, all the products of Montana mines."[26]

Despite its dreamlike facade, the White City was still made of matter, with infrastructure for a city of 300,000 people, including water, sewage, gas, and the electrical systems that powered the 95,000 incandescent and arc lamps. This fixed infrastructure, like the dead work of the Butte underground, lasted long after the fair functioned as a site for producing and circulating value. And just like under Butte, corners were cut where possible. Perhaps nothing better-illustrated the gilded contradictions of dead work than the cheap and flammable faux-marble plaster called staff. Staff wasn't marble; but it could be sold as such, and there was an increasing understanding among the salesmen and showmen of the time that selling was all that mattered, quality and safety be damned.[27]

In many respects, the Dream City amounted to a colossal metastasis of the spectacle that had characterized the opening of the Bijou a decade earlier. Nowhere was this more apparent than in the Electricity Building, located at the center of the fairgrounds. The Edison Electric Tower was among the most iconic symbols of the exposition. But the most popular exhibit belonged to the Western Electric Company of Chicago. "The Western Electric company of this city," the *Chicago Daily Tribune* reported, "have installed an Electric Scenic theater in the southeast corner of Electricity Building that is proving by far the most interesting exhibit in the building, if not the entire Fair." The theater was designed "to re-produce in as realistic a manner possible the natural transitions of light during a period of twenty-four consecutive hours of night and day." Facing audiences with a stage ten feet wide by nine feet high, the electric theater evoked the Swiss Alps. The scene included a village, a church, a castle, and a bridge over a stream connecting a waterfall and lake with real flowing water. Snow-covered mountains extended into the distance behind the alpine scene; miniature streetlamps and house lamps illuminated the scene at "night." Starting in "evening" and ending in the moonlight of the following night, in the span of about twenty minutes audiences watched night fall and sun rise, then a miniature military procession followed by villagers, a storm, a rainbow, sunset, and moonrise. The room seated around 175. According to the author of *Electricity at the Columbian Exposition*, over 3,000 people viewed this theater each day, and the "demand for seats was so great that it was finally decided to give consecutive performances from 10 o'clock in the morning until 9 o'clock in the evening, in order to accommodate all who might desire to see these effects." But most importantly, "all these effects are produced by automatic machinery, requiring the attendance of but one man." The whole automatic reproduction of alpine nature "was designed and carried out by Mr. A. L. Tucker . . . who has entire charge and control of the Scenic Theater."[28] Not only did the theater explicitly illustrate middle-class dreams of order, harmony, and precision. It symbolically allowed "but one man," through the power of electricity, to exercise mastery over both nature and society.

Despite the confidence projected by the Dream City's exhibits and boosters, the many tensions underlying the Gilded Age economy could not be permanently repressed or wished away. Following the economic crisis of 1893, the fair largely shut down, not even making it through October. And the labor tensions so artfully disguised in the White City boiled over there and nationally as the illusions of the fair went up in flames. During the winter of 1893 and the spring of 1894, fires routinely swept across the staff-coated fairgrounds, a phenomenon that Chicago officials blamed on the now-unemployed "tramps" squatting nearby. The final conflagration took place in the midst of the famous Pullman strike. The fairgrounds, situated "halfway between the town of Pullman and downtown Chicago, . . . became a nexus for Chicago's unemployed, striking workers, and police officers." On July 5, 1894, strikers "stopped several trains within blocks of the fairgrounds during the afternoon," and after "the Court of Honor caught flame later that night, suspicion immediately turned to homeless residents of the fairgrounds and to the strikers, who had been known in recent weeks to set railroad cars afire." Meanwhile, crowds gathered and watched as the dreamscape of the White City was consumed by the fire of labor struggles.[29]

O NLY in the distorting lens of retrospection do the electric futures so artfully staged in the 1880s and 1890s seem reasonable or peaceful events in a march of progress. The utopian visions of Gilded Age boosters — which to the modern observer seem natural, albeit quaint in their naiveté — were, in fact, indelibly shaped by the contemporary simmering threat of social conflict and were articulated to wish it away. The shock and fear with which white, native-born, middle-class Americans continually responded to the Knights of Labor, the Western Federation of Miners, the industrial strikes, and economic depressions were real. They were rooted in a blindness as deep as it was ideological, and it is impossible to understand either the nativism of middle-class Americans and their support for brutal state-sponsored violence against workers or the growing militancy of those workers without also understanding the enduring tensions between dead work and the selling of electric utopias.

The spell that the Wizard of Menlo Park first began to weave in a Boston theater in 1882 continues to shroud popular imaginings of what technology is, what it does, and where it comes from. As a result, those of us under the spell have blinded ourselves to a history of light that needs to be seen. It is the history I've tried to tell in this book. When Edison and Senator Platt made pronouncements on electric progress, mastery, cleanliness, and freedom from work, they were entangled in the history that whale oil began and camphene made.

The past was not even past. Shortly after Platt declared men had become electric gods, a kerosene lamp exploded in a New York tenement housing ten

families of "Italians, negroes, and Polish Hebrews chiefly," the flames bursting out of the windows and setting fire to neighboring buildings, killing at
least one "German woman." Before, during, and after the "age of electrification," thousands of young people continued to make trillions of lucifer
matches, and despite labor laws, increased mechanization, strikes, and moral
campaigns to improve conditions and to ban white phosphorus, not until
the First World War (over seventy years after the discovery that phosphorus was disfiguring and killing match workers) did American and British
match manufacturers, some of the richest men in the world, finally replace
the toxic white phosphorus with nontoxic forms. Moreover, electricity itself
was still rooted in rock and grease and flesh. Almost all the energy powering
electric lights could be traced back to coal-fired engines transforming that
dirty, fiery heat into electricity through dynamos. And both coal and copper
miners continued to mine by the lights of millions of candles made through
the deaths of millions of hogs and cattle — animals that, until the 1930s, were
mostly raised by farmers using kerosene lamps. In 2010, Truman Young, a
Vermont dairy farmer for ninety-two years, recalled how his father, out of
fear of barn fires, "made a rule in the Fall when we'd begin to take the cattle
in if we were gonna feed any hay, get up there and throw it down, before it
gets dark. . . . 'Cause you see, they'd have to lug those kerosene lanterns up
in the hay mound." As the Youngs well knew, the Great Chicago Fire "was
caused with a lady milking a cow with a kerosene lantern."[30] Oral traditions
warning against the dangers of exploding and volatile lights reached back
further than kerosene and Mrs. O'Leary's cow, but not into the deep past.
Such warnings and such technologies were decidedly modern, nineteenth-
century creations.

For most of the years between 1880 and 1920, not only did electricity
remain out of reach for most people, but new kinds of gaslights actually
spread much more rapidly than electric lamps, gaslights that were often
used to extend and further capital's control over labor. The Italian immigrant
Leonard Costello described how this process shaped life for working families in turn-of-the-century New York, when "those of us of the Aviglianese
colony moved to tenements several blocks away. Instead of kerosene lamps,
we now had gas light and a gas stove and a meter which kept us constantly
scurrying for quarters." They may not have exploded like kerosene, but for
the cash-starved working families that chose or were forced to live around
them, these coin-operated gaslights reorganized spaces of consumption and
monetized domestic practices that had previously functioned largely parallel to the money economy. "In the middle of a meal or at night while I was

reading," Costello recalled, "the gas would lower under a boiling pot of spaghetti or the light would dim, and the meter would have to be fed. My father said it was like having an extra mouth in the family."[31] These extra "mouths" also meant extra waste, and here the nature of lights mattered greatly. One of the most celebrated qualities of electric lights was that, unlike gaslights, they did not consume air, produce fumes, or raise temperatures in the rooms in which they shone. The late-nineteenth-century class divisions between gaslights and the more expensive electric lights, then, were more than markers of status. They actively produced ecological inequalities, with gaslights and kerosene poisoning working-class air and continually threatening explosion while electric bulbs merely illuminated the spaces of the wealthy.

How luminous ecologies lived was critically important. So too was how they died and were recycled by capital and nature. Unlike spaces made of whale oil, the landscapes of copper and coal exhausted the land and poisoned the earth with toxic chemicals that could and did accumulate in living plants, animals, and human bodies. It's no accident that, today, the abandoned copper mines of Butte form the largest superfund site in the United States, and copper smelting continues to be one of the most toxic processes in an electric ecology.[32]

T HE first century of the industrialization of light was a story of a hidden relationship between industrial slavery, industrial captivity, the exploitation of children and outworking free women, and the democratization of artificial light. It's easy to miss, especially if camphene is overlooked. Even in the antebellum decades, when camphene was the most widely used illuminant in the urban United States, people never imbued it with the kind of cultural meaning they attached to whale oil, coal gas, candles, matches, or kerosene. Camphene was never the star of its own drama, but the struggles surrounding its journey from turpentine camps in North Carolina piney woods to explosive lamps in New York tenements did dramatically, if often surreptitiously, transform American debates and conditions concerning slavery, industry, and gender. And when the means of light are recognized to be the material expressions of social struggles, it is suddenly harder to ignore how, through the antebellum making and use of this intemperate piney light, tens of thousands of white women laboring in household workshops and thousands of enslaved black men tapping pines far from plantations endured risks rendered nearly invisible to underwrite the increasingly public worlds of millions of northern and southern white men.

Indeed, when the complicated relations and struggles of workers with one another and with those who sought to exploit them are given proper consideration, new processes, turning points, and continuities in the history of light become apparent, histories that decenter electricity. The importance of sugar slavery, the slave trade, and cotton textiles for the whale fishery become clearer. The industrial slaveries of southern gasworks and of the Richmond coal mines supplying eastern gasworks take on added importance. So, too, does the expansion of slavery to the vast coalfields of western Virginia, and how a surging coal-oil industry and petroleum rush ran headlong into secession, the Civil War, and the destruction of slavery. The Cincinnati mass slaughter of Ohio Valley hogs and mass manufacture of lard-based candles and lamp oils also take on new meaning when seen in relation to an industrializing slave South rather than solely a free-labor North. When the questions are about work and energy and power, it becomes easier to see how the lucifer matches tying together the expansion of all these antebellum lights entangled one of the world's most toxic industries — an industry structured by the same gendered and ageist internal struggles of working-class families organizing the use of camphene — with the extraordinarily deadly saladeros of the Plata, and to see how they made possible the democratization of flame that at once trapped increasing numbers of workers in night labor and placed the means of instant fire in the dissatisfied hands of the young, the enslaved, and the revolutionary.

The Civil War destroyed the industrial slavery upon which camphene and much of eastern and southern coal gas depended, but the emancipation of some American lucifers did little for others. The coincidental discovery of petroleum in western Pennsylvania and the cheap kerosene that rushed in to illuminate the spaces and social relations first produced around camphene and matches meant that little changed for the women and children who had been forced to work deep into the night around camphene lamps and now found themselves sewing or making matchboxes at the same hours for even lower wages around kerosene lamps that were just as, if not more, likely to explode. And the only things promising to liberate the thousands of children poisoned by the work of making matches were manufacturers' threats to replace them with either machines or convict labor. Nor did the hogs and cattle trapped in the deadly production of candles and matches find any freedom in the age of emancipation. As today's industrial agriculture and slaughter should remind us, so long as they haven't seen, heard, or smelled it, most consumers have been willing to tolerate pretty much any level of exploitation of those they've called animals. Kerosene, free labor, monopoly capitalism, and lights freed and freeing from toil were not the future. They were one future, largely imagined, and written into being by the Civil War. Only later were they rewritten into progressive destiny by men trying to sell oil, electricity, and American Exceptionalism.

We have to stop swallowing the sales pitch. In today's increasingly automated world—where for many, especially the powerful and privileged, physical labor is something that happens out of sight and out of mind—it is more important than ever to remember the backstage worlds and workers behind the automation. A collective failure of imagination has allowed technologies to mask histories and social relations, to allow lights to appear as "the expressed thought of man," to convince ourselves that we're better than mud. To misquote Melville, the next time you flip a switch to turn on a light, for God's sake, be economical with your electricity. It may not even flicker, but behind that glow lie hidden worlds of work layered beneath deep histories of struggle. Struggles that made that electric action effortlessly possible, the workers invisible. That made some people powerful and others weak. That let there be light.

Acknowledgments

This project began with a desire to challenge a foundational myth of invention, individualism, and the self-directed Progress of Man. Whether or not I have succeeded in doing so in these pages is up to the reader, but my own experiences have taught me one thing for sure. It was not just Thomas Edison who benefited from the kindness and sacrifices of others. No one builds anything alone, and I hope to be more gracious in recognizing that assistance than the Wizard of Menlo Park. This project would not have been possible without the extraordinary generosity of friends, family, mentors, and strangers. I am deeply in their debt, and ever will be.

Thanks to Eric Olsen, Joe McCoy, and Helene Lerner for changing how I thought about school and for teaching me to love to write. To Paul Olson, Priya Satia, and Richard White for inspiring me to pursue scholarship as a career. To my friends and family for all the joy, all the support, and all the perspective. To Richard White, Zephyr Frank, Jon Christensen, and Erik Steiner for the opportunity to work in the Spatial History Lab and for convincing me of the importance of space. Thank you to the departments of history at Stanford and Harvard for the education, the institutional support, and giving me the chance to pursue academic history.

I would like to extend a special thanks to my advisor Walter Johnson, who inspired, supported, and helped bring moral clarity to my scholarship during graduate school. I would not be where I am today without the guidance, the institutional support, and the academic community Walter provided. I am also especially indebted to my other readers and advisors: Rachel St. John, Vince Brown, and Emma Rothschild. Rachel spent hours above and beyond what I could have expected reading and discussing my work with me. She has also been a moral guide and friend. Vince has supported and pushed the limits of my scholarship through example, discussion, and collaboration on digital and visual history. Emma has been a guide, a patron, and a clear and incisive critic. Without the weekly teas and the community provided by the CHE, I might never have kept my sanity through graduate school. I also owe many thanks to Sven Beckert, Laurel Thatcher Ulrich, Ivan Gaskell, Jill Lepore, Vernie Oliveiro, Emmanuel Akyeampong, David Armitage, Joyce Chaplin, Lizabeth Cohen, Caroline Elkins, Alison Frank Johnson, Ian Miller, James Kloppenberg, and Dan Smail for their thoughtful comments in meetings, workshops, and presentations, for teaching me the craft of history, and for teaching me how to teach.

The Program for the Study of Capitalism, the Energy History Project, the Charles Warren Center, the American Society for Environmental History, the Southern Historical Association, the Society for Historians of the Early American Republic, and the Agricultural History Society all generously provided opportunities to present and discuss my work with incredible scholars from around the world. Generous fellowships and financial support from Harvard University, the Charles Warren Center for the Study of American History, the Project on Justice, Welfare, and Economics, the Center for History and Economics, Lafayette College, and the McNeil Center for Early American Studies provided the time, space, and resources necessary for researching and writing this project. And none of this would have been possible without the staff who actually make the world go round. Special thanks to Matthew Corcoran, Emily Gauthier, Jessica Barnard, Arthur Patton-Hock, Cory Paulsen, Liana DeMarco, Tammy Yeakel, Barbara Natello, and Amy Baxter Bellamy.

I am also greatly in debt to the many archivists and librarians who assisted with this project. Thank you to the research staff at Houghton Library and Baker Library; to Ben Simons and Libby Oldham at the Nantucket Historical Association; to Laura Pereira at the New Bedford Whaling Museum Research Library; to Rachel Lilley, Brian Shovers, Zoe Ann Stoltz, and Ellen Arguimban at the Montana Historical Society Archives; and to Penny Pugh and Kevin Fredette at the West Virginia & Regional History Collection. And I could never have written Chapter 5 without the help of Tyler Schwartz, my amazing undergraduate research assistant. Thank you.

I owe particular thanks to those who warmly welcomed me into their homes as I traveled for research. Thanks to Jan and Ken Krantz for everything, and for introducing me to Truman Young. Thank you to Greg Krantz for hosting, driving, and even assisting me in the dusty archives of the Butte-Silver Bow Courthouse. To Erica and David Greeley for generously housing and feeding me, and to Thomas, Jack, and Luke for the much-needed breaks from studying. To Vince Brown and Ajantha Subramanian for sharing their home with me, and to Zareen and Anisa for putting up with a strange boy who still wears an uncool backpack. I'm sorry I never got the chance to read

Redwall to you. Thank you also to my parents for putting up with me for a whole year after they'd thought I'd moved out forever.

While too many people read parts of this manuscript in one form or another for me to thank them all, I will do my best to thank those who did truly heroic work. Any errors or mistakes contained in this work are entirely my own. Walter Johnson, Rachel St. John, Emma Rothschild, Vincent Brown, Ian Miller, Joshua Specht, Jennifer Pandiscio Zallen, Robi Zallen, Barry Zallen, Richard White, Scott Nelson, Christopher Jones, Kathy Brown, Ann Norton Greene, Chris Heaney, Chris Phillips, Josh Sanborn, Ben Cohen, Jess Roney, Judy Ridner, Nancy Gallman, Andrew Dial, Andrew Ferris, Aaron Hall, Elaine LaFay, Alicia Maggard, Jordan Smith, Sam Sommers, and Andrew Zonderman all read the entire manuscript, and some read and commented on multiple drafts. Laurel Thatcher Ulrich, Jill Lepore, Alison Frank Johnson, Gunther Peck, Joshua Guild, Kathryn Morse, Jennifer Seltz, Seth Rockman, John McNeill, Adam Rothman, Dan Richter, Fred Anderson, Alex Lichtenstein, and Susan O'Donovan all took time out of their busy faculty schedules to generously read, comment on, and discuss drafts of chapters or sections of the manuscript. Thank you to Cemal Kafadar for inviting me to Florence to participate in the fascinating exploratory seminar on a history of the nighttime in the early modern world. Lynda Yankaskas, Scott Gordon, Jeff Hotz, Michelle LeMaster, Monica Najar, Chris Phillips, and Frederick Staidum generously welcomed me into their Lehigh Valley writers group. Their insights and critiques have been invaluable. Molly Holz and the editorial staff at *Montana* worked tirelessly and patiently to turn a dissertation chapter into a published article, parts of which appear here in the epilogue. I would also like to thank James Campbell, Gregg Mitman, and the other search committee members at Stanford University and the University of Wisconsin-Madison for their comments and consideration. I owe special thanks to everyone at the University of North Carolina Press for actually making this into a book, particularly Brandon Proia and his team, and Stephanie Wenzel, for being models of what editors should be.

My colleagues at Lafayette College, who have been so welcoming, supportive, and inspiring, also deserve special thanks, and my unending gratitude. Thank you to Joshua Sanborn, Rachel Goshgarian, Don Miller, Steve Belletto, Wendy Wilson-Fall, Paul Barclay, Leigh Campoamor, Tamara Carley, Nathan Carpenter, Jessica Carr, Andrew Fix, Chris Lee, D. C. Jackson, Hafsa Kanjwal, Rebekah Pite, Deborah Rosen, Caroline Séquin, Kathleen Vongsathorn, and Bob Weiner. I am forever in your debt.

In addition to faculty readers, I benefited from the invaluable feedback, scholarship, company, and friendship of an amazing group of fellow graduate students. Thanks especially to Joshua Specht, who not only challenged and helped me to grow academically but has been a world-class friend. Special thanks also to Branden Adams, Bina Arch, Rudi Batzell, Niko Bowie, Eli Cook, Nick Crawford, Rowan Dorin, Erin Dwyer, Philippa Hetherington, Kristen Keerma, Craig Kinnear, Philipp Lehman, Yael Merkin, Ross Mulcare, Graham Pitts, Caitlin Rosenthal, Josh Segal, Victor Seow, Sarah Shortall, Ben Siegel, David Singerman, Katherine Stevens, Gloria Whiting, Tsione Wolde-Michael, and all the organizers and participants in the following invaluable workshops: the Center for History and Economics graduate student workshop, the Program on the Study of Capitalism workshop, the 19th Century U.S. dissertation workshop, the Early American dissertation workshop, and Georgetown's Environmental History workshop. Thanks also to the wonderful graduate student fellows at the McNeil Center: Emilie Connolly, Andrew Dial, Nicole Dressler, Andrew Ferris, Aaron Hall, Elaine LaFay, Alicia Maggard, Alexandra Montgomery, Hayley Negrin, Alexis Neumann, Jordan Smith, Sam Sommers, and Andrew Zonderman.

I would be remiss if I did not also acknowledge the work done by today's American lucifers who made this project visually possible. I have been especially dependent on the work of the employees of NSTAR, Pepco, and PECO for providing and maintaining the means of electric lighting. I am also in debt to the underpaid Foxconn workers who made the interactive lights of my computer, tablet, and phone through which most of this manuscript was produced and much of it researched. I am also grateful to the work of the copper miners, bulb makers, and other electrical manufacturers, engineers, and workers who made and continually remake our electric worlds.

Finally, to Jenny, to whom I owe everything. You've been my best reader, my best critic, and my best friend. I don't know how you made time for me or for the history of light, but you did. Thank you. You are amazing.

ℕotes

PROLOGUE

1. "Martha Ballard's Diary Online," February 22, 1797. Because Ballard made candles almost always by the dozen, and very occasionally by the half-dozen, but in no other numbers, it is most likely that she owned candle dipping rods that she could use to dip three pairs (wicks draped over rod, making two candles each). Between 1786 (the first year she recorded making candles) and 1809 (the last year she made such a note, at age seventy-four), Ballard made 694 dozen candles (8,328). For more on Martha Ballard, see Ulrich, *Midwife's Tale.* For more on tallow candle making, an old art, long practiced by women and girls in northern Europe and carried into colonial landscapes like Ballard's New England

frontier community, see Ekirch, *At Day's Close*, 104–7; Brox, *Brilliant*, 12–15; and Strasser, *Never Done*, 58–61.

2. "Martha Ballard's Diary Online," February 21, 22, 1797. For a popular guide on domestic candle making, see Beecher, *Treatise on Domestic Economy*, 283–84. On November 15, 1797, Ballard and Lambart made another 20 dozen candles. On January 9, 1798, Ballard made 26½ dozen more, noting that "it is the last Tallow we have that came out of the cow we killd a year ago." Eighty-nine and a half dozen (1,074) candles from one cow, when considered with her other entries, suggests about 1½ dozen candles per pound of tallow, meaning each candle was slightly less than an ounce.

3. "Martha Ballard's Diary Online," February 22, 1797, December 8, 1796.

4. "Martha Ballard's Diary Online," February 22, 1797.

5. Orpen, *Memories of Old Emigrant Days in Kansas*, 95–96.

6. Platt, "Invention and Advancement," 63–64.

7. 2014 Mazda CX-5 TV Spot, "Edison," https://www.ispot.tv/ad/7n8j/2014-mazda-cx -5-edison; Strasser, *Never Done*, 50–66. The myth of electric light crossed even the most militantly policed borders among the political and economic regimes vying for control of the twentieth century. For American capitalists, Edison's light was proof—pure, beautiful proof—that the American system of private enterprise was mastering nature and uplifting civilization, that more electricity meant more freedom; for Lenin, electricity was the key to a workers' state; for European imperialists, electricity would secure colonies and fulfill the civilizing mission; European fascists would ride hydroelectricity into fascist modernity and expanding empire. Americans, Soviets, Fascists, and Colonialists may have envisioned and fought for sometimes radically different futures, but those futures all included electric light as a *liberating* force. See White, *Organic Machine*; Mitchell, *Rule of Experts*; McNeil, *Something New under the Sun*, 175–76; and Blackbourn, *Conquest of Nature*.

8. I take this way of viewing history and society and energy from the field of environmental history. See particularly Cronon, *Nature's Metropolis*; White, *Organic Machine*; Pyne, *Vestal Fire*; West, *Contested Plains*; Morse, *Nature of Gold*; and Andrews, *Killing for Coal*.

9. Kopytoff, "Cultural Biography of Things." For more on the environmental history of New England, see Cronon, *Changes in the Land*.

10. Tandeter, "Forced and Free Labour in Late Colonial Potosí"; Moore, "'Modern World-System' as Environmental History?"; Fischer, *Cattle Colonialism*, 100–164; Reséndez, *Other Slavery*, 100–124; Tutino, *Making a New World*.

11. Melville, *Moby-Dick*, 206.

12. Strasser, *Never Done*, 76–77.

13. Indeed, new kinds of gaslights, called mantles, which worked through incandescence, actually spread much more rapidly than electric lights from about 1880 to 1920. Observers described gas mantles as electric lights without electricity. It was no wonder that, at less than one-fifth the cost of electric lamps and with no new infrastructure required, gas mantles outmatched electric incandescents until well into the twentieth century, when increased public spending finally allowed electricity to achieve economies of scale. See Baldwin, *In the Watches of the Night*, 158–59, and Schivelbusch, *Disenchanted Night*, 47–49.

14. Schivelbusch's *Disenchanted Night* still stands as an indispensable account. See also Melbin, *Night as Frontier*; Jakle, *City Lights*; Ekirch, *At Day's Close*; and Baldwin, *In the Watches of the Night*. For some comprehensive illustrated chronicles of the history of lighting technologies and use, see O'Dea, *Social History of Lighting*; Bowers, *Lengthening*

the Day; and Dillon, *Artificial Sunshine*. For recent popular histories that try to reevaluate whether all this light has been as inevitable or positive as we thought, see Jonnes, *Empires of Light*; Crosby, *Children of the Sun*; and Brox, *Brilliant*. David Nye wrote a history of electrification in the United States in which he took as his underlying premise the notion that "in the United States electrification was not a 'thing' that came from outside society and had an 'impact'; rather, it was an internal development shaped by its social context. Put another way, each technology is an extension of human lives: someone makes it, someone owns it, some oppose it, many use it, and all interpret it" (Nye, *Electrifying America*, ix). For important works and discussions on the history of technology, see Hughes, *Networks of Power*; Smith and Marx, *Does Technology Drive History?*; and Edgerton, *Shock of the Old*. Many cultural historians agree with Nye that the history of light is a liberal one, but they try to explain (rather than presume) the formation of liberal subjects. Yet because they cannot see the life-consuming and community-shattering coal mines or the uprooted armies of "free" migrant copper and coal miners as connected to a "political" history of light and vision in the Victorian city, they can safely say that gas and electric lighting systems in nineteenth-century London "can be seen as broadly liberal, in that they were invariably designed with certain aspects of human freedom in mind" (Otter, *Victorian Eye*, 258). See also Kern, *Culture of Time and Space*; Crary, "Techniques of the Observer"; and Joyce, *Rule of Freedom*. The recent history of open-pit copper mining, LeCain, *Mass Destruction*, has also been an indispensable work for thinking about the historical intersections of social, environmental, and technological processes.

15. Zakim and Kornblith, *Capitalism Takes Command*. We should remember Peter Way's entreaty to recognize that the historiographies of labor and capitalism leave the politics and work of the vast majority of people—the enslaved, women, children, native peoples, migrants, and unskilled workers—out of the picture. "It is time," he writes, "that a more Malthusian rendering of the past should once again be allowed to creep into our interpretations. This does not mean treating people as victims, but as merely human" (Way, *Common Labour*, 12). For a recent and thorough review of the state of the field, see Rockman, "What Makes the History of Capitalism Newsworthy?"

16. Le Guin, "Day before the Revolution," 300. There's wisdom for historians in science fiction, if we have the courage to read and report back.

17. See especially Schivelbusch, *Disenchanted Night*; Nye, *Electrifying America* and *American Illuminations*; Otter, *Victorian Eye*; LeCain, *Mass Destruction*; and Baldwin, *In the Watches of the Night*. I have written elsewhere in more depth on the early social history of electric lighting: Zallen, "'Dead Work,' Electric Futures, and the Hidden History of the Gilded Age."

CHAPTER ONE

1. Smith, *Story of Boston Light*, 66; Noble, *Lighthouses and Keepers*, 1–27; O'Dea, *Social History of Lighting*, 199–211; Brox, *Brilliant*, 48–57.

2. Hixson Journal, September 7, 1832.

3. Hixson Journal, March 2, 3, 1833.

4. Dolin, *Leviathan*, 265–67; Melville, "Cutting In," in *Moby-Dick*, 296.

5. Melville, "Cutting In," in *Moby-Dick*, 297; Davis, Gallman, and Gleiter, *In Pursuit of Leviathan*, 343.

6. Melville, "The Try-Works," in *Moby-Dick*, 401–2.

7. Hixson Journal, March 2–4, June 22, 1833; Melville, "The Try-Works," in *Moby-Dick*, 402–3; "On the Water."

8. Smith, *Story of Boston Light*, 24; Ellis, *Men and Whales*, 144; Dolin, *Leviathan*, 108–35; Ekirch, *At Day's Close*, 67–74, 330–37; Brox, *Brilliant*, 20–57; Kugler, *Whale Oil Trade*, 165; Beattie, *Policing and Punishment in London*, 221–25; Linebaugh, *London Hanged*; Davis, Gallman, and Gleiter, *In Pursuit of Leviathan*, 342–68; Bailey, "Slave(ry) Trade and the Development of Capitalism in the United States," 375, 381–82, 386, 392–93; Bailey, "Other Side of Slavery," 46–47; Sohn, "Other Side of the Story," 6–12.

9. Richards, *Unending Frontier*, 585–88; Ellis, *Men and Whales*, 33–47; Weaver, *Red Atlantic*, 87–98; Lipman, *Saltwater Frontier*, 222–35; Shoemaker, *Native American Whalemen and the World*; Smith, *Story of Boston Light*, 12–13; Beattie, *Policing and Punishment in London*, 221; Huang, "Franklin's Father Josiah."

10. Rediker, *Villains of All Nations*; Smith, *Story of Boston Light*, 44.

11. Norling, *Captain Ahab Had a Wife*; Beattie, *Policing and Punishment in London*, 169–225; Koslofsky, *Evening's Empire*, 128–97; Baldwin, *In the Watches of the Night*, 1–118.

12. For recent examples, see Baldwin, *In the Watches of the Night*, 14–33; Ekirch, *At Day's Close*, 330–32; Koslofsky, *Evening's Empire*, 130–66; and Brox, *Brilliant*, 20–36.

13. OBP, September 11, 1751, trial of William Newman and James March (emphasis added).

14. OBP, *Ordinary of Newgate's Account*, October 23, 1751.

15. OBP, September 11, 1751, trial of David Brown, and *Ordinary of Newgate's Account*, October 23, 1751. Many of the thefts had involved some kind of assault, so it is not to say they were not violent; but the cases were far more about protecting property than protecting life.

16. OBP, *Ordinary of Newgate's Account*, October 23, 1751; Linebaugh, *London Hanged*, 7–41, 74–118. The Ordinary's accounts of Newgate are a remarkable resource for exploring the production of crime and criminals at the heart of the British Empire.

17. OBP, *Ordinary of Newgate's Account*, October 23, 1751.

18. Stackpole, "Nantucket Whale Oil and Street Lighting," 25.

19. Parnell, *Applied Chemistry*, 50; Stackpole, "Nantucket Whale Oil and Street Lighting," 26; Dolin, *Leviathan*, 105; de Beer, "Early History of London Street-Lighting"; www.oldbaileyonline.org/forms/formStats.jsp, June 2018.

20. From 1736 to 1751, the state killed 711 men and women in London. From 1751 to 1766, it killed 748. The next fifteen years of expanding lights saw 1,191 hanged, and from 1781 to 1796, the figure climbed to 1,486. Even accounting for population growth (from about 700,000 in 1730 to 1 million in 1801), that represents an approximately 50 percent increase in hangings per capita from 1736 to 1796. See www.oldbaileyonline.org/forms/formStats .jsp, March 2011. For lamp smashing, see Schivelbusch, *Disenchanted Night*, 97–114; Ekirch, *At Day's Close*, 74, 246, 336–37; and Baldwin, *In the Watches of the Night*, 10, 14, 17, 30.

21. That London launched its all-night street-lighting campaign in 1736 was no accident of timing. Following the end of the War of Spanish Succession in 1714, a practice of gin drinking that had been growing for decades among the city's poor suddenly became, in the minds of elites, a "craze" and a threat to order. In 1736, Parliament passed not only the most infamous and hated of the Gin Acts, but the act ordering the construction of nearly 5,000 streetlamps to be lit every night, all night, all year. See Warner, *Craze*; Rogers, *Mayhem*; and de Beer, "Early History of London Street-Lighting," 322–24.

22. Hutchinson, *History of the Province of Massachusets-Bay*, 445; Dolin, *Leviathan*, 120.

23. Kugler, *Whale Oil Trade*, 15; Bailey, "Slave(ry) Trade and the Development of Capitalism in the United States," 373–414; Bailey, "Other Side of Slavery," 35–50; Sohn, "Other Side of the Story," 6–12; Ellis, *Men and Whales*, 144; Dolin, *Leviathan*, 113; Foster, "Hadwen & Barney Candle Factory," 9.

24. Brown, "Eating the Dead," 119. For a particularly insightful and invaluable contemporary critique, see Stephen, "Of the Excess of forced Labour in point of Time," in *Slavery of the British West India Colonies*, 2:82–160. For the history and meaning of racial capitalism, see Robinson, *Black Marxism*. For some other works challenging and critiquing the Marxist inability to see slavery as central to "primitive accumulation" and to the production of surplus value in the eighteenth and nineteenth centuries, see Williams, *Capitalism and Slavery*; Hymer, "Robinson Crusoe and the Secret of Primitive Accumulation"; Johnson, "Pedestal and the Veil"; Beckert and Rockman, *Slavery's Capitalism*; and Rood, *Reinvention of Atlantic Slavery*.

25. In the British West Indies it was illegal for the enslaved to work longer than ten-hour days. But as this legal barrier was understood only to apply to the cane fields, planters circumvented it by moving the enslaved back and forth from field to boiling house, thus exploiting even more totally the time, labor, and life of the people they enslaved, while narrowly adhering to a literal interpretation of the law. See House of Commons, *Protectors of Slaves Reports*, and Stephen, *Slavery of the British West India Colonies*, 2:82–160.

26. The methods sugar planters used to illuminate night work were only rarely mentioned by contemporaries and still less often by historians. For evidence that planters used whale-oil lamps to light the night shifts in boiling houses, see Thistlewood Diary, Friday, February 17, 1769, 216; Dalby Thomas, "An Historical Account of the Rise and Growth of the West-Indian Colonies," *Harleian Miscellany* 2 (1774): 349; Clark, "History of an Aneurism of the Crural Artery," 327; Long, *History of Jamaica*, 1:414–15, 462, 492, 501, 505, 541, 551, 2:168; Roughley, *Jamaica Planter's Guide*, 199–200; and Dod, "Stray Glimpses of the Cuban Sugar Industry," 93. For discussions and descriptions of night work in sugar plantations more generally, see Ramsay, *Essay on the Treatment and Conversion of African Slaves*, 75–77; House of Commons, *Abridgement of the Minutes of the Evidence (1790)*, 135; House of Commons, *Abstract of the Evidence Delivered before a Select Committee (1791)*, 56–57; House of Commons, *Protectors of Slaves Reports (1829)*, 17–21; House of Commons, *Report from the Select Committee on the Extinction of Slavery (1832)*, 33–39; Stephen, *Slavery of the British West India Colonies*, 2:137–60, 174–80; Mintz, *Sweetness and Power*, 46–52; and Follett, *Sugar Masters*, 106–7. For more on the dynamism of the sugar industry and slavery, especially when it came to technology, see Rood, *Reinvention of Atlantic Slavery*, and Smith, "Invention of Rum."

27. Clark, "History of an Aneurism of the Crural Artery," 327; House of Commons, *Abridgement of the Minutes of the Evidence (1790)*, 135; Thistlewood Diary, Friday, February 17, 1769, 216.

28. Dod, "Stray Glimpses of the Cuban Sugar Industry," 93; Roughley, *Jamaica Planter's Guide*, 199–200.

29. Mintz, *Sweetness and Power*, 74–150; Smallwood, *Saltwater Slavery*. Sugar was also part of a related temporal process involving the seventeenth- and eighteenth-century proliferation of coffeehouses, which were primarily patronized at night, in European and American cities. Recent scholarship has shown that these coffeehouses were integral

spaces in the formation of bourgeois culture and politics and were part of the middle- and upper-class colonization of urban nights that oil streetlamps were designed to help secure. See especially Koslofsky, *Evening's Empire*, 174–85, and Cowan, *Social Life of Coffee*.

30. De Crèvecoeur, *Letters from an American Farmer*, 179, 124.

31. Dolin, *Leviathan*, 156, 163.

32. Ellis, *Men and Whales*, 159–60; Dolin, *Leviathan*, 206; Davis, Gallman, and Gleiter, *In Pursuit of Leviathan*, 4.

33. Hixson Journal, September 10, 1832; Rosenberg, *Cholera Years*, 13–98; Hobsbawm, *Industry and Empire*; Beckert, *Empire of Cotton*.

34. *Maxwell v. Eason*, 514.

35. *Clark v. Manufacturers' Ins. Co.*, 896.

36. Massachusetts Historical Commission, "Mann's Cotton Mill Double Worker Housing"; Davis, Gallman, and Gleiter, *In Pursuit of Leviathan*, 344.

37. Sohn, "Other Side of the Story," 8–9. See also Bailey, "Slave(ry) Trade and the Development of Capitalism in the United States," 373–414, and "Other Side of Slavery," 35–50. Opening and operating new palm plantations in West Africa may have somewhat reduced the incentive for selling slaves into an illegal overseas trade, but if anything, it invigorated the need for and use of plantation slave labor in Africa itself. See Lovejoy, "Slavery and 'Legitimate Trade' on the West African Coast."

38. Hixson Journal, September 12–14, 1832.

39. Melville's first and most popular (during his lifetime) novel, *Typee: A Peep at Polynesian Life* (1846), combined the adventure and eroticism of Pacific whaling and was based in large part on his own experiences living among a group of Typee natives in the Marquesas Islands after he deserted from a New Bedford whaleship.

40. Thompson, *Life of John Thompson*, 107–32; Dolin, *Leviathan*, 123, 224–25; Bolster, "'To Feel like a Man,'" and *Black Jacks*, 176–80. The whale fishery must be considered a maroon geography: a haven for escaped slaves and free blacks and an incubator of antislavery politics, even as it produced the material and social means of continued, expanding enslavement through the circulation and consumption of whale oils for light and lubrication. For more on reconsidering the definition of marronage and expanding our understanding of maroon spaces, see Hahn, *Political Worlds of Slavery and Freedom*, 1–53.

41. Hixson Journal, October 19, 1834, and "Names of the Maria's crew." Thank you to Richard White for pointing out that in the New England census, "Coloured" also included Indians. See also Weaver, *Red Atlantic*, 87–98; Lipman, *Saltwater Frontier*, 222–35; and Shoemaker, *Native American Whalemen and the World*.

42. Thompson, *Life of John Thompson*, 120–27.

43. SRJD, *Documents Relative to the House of Refuge*, 160–61, 210; SRJD, *Thirteenth Annual Report* (1838), 28, 35; SRJD, *Fourteenth Annual Report* (1839), 4, 20, 35; SRJD, *Fifteenth Annual Report* (1840), 23, 40, 42; SRJD, *Sixteenth Annual Report* (1841), 36; SRJD, *Twentieth Annual Report* (1845), 25; SRJD, *Twenty-First Annual Report* (1846); 26, 32–33; SRJD, *Twenty-Second Annual Report* (1847), 23; SRJD, *Twenty-Third Annual Report* (1848), 21, 26, 29; SRJD, *Twenty-Fourth Annual Report* (1849), 25; SRJD, *Twenty-Fifth Annual Report* (1850), 33; and SRJD, *Twenty-Sixth Annual Report* (1851), 29. Thanks to Seth Rockman for pointing me toward the incredible archive of the New York House of Refuge. For studies on maritime antislavery and countermastery political cultures, see James, *Mariners, Renegades, and Castaways* (1953); Scott, "Common Wind"; Rediker, *Between*

the *Devil and the Deep Blue Sea*, *Villains of All Nations*, and *Slave Ship*; Bolster, *Black Jacks*; Linebaugh and Rediker, *Many-Headed Hydra* (2000); and Cecelski, *Waterman's Song*.

44. Hixson Journal, September 14, 15, 21, 1832. Captain Macy had followed paths similar to the *Maria*'s. He first became a whaling captain in 1821. In the following decade he captained three Pacific voyages responsible for accumulating 6,245 barrels of sperm oil. Figures calculated from "American Offshore Whaling Voyages."

45. Browne, *Etchings of a Whaling Cruise*, 24, 43.

46. Ellis, *Men and Whales*, 174.

47. Hixson Journal, September 24, October 8, 9, 13, 19, 1832, May 16, 1833.

48. Hixson Journal, October 19, November 26, 1832. Whaleship captains were expected to serve as medical officers, but most lacked both training and experience. See Ellis, *Men and Whales*, 180.

49. Hixson Journal, November 21, 27, 23, 1832.

50. Hixson Journal, April 6, 1833; Dolin, *Leviathan*, 195–96.

51. Hixson Journal, April 6, 1833; Charles Murphey, "Logbook of Maria of Nantucket" (October 13, 1832–March 21, 1836), Ships' Log Collection, Nantucket Historical Association (material accessed from association's online database), April 7, 1833. Murphey, not a terribly pious individual—he complained about the captain's Sunday Bible readings (see Murphey, "Logbook of Maria of Nantucket," August 13, 1833)—was almost certainly being sarcastic, although it is possible he did not participate in the heavy labor and enjoyed his relative ease.

52. Hixson Journal, November 20, 1833, May 23, 13, 1834.

53. Ellis, *Men and Whales*, 178; Davis, *Nimrod of the Sea*, 326.

54. Hixson Journal, December 30, 1834.

55. Thompson, *Life of John Thompson*, 112–14; Hixson Journal, April 26, 1833.

56. Hixson Journal, July 12, 1833.

57. Hixson Journal, February 11, 1834.

58. Hixson Journal, July 31–August 3, 1834.

59. Melville, *Moby-Dick*, 437. According to modern estimates, there were approximately 1.1 million sperm whales in the world's oceans in 1750 and around 800,000 by 1880. This was a biomass reservoir replenished by the sperm whales eating close to 300 million tons of squid each year. Damaging though their activities were to populations, American whalers only tapped into a fraction of this energy, killing around 200,000 sperm whales over the nineteenth century. See especially Whitehead, "Estimates of the Current Global Population Size"; Jackson, "When Ecological Pyramids Were Upside Down"; Davis, Gallman, and Gleiter, *In Pursuit of Leviathan*, 133–49; Ellis, *Monsters of the Sea*, 245; and Burnett, *Sounding of the Whale*.

60. Hixson Journal, April 25–26, June 23–26, July 5, 16–18, December 10–12, 1833, February 18, March 17–22, 30, April 3–5, May 13, June 29, August 22, 26, September 11, October 7, 9, 1834. For more on coopering and whaling, see Howard, "Coopers and Casks."

61. Hixson Journal, March 17, 24, 1834.

62. Dolin, *Leviathan*, 255–74; Davis, Gallman, and Gleiter, *In Pursuit of Leviathan*, 150–213; Moment, "Business of Whaling in America."

63. Starbuck, "Returns of Whaling Vessels, Sailing from American Ports, Since the Year 1715," in *History of the American Whale Fishery*, 168–659. See also Nantucket Historical Association online logbook search (www.nha.org/library/librarydatabases.html), and

Eliza Brock, "Eliza Brock Journal," March 17–18, 1856, Ships' Log Collection, Nantucket Historical Association.

64. Hixson Journal, May 24, 1834; Ellis, *Men and Whales*, 163.

65. Hixson Journal, November 17, 1833. Between April 27 and July 23, 1833, Hixson counted that the *Maria* spoke with ships fifty-two times in a period of only eighty-seven days. See Hixson Journal, July 9, 1834.

66. Hixson Journal, July 6, 1833, January 5, 1834.

67. Dolin, *Leviathan*, 256–59; Hixson Journal, January 25, August 13, 1834. Thanks to Jennifer Seltz for bringing the significance of the "watch below" to my attention. See also Dening, *Mr. Bligh's Bad Language*.

68. Davis, Gallman, and Gleiter, "Product Markets," in *In Pursuit of Leviathan*, 342–80; "American Offshore Whaling Voyages"; Charles Murphey, "Logbook of Maria of Nantucket" (October 13, 1832–March 21, 1836), Ships' Log Collection, Nantucket Historical Association.

69. Christopher Mitchell & Co. Papers, Letter Book 1, 39–44, Nantucket Historical Association.

70. Gardner, *Three Bricks and Three Brothers*; Foster, "Hadwen & Barney Candle Factory," 5.

71. Gardner, *Three Bricks and Three Brothers*, 34, 36.

72. Davis, Gallman, and Gleiter, *In Pursuit of Leviathan*, 344; Gardner, *Three Bricks and Three Brothers*, 36; Foster, "Hadwen & Barney Candle Factory," 6.

73. Gardner, *Three Bricks and Three Brothers*, 37–38.

74. Melville, "The Advocate," in *Moby-Dick*, 116–20.

75. Quoted from Davis, Gallman, and Gleiter, *In Pursuit of Leviathan*, 345.

76. Christopher Mitchell & Co. Papers, Letter Book 2, "Nantucket, 6th June, 1838. Saml. W. Sevete, Est. National Ins. Office, Boston," Nantucket Historical Association; The Warder, "Great Fire at Nantucket: Awful Calamity," *Boston Daily Atlas*, July 15, 1846; Nantucket Historical Association, "Island in Time," 26.

77. Davis, Gallman, and Gleiter, *In Pursuit of Leviathan*, 346, 354–55.

78. Kaye, "Second Slavery"; Consul Campbell to the Earl of Clarendon, September 15, 1857, in House of Commons, *Class B., Correspondence with British Ministers and Agents*, 240–38.

79. Dolin, *Leviathan*, 211–12; "The Junior Mutineer," *Whalemen's Shipping List, and Merchant's Transcript*, July 12, 1859.

CHAPTER TWO

1. "A Double Casualty," *New York Commercial Advertiser*, June 17, 1858; "Chapter of Accidents—Woman Burned and Child Fatally Injured," *New York Evening Post*, June 17, 1858; "Accident from the Use of Burning Fluid," *New York Daily Tribune*, June 18, 1858. For descriptions of seamstresses' labor, especially their night work, see "Labor in New-York: Its Circumstances, Conditions and Rewards. No. 1.—The Seamstresses," *New York Daily Tribune*, August 14, 1845; Methodist Protestant, "The Seamstress, or the Value of Labor," *Boston Recorder*, October 15, 1846; C. L., "The New York Needle Woman," *Advocate of Moral Reform and Family Guardian* 18 (September 1, 1852): 135; C. L. B., "Efforts for the New-York Poor," *NYT*, November 2, 1852; *Boston Herald*, February 8, 1853; "Shirt Case. Additional Charges against Davis & Co.—More Human Sewing Machines in the Field,"

NYT, February 27, 1855; "The Case of Margaret Bryne," *NYT*, March 1, 1855; "Sewing Women," *NYT*, March 27, 1857; "Sewing and Starving," *New York Ledger*, December 12, 1863; Penny, *Employments of Women*, 296–98, 351; "The Sewing Women and Their Employers," *New York Evening Post*, March 22, 1864; "The Sewing Women of New York. How Northern Philanthropy Is Supported and Miscegenation Encouraged," *Philadelphia Daily Age*, March 23, 1864; and "Working Women," *Boston Post*, March 24, 1864. For recent social histories of women in the antebellum clothing industry, see Zakim, *Ready-Made Democracy*, 138–44; Stansell, *City of Women*, 106–19; and Boydston, *Home and Work*, 82–83.

2. "Labor in New-York," *New York Daily Tribune*, August 14, 1845; "Shirt Case," *NYT*, February 27, 1855; "The Case of Margaret Bryne," *NYT*, March 1, 1855; "Sewing Women," *NYT*, March 27, 1857; "Sewing and Starving," *New York Ledger*, December 12, 1863; "The Sewing Women and Their Employers," *New York Evening Post*, March 22, 1864; "The Sewing Women of New York," *Philadelphia Daily Age*, March 23, 1864; Zakim, *Ready-Made Democracy*, 138–44, 197–200; Stansell, *City of Women*, 106–19; Boydston, *Home and Work*, 82–83.

3. "Artificial Illumination.—Burning Fluids," *Scientific American*, January 2, 1858, 133. Camphene, also called "burning fluid" or "spirit gas," was defined as a liquid mixture of one part spirits of turpentine and four parts highly distilled alcohol (88 to 95 percent); see U.S. Congress, *Alcohol in the Manufactures and Arts*, 375–81, and "Camphene, Burning Fluids, &c," *Scientific American*, October 8, 1853, 26. For camphene market prices, see "Pure Camphene," *New Hampshire Patriot and State Gazette*, February 15, 1849 (50 cents/gal.); "Our State Prisons," *NYT*, August 30, 1854 (55–65 cents); "Weekly Review of the Chicago Wholesale Market," *Chicago Press and Tribune*, November 24, 1858 (48 cents); "Kerosene Oil: Great Reduction in Price," *New York Herald*, September 25, 1859 (63 cents); "Arson Case in Greenwich Street," *New York Herald*, April 6, 1860 (56 cents); U.S. Congress, *Alcohol in the Manufactures and Arts*, 376, and *Reports of a Commission*, 161–62; and Williamson and Daum, *American Petroleum Industry*, 33–37.

4. "A Double Casualty," *New York Commercial Advertiser*, June 17, 1858, 2; "Chapter of Accidents—Woman Burned and Child Fatally Injured," *New York Evening Post*, June 17, 1858, 2; "Accident from the Use of Burning Fluid," *New York Daily Tribune*, June 18, 1858, 5; "Another Camphene Accident," *NYT*, July 24, 1858, 1; "Fatal Camphene Accident," *New York Daily Tribune*, July 24, 1858, 7.

5. James R. Grist to Allen Grist, "At Ben's," September 17, 1852, box 2, CS 1845–1852, JRGBR.

6. The most recent histories of light and night in America barely mention camphene at all. Brox, *Brilliant*, devotes less than one page (78–79) to camphene. Baldwin, *In the Watches of the Night*, gives camphene one sentence (18). Dillon, *Artificial Sunshine*, which, as a "social history of domestic lighting," would seem particularly well-aimed for a discussion of camphene, has no index entry and gives only a short paragraph (99).

7. Meriam, "Deaths and Injuries from the Use of Camphene."

8. "Camphene," *Cleveland Daily Herald*, April 28, 1855; "Another Camphene Murder," *Cleveland Daily Herald*, January 12, 1859.

9. For the most comprehensive study of the South's turpentine industry, see Outland, *Tapping the Pines*.

10. Zakim, *Ready-Made Democracy*, 38, 127–56; Blackmar, *Manhattan for Rent*, 44–71, 183–249. See also Stansell, *City of Women*; Boydston, *Home and Work*; and Rockman, *Scraping By*.

11. "Sewing Women," *NYT*, March 27, 1857.

12. "Shirt Case. Additional Charges against Davis & Co.—More Human Sewing Machines in the Field," *NYT*, February 27, 1855; Methodist Protestant, "The Seamstress, or the Value of Labor," *Boston Recorder*, October 15, 1846; "Labor in New-York," *New York Daily Tribune*, August 14, 1845; *Boston Herald*, February 8, 1853; Zakim, *Ready-Made Democracy*, 129, 143–51.

13. "Labor in New-York," *New York Daily Tribune*, August 14, 1845; C.L., "The New York Needle Woman," *Advocate of Moral Reform and Family Guardian* 18 (September 1, 1852): 135.

14. "Sewing Women," *NYT*, March 27, 1857; "The Case of Margaret Bryne," *NYT*, March 1, 1855.

15. Blackmar, *Manhattan for Rent*; Beckert, *Monied Metropolis*; Baldwin, "Lighting the Heart of Darkness," in *In the Watches of the Night*, 14–33.

16. Zakim, *Ready-Made Democracy*, 100; "Working Women," *Boston Post*, March 24, 1864.

17. Boydston, *Home and Work*, 88–92, 113; Blackmar, *Manhattan for Rent*, 115, 123, 136–37; Zakim, *Ready-Made Democracy*, 127–84; C. L., "The New York Needle Woman," *Advocate of Moral Reform and Family Guardian* 18 (September 1, 1852): 135; Penny, *Employments of Women*, 351; Williamson and Daum, *American Petroleum Industry*, 34–36.

18. Outland, *Tapping the Pines*, 3, 8, 37–121. Rosin, the thick carbon-rich material left after the spirits had been distilled, also found new uses as a lubricating oil for the heavy machinery of New England cotton mills and as the source of illuminating gas in urban gasworks.

19. Vollmers, "Industrial Slavery in the United States," 374; Outland, *Tapping the Pines*, 40–41; Perry, "Naval-Stores Industry in the Old South," 524.

20. "The Turpentine Region," *FO*, January 27, 1846; "The Pine Forests of North Carolina," *Raleigh Register*, August 29, 1849; "Negro Hiring," *FO*, January 6, 1853; "The Tide Turned," *FO*, January 25, 1853; "High Prices," *FO*, February 1, 1853; Turpentine Maker, letter to the editor, *FO*, February 10, 1853; Tarboro' Southerner, *FO*, December 22, 1853; "Negro Hire," *FO*, January 2, 1854; "Bad Policy," *FO*, January 1, 1855; "A North Carolina Farmer," *FO*, March 26, 1855; "The Cost of Living," *FO*, April 12, 1855; "Shameful," *FO*, April 30, 1855; "Hired Negroes," *FO*, January 9, 1860; "High Prices for Turpentine Hands," *FO*, January 9, 1860; Outland, *Tapping the Pines*, 37–59; de Boer, *Nature, Business, and Community*, 55–86.

21. Outland, *Tapping the Pines*, 41–42, 45, 69; de Boer, *Nature, Business, and Community*, 75. For studies of the political geographies of turpentine slavery, see de Boer, *Nature, Business, and Community*, 64–69; Outland, *Tapping the Pines*, 60–97; and Cecelski, *Waterman's Song*, 128–33. For slaveholders' correspondence about turpentine camps, see especially JRGBR. Other valuable manuscript collections include the following at RLDU: Tillinghast Family Papers; Daniel W. Jordan Papers; A. J. Turlington Papers; William H. Turlington Papers; William R. Smith, Memorandum Book, 1852–1855; and Francis Harper Papers. See also the following collections at the Southern Historical Collection, Wilson Library, University of North Carolina at Chapel Hill: Mercer Family Papers #2990-z; Grimes Family Papers #3357; and Thomas David Smith McDowell Papers #460.

22. In 1856, twenty-five slaves labored at the turpentine camp at Grist Depot (located along the Wilmington and Manchester Railroad) to dip and scrape 7,709 barrels of resin, distilling 900 barrels of turpentine and 4,323 barrels of rosin and shipping them east to Wilmington. To power and supply the means of this labor and production, the Grist family

shipped to the camp 44 barrels of pork, 175 sacks of meal, 100 pairs of shoes, 26 bales of hay (for the mules to haul the turpentine), 4 boxes of hats, 3 barrels of glue, 4 kegs of nails, 2 boxes of dippers, 1 box of axes, 38 files, and 1 cooper's adze. See M. Jones, "1856—Receved from Wilmington," in "Account Book, 1856–1859," RLDU.

23. Cecelski, *Waterman's Song*, 109. For more on the politics and labor of early republican canals, see Way, *Common Labour*, and Williams, *Americans and their Forests*.

24. T. Banks to James R. Grist, Fayetteville, August 15, 1854, box 3, CS 1853–1854, JRGBR; "Steamer James R. Grist," *FO*, August 28, 1854. The owners of the thirty-four stills in Wilmington in 1845, capitalized at $87,000, held a slave labor force worth $66,000, paid $6,000 in overseers' wages, and spent an additional $83,750 to run the distilleries as they consumed over 200,000 barrels of resin to produce spirits of turpentine and rosin valued at more than $400,000. See Wilmington Chronicle, "Statistics of Wilmington," *FO*, September 24, 1845, and Outland, *Tapping the Pines*, 49–54.

25. For Wilmington fires, see "Fires," *FO*, February 2, 1842; "And Yet Another!," *Raleigh Register*, May 24, 1842; "Fire in Wilmington," *FO*, April 24, 1844; "Fire," *Raleigh Register*, April 30, 1844; "Fire," *FO*, May 15, 1844; Wilmington Chronicle, *Raleigh Register*, January 13, 1846; *FO*, November 24, 1846; *FO*, June 1, 1847; "Fires," *FO*, April 9, 1850; "Fire," *FO*, May 27, 1851; *FO*, July 31, 1851; "Fire," *FO*, November 11, 1852; "Fire," *FO*, January 27, 1853; "Fire," *FO*, August 18, 1856; and "Fires in Wilmington," *FO*, September 17, 1857. For fires in the piney woods, see "Fire in the Dismal Swamp," *FO*, April 23, 1845; "Destructive Fires," *FO*, March 12, 1855; "Fires," *FO*, March 19, 1855; "Terrific Fires in the Pineries of Carolina," *FO*, March 22, 1855; "Fire in Bladen County," *FO*, April 16, 1855; "The Fires in the Woods," *FO*, April 19, 1855; "The Fires in the Woods," *FO*, April 26, 1855; "The Fires," *FO*, April 30, 1855; and "Forest Fires," *FO*, May 3, 1855. For examples of railroad and steamship fires involved in transporting turpentine, see "Burning of Freight Cars on the Manchester Railroad," *FO*, June 24, 1858, and "Steamer Rowan Burnt," *FO*, September 8, 1859.

26. Benjamin Grist to Allen Grist, St Pauls P.O., January 21, 1851, box 2, CS 1845–1852, JRGBR.

27. Olmsted, *Journey in the Seaboard Slave States*, 339; "The Manufacture of Turpentine in the South," *DeBow's Review*, May 8, 1850, 452.

28. Benjamin Grist to Allen Grist, St Pauls P.O., January 21, 1851, box 2, CS 1845–1852, JRGBR.

29. Olmsted, *Journey in the Seaboard Slave States*, 339–40.

30. "The Manufacture of Turpentine in the South," *DeBow's Review*, May 8, 1850, 453; Olmsted, *Journey in the Seaboard Slave States*, 339–40.

31. M. Jones, "Account Book, 1856–1859," RLDU; Vollmers, "Industrial Slavery in the United States," 382–83.

32. James R. Grist to Mr. Perry, Robeson P.O., September 9, 1850, box 2, CS 1845–1852, JRGBR.

33. "Turpentine. Hints for Those about to Engage in Its Manufacture," *DeBow's Review*, October 19, 1855, 486–87; "Product of Turpentine at the South," *DeBow's Review*, September 11, 1851, 303.

34. Outland, *Tapping the Pines*, 67.

35. "Product of Turpentine at the South," *DeBow's Review*, September 11, 1851, 304; Olmsted, *Journey in the Seaboard Slave States*, 342; Outland, *Tapping the Pines*, 49, 68–69, 72–73, 86, 98.

36. Olmsted, *Journey in the Seaboard Slave States*, 342.

37. "The Pine Forests of the South," *DeBow's Review*, February 3, 1867, 197.

38. Olmsted, *Journey in the Seaboard Slave States*, 342; M. Jones to James R. Grist, October 30, 1855, box 3, CS 1855–1856, JRGBR.

39. At Grist Depot, a turpentine camp located not on a river but near the South Carolina border along the Wilmington and Manchester Railroad, M. Jones, the operator, always labeled a crop by the name of the enslaved man who boxed it, even for measuring dipping: "Dick crop dipt 48, Squear crop dipt 46, Jim Turdin crop 44," and so on (M. Jones to James R. Grist, Grist Depot, August 22, 1858, box 3, CS 1857–August 1858, JRGBR). None of the other men escaping and returning admitted to seeing Miles Clark (they did mention others who had not returned). After that, there is a gap in the record, but no mention is ever made of him again, suggesting one of four possibilities: he was eventually caught and sold; he escaped to the North; he remained living as a maroon in one of the region's many swamps; or he died somewhere in the woods.

40. James R. Grist to Mr. Perry, "Roberson P.O.," September 9, 1850, box 2, CS 1845–1852, JRGBR.

41. Outland, *Tapping the Pines*, 82–84; Perry, *Treatise on Turpentine Farming*, 98; James R. Grist to Mr. Perry, "Roberson P.O.," September 9, 1850, box 2, CS 1845–1852, JRGBR (emphasis mine).

42. M. Jones to James R. Grist, Grist Depot, August 29, 1858, box 3, CS 1857–August 1858, JRGBR; Outland, *Tapping the Pines*, 82.

43. M. Jones to James R. Grist, October 24, 1855, box 3, CS 1855–1856, JRGBR; Outland, *Tapping the Pines*, 89; de Boer, *Nature, Business, and Community*, 26–30.

44. R. M. Wadsworth to James R. Grist, Lynches Creek PO, August 11, 1854, box 3, CS 1853–1854, and Benjamin Grist to James R. Grist, Fish River, May 29, 1860, box 3, CS November 1859–June 1860, JRGBR; quoted from Outland, *Tapping the Pines*, 90.

45. Of a sample of seventy-seven letters from the JRGBR written from North Carolina between 1850 and 1860, there are sixty-one separate mentions of runaway turpentine slaves. Only four fell outside the April–October dipping season.

46. Quoted in Cecelski, *Waterman's Song*, 124, 128–29; de Boer, *Nature, Business, and Community*, 67–68.

47. John T. Council to James R. Grist, April 21, 1857, box 3, CS 1857–August 1858, JRGBR.

48. John T. Council to James R. Grist, April 21, 1857, box 3, CS 1857–August 1858, JRGBR; M. Jones, "Account Book, 1856–1859," RLDU. The nine enslaved Clark men listed in Jones's account book were Miles Clark, Tom Clark, Prince Clark, John Clark, Ben Clark, Moses Clark, Clem Clark, Prive Clark, and Jack Clark.

49. John T. Council to Allen & James R. Grist, May 7, 1858, box 3, CS 1857–August 1858, JRGBR.

50. M. Jones to James R. Grist, Grist Depot, July 4, 11, 1858, box 3, CS 1857–August 1858, and September 20, 12, 1858, box 3, CS September 1858–April 1859, JRGBR. For discussion of carceral landscapes, incarceration, and excarceration, see Johnson, *River of Dark Dreams*, 209–43, and Linebaugh, *London Hanged*, 7–41.

51. M. Jones to James R. Grist, Grist Depot, September 12, 1858; John T. Council to James R. Grist, October 13, 1858; and John T. Council to James R. Grist, Prospect Hall P.O., October 25, 1858, all in box 3, CS September 1858–April 1859, JRGBR.

52. Benjamin Grist to James R. Grist, Gristville, July 5, 1857, and Benjamin Grist to James R. Grist, Danley Mills, April 6, 1858, box 3, CS 1857–August 1858, JRGBR; Benjamin

Grist to James R. Grist, Dannelley Mill, February 27, 1859, box 3, CS September 1858–April 1859, JRGBR. The written record falls off before the conflict was resolved, but by 1860, all the Clark men except Miles were back at work at Grist Depot with Selvester, Griffin, Lewis, and Clem. M. Jones makes no mention of any more runaways after 1858, and it is not clear what changed. For more examples of how slaveholders used the internal slave trade and the network of jails to discipline and dislocate those who caused them trouble, see William T. Parham to Mr. James Grice, Jarratts, Sussex Co. Va, May 1, 1854, box 3, CS 1853–1854, and Henry S. Clark to James R. Grist, Greenville, N.C., March 26, 1860, box 3, CS November 1859–June 1860, JRGBR.

53. "Turpentine. Hints for Those about to Engage in Its Manufacture," *DeBow's Review*, October 19, 1855, 487–88.

54. Benjamin Grist to James R. Grist, Gristville, May 19, 1852, box 2, CS 1845–1852, JRGBR; Perry, *Treatise on Turpentine Farming*, 119.

55. Absalom Davis to Allen Grist, July 28, 1851, box 2, CS 1845–1852, JRGBR.

56. John T. Council to James R. Grist, August 9, 1859, box 3, CS May–October 1859, JRGBR.

57. Perry, *Treatise on Turpentine Farming*, 110; Benjamin Grist to James R. Grist, Dannelley Mill, April 3, 1859, and Allen Grist Jr. to James R. Grist, Fish River, April 27, 1859, box 3, CS September 1858–April 1859, JRGBR.

58. Benjamin Grist to James R. Grist, Gristville, April 22, 1852, and James R. Grist to Allen Grist, "At Ben's," September 17, 1852, box 2, CS 1845–1852, JRGBR; Olmsted, *Journey in the Seaboard Slave States*, 344–45.

59. Olmsted, *Journey in the Seaboard Slave States*, 344–45; Outland, *Tapping the Pines*, 75–77.

60. "Turpentine. Hints for Those about to Engage in Its Manufacture," *DeBow's Review*, October 19, 1855, 489.

61. Council James to James R. Grist, August 23, 1853, box 3, CS 1853–1854, JRGBR.

62. Outland, "Suicidal Harvest on the Move," in *Tapping the Pines*, 98–121.

63. Benjamin Grist to James R. Grist, Gristville, July 5, 1857, box 3, CS 1857–August 1858, JRGBR.

64. Benjamin Grist to James R. Grist, Dannelley Mill, May 9, 1859, box 3, CS May–October 1859, and J. H. Farmer to James R. Grist, Mobile, Alabama, March 23, 1854, box 3, CS 1853–1854, JRGBR.

65. Benjamin Grist to James R. Grist, Danley Mills, April 6, 1858, box 3, CS 1857–August 1858, JRGBR.

66. Benjamin Grist to James R. Grist, Dannlley Mills, June 1, 1858; W. G. Whitfield to James R. Grist, Alabama, Balddins County, Danleys Mills P.O., July 27, 1858; Benjamin Grist to James R. Grist, Dannlley Mill, August 1, 1858; W. G. Whitfield to James R. Grist, Alabamma, Balddinge County, Danley Mills Post Office, August 3, 1858, all in box 3, CS 1857–August 1858, JRGBR.

67. Benjamin Grist to James R. Grist, Dannelley Mill, February 13, 27, 1859; Benjamin Grist to Allen Grist, Dannelley Mill, February 27, 1859, all in box 3, CS September 1858–April 1859, JRGBR.

68. Hussey & Murray to Allen and James R. Grist, New York, December 28, 1850, box 2, CS 1845–1852, JRGBR.

69. Beckert, *Monied Metropolis*, 53; Classified Ad 5— "The Doric Lamp," *New York Daily Tribune*, February 24, 1844, 6; Dietz, *Leaf from the Past*, 74.

70. "Fire in Williamsburgh," *Albany Evening Journal*, November 18, 1848; Ripley and Dana, *New American Cyclopaedia*, 3:737–38; Freedley, *Philadelphia and Its Manufactures*, 146–49; U.S. Congress, "Reduction, since 1860, in the Production and Consumption of Distilled Spirits in the United States," 161–62, and *Alcohol in the Manufactures and Arts*, 375–81; Herrick, *Denatured or Industrial Alcohol*, 207–9.

71. Adams, *Home Fires*.

72. "Camphene," *Scientific American*, March 2, 1850, 189; "What Is Camphene Doing," *Plattsburgh Republican*, November 11, 1854; Newburyport Herald, "Camphene Statistics," *Lowell Daily Citizen and News*, May 31, 1859; Manuel, "Original Bridge Fuel." For a good example of the equation of temperance and refraining from using camphene, see Hinckley, *Camphene Lamp*.

73. "Camphene," *Scientific American*, March 2, 1850, 189; "Camphene, Burning Fluids, &c.," *Scientific American*, October 8, 1853, 26.

CHAPTER THREE

1. "Terrible Coal Pit Explosion," *Richmond Daily Dispatch*, March 21, 1855; "The Recent Explosion at the Coal Pits," *Richmond Daily Dispatch*, May 25, 1854; "Four More Deaths from the Midlothian Coal Pits Explosion," *NYT*, March 24, 1855.

2. For more on Davy lamps, see Pohs, *Miner's Flame Light Book*, 279–93.

3. "At What Price Gas May Be Produced," *Scientific American*, January 24, 1852, 148.

4. Political economy offers an explanation for why the vast coalfields of western Virginia remained undeveloped relative to adjacent western Pennsylvania deposits, but it can overstate the success of Pittsburgh and the demise of Richmond. See Adams, *Old Dominion, Industrial Commonwealth*. In the United States overall for 1853, at least 113,000 tons of bituminous coal were consumed by municipal gasworks to produce over 1 billion cubic feet of gas. See Bruce, *Virginia Iron Manufacture in the Slave Era*, 80–109; Williamson and Daum, *American Petroleum Industry*, 39; Lewis, *Coal, Iron, and Slaves*, 6–7; Brox, *Brilliant*, 58–76; and Wilson, *New York Industrial Exhibition*, 29–32.

5. Howe, *Historical Collections of Virginia*, 230; Pohs, *Miner's Flame Light Book*, 215–54; "Account Book, 1841–1843," Jones Papers, RLDU.

6. Long, *Where the Sun Never Shines*, 24–51; Simonin, *Underground Life*, 114–79, 204–29.

7. "Description of the Explosion Inside," *Pittsburgh Gazette*, March 24, 1855; "Terrible Coal Pit Explosion," *Richmond Daily Dispatch*, March 21, 1855.

8. Richmond Enquirer, "By Yesterday's Southern Mail: Explosion of the Midlothian Coal Pits—Shocking Loss of Life," *Alexandria Gazette*, March 22, 1855; "Terrible Coal Pit Explosion," *Richmond Daily Dispatch*, March 21, 1855; "Description of the Explosion Inside," *Pittsburgh Gazette*, March 24, 1855.

9. "Terrible Coal Pit Explosion," *Richmond Daily Dispatch*, March 21, 1855.

10. Long, *Where the Sun Never Shines*, 38; "Account Book, 1841–1843," Jones Papers, RLDU; Adams, "Dark as a Dungeon."

11. "Terrible Accident at the Midlothian Coal Pits in Chesterfield—Eleven Lives Lost," *Richmond Daily Dispatch*, December 13, 1856.

12. "Account Book, 1841–1843," Jones Papers, RLDU; "Terrible Coal Pit Explosion," *Richmond Daily Dispatch*, March 21, 1855.

13. Frantel, *Chesterfield County Virginia Uncovered*, 7, and Appendix 7; Richmond Compiler, "The Black Heath Coal Mine," *Richmond Enquirer*, March 23, 1839; "The

Disastrous Accident," *Richmond Enquirer*, March 21, 1839; "The Explosion at the Black Heath Pits," *Alexandria Gazette*, March 22, 1839.

14. "Account Book, 1841–1843," Jones Papers, RLDU; "Manoeuvres of the Whigs," *Richmond Enquirer*, April 12, 1844; Olmsted, *Journey in the Seaboard Slave States*, 47–48; Richmond Compiler, "The Black Heath Coal Mine," *Richmond Enquirer*, March 23, 1839.

15. Richmond Compiler, "The Black Heath Coal Mine," *Richmond Enquirer*, March 23, 1839.

16. "Deep Coal Pit—Mid Lothian," *Richmond Enquirer*, September 13, 1839; Lewis, *Black Coal Miners in America*, 4–5.

17. "Mid-Lothian Notice," *Richmond Enquirer*, January 9, 1840.

18. Lewis, *Black Coal Miners in America*, 6; Adams, *Old Dominion, Industrial Commonwealth*, 208.

19. "Pit Hands Wanted," *Richmond Whig*, January 3, 1843; "Account of the Burning of the Midlothian Coal Mines," *Daily National Intelligencer*, July 28, 1843; Howe, *Historical Collections of Virginia*, 230–32; Richmond Compiler, "Visit to a Coal Mine," *Daily National Intelligencer*, November 2, 1843; "Explosion and Loss of Life at the Black Heath Coal Mines," *Richmond Daily Dispatch*, November 28, 1855. Richmond companies likely responded by trying to substitute such lamps with newly popular locked models. Not trusting miners or their knowledge of the mines, owners tried to force scientific expertise into work practices by controlling access to flame and light with lock and key. See Pohs, *Miner's Flame Light Book*, 369–70.

20. "Latest from the Midlothian Pits," *Richmond Daily Dispatch*, March 22, 1855. For more on slave life insurance and industrial slavery, see Starobin, *Industrial Slavery in the Old South*, 71–74; Murphy, "Securing Human Property" and *Investing in Life*, 184–206; Baptist, "Toxic Debt, Liar Loans," 69–92; Ryder, "'Permanent Property'"; and Ralph, "Price of Life."

21. These figures are based on analysis of the account books for 1841–43 at nearby pits operated by David Watkins & Co., found in "Account Book, 1841–1843," Jones Papers, RLDU. See also "Latest from the Midlothian Pits," *Richmond Daily Dispatch*, March 22, 1855; Richmond Enquirer, "The Midlothian Explosion," *Alexandria Gazette*, March 24, 1855; P. B. Price to A. B. Coulter, Richmond, February 5, 1857, quoted in Frantel, *Chesterfield County Virginia Uncovered*, 178.

22. Murphy, *Investing in Life*, 184–206.

23. See especially Johnson, *Soul by Soul*.

24. "Latest from the Midlothian Pits," *Richmond Daily Dispatch*, March 22, 1855; Richmond Enquirer, "The Midlothian Explosion," *Alexandria Gazette*, March 24, 1855.

25. Richmond Enquirer, "The Midlothian Explosion," *Alexandria Gazette*, March 24, 1855.

26. "State and City News.: Life Insurance," *Richmond Whig*, April 17, 1855; "Virginia: Midlothian Pits," *Richmond Daily Dispatch*, April 11, 1855.

27. "Gas," *Baltimore Gazette and Daily Advertiser*, June 25, 1834; "New Orleans, June 10," *Southern Patriot*, June 20, 1834.

28. "From the New Orleans Bulletin of the 9th. Gas," *Daily National Intelligencer*, August 25, 1834; "New Orleans, June 10," *Southern Patriot*, June 20, 1834.

29. "The Gas Works," *New Orleans Daily Picayune*, September 9, 1840. A few years later reporters repeated these incredible claims in "The New Orleans Gas Works," *New Orleans Daily Picayune*, March 23, 1842.

30. "The New Orleans Gas Works," *New Orleans Daily Picayune*, March 23, 1842; Baldwin, *In the Watches of the Night*, 30; Brox, *Brilliant*, 70.

31. "A Glance at the Gas Works," *New Orleans Daily Picayune*, April 4, 1850.

32. "The New Orleans Gas Works," *New Orleans Daily Picayune*, March 23, 1842; "A Glance at the Gas Works," *New Orleans Daily Picayune*, April 4, 1850; David Biggs, "Gas and Gas-Making," *Harper's New Monthly Magazine*, December 1862, 20.

33. Sinclair, *Port of New Orleans*, 191.

34. Quoted in "Echoes of Slavery Days," *NYT*, June 29, 1901, 10.

35. *The New Orleans Gas Light and Banking Company v. George R. Botts*, 9 La. 305 (1844). For more on how slaveholders, slave traders, and the enslaved tried to shape the moment of sale, see Johnson, *Soul by Soul*.

36. Baptist, "Toxic Debt, Liar Loans," 85; "An Act to Incorporate the New-Orleans Gas Light and Banking Company," *Acts Passed at the First Session of the Twelfth Legislature of the State of Louisiana* (New Orleans: Jerome Bayon, 1835), 92–112. On June 10, 1840, as the company was attempting to transition solely to producing gas and stopping its banking activities, it burned $2,178,000 worth of notes that it had called in and redeemed with specie, leaving only $29,000 of paper Gas Bank notes still in circulation. In 1845, when the charter of the company was amended, its capital was reduced to $1,200,000. See "Extensive Conflagration," *New Orleans Daily Picayune*, June 11, 1840, and "No. 100 — An Act to Amend the Charter of the New Orleans Gas Light and Banking Company," *Acts Passed at the First Session of the Seventeenth Legislature of the State of Louisiana* (New Orleans: Magne & Weisse, 1845), 59–60.

37. "Echoes of Slavery Days," *NYT*, June 29, 1901, 10.

38. W. Newton Mercer to Genl. Butler, September 9, 1862, #506 1862, Letters Received, ser. 1756, Dept. of the Gulf, RG 393 Pt. 1 [C-508], in Berlin, Fields, Glymph, Reidy, and Rowland, *Freedom*, 222–23.

39. Berlin, Fields, Glymph, Reidy, and Rowland, *Freedom*, 193; Brigr. Genl. J. W. Phelps to Captain R. S. Davis, September 6, 1862, #26 1862, Letters Received, ser. 1756, Dept. of the Gulf, RG 393 Pt. 1 [C-508], 221–22; Capt. R. S. Davis to Brg. Gen. J. W. Phelps, September 8, 1862, vol. 2 DG, p. 312, Letters Sent, ser. 1738, Dept. of the Gulf, RG 393 Pt. 1 [C-508], 222; J. W. Phelps, Report of service, January 16, 1873, vol. 6, Generals' Reports of Service, ser. 160, RG 94 [JJ-5], 222; "Former Slaves of the New Orleans Gas Works to the Commander of the Department of the Gulf," Isaac White et al. to Major General N. P. Banks, February 23, 1863, W-43 1863, Letters Received, ser. 1920, Civil Affairs, Dept. of the Gulf, RG 393 Pt. 1 [C-508], 223–24, all in Berlin, Fields, Glymph, Reidy, and Rowland, *Freedom*.

40. "Gas in Mobile," *New York Commercial Advertiser*, September 26, 1836; *Baltimore Gazette and Daily Advertiser*, October 25, 1836; *Public Ledger*, March 23, 1837; "A Glance at the Gas Works," *New Orleans Daily Picayune*, April 4, 1850.

41. "At What Price Gas May Be Produced," *Scientific American*, January 24, 1852, 148; "Give Us Cheap Gas," *Scientific American*, December 11, 1852, 101.

42. "Give Us Cheap Gas," *Scientific American*, December 11, 1852, 101.

43. Commissioners for inquiring into the Employment and Condition of Children in Mines and Manufactories, *The Condition and Treatment of the Children Employed in the Mines and Collieries of the United Kingdom* (1842).

44. Boston Gas Light Company, *Trial*, 70.

45. Wilson, *New York Industrial Exhibition*, 29–32.

46. Philadelphia Common Council, *Journal of the Common Council*, 86, 372, 359, 361–62, 373, 352–56, 364–65; Wilson, *New York Industrial Exhibition*, 32.

47. Philadelphia Common Council, *Journal of the Common Council*, 382–85.

48. Boston City Council, *Report of a Committee*, 4, 7; Boston Gas Light Company, *Trial*, 71–72.

49. Running water, for those fortunate enough to have access to it through private or public waterworks, was not metered until the end of the century, and even then metering was unpopular. See Melosi, *Sanitary City*, 86–87.

50. Boston Gas Light Company, *Trial*, 37–38, 18.

51. Boston City Council, *Report on the Erection of a Gasometer*; Boston Gas Light Company, *Report of the Hearings*, 20.

52. Boston City Council, *Report on the Erection of a Gasometer*, 11. For more on the antebellum politics and law of public good, public health, *salus populi*, "noxious trades," and the "well-regulated society," see Novak, *People's Welfare*.

53. "Explosion in Hanover Street," *Daily Evening Transcript*, February 21, 1852; Boston City Council, *Report on the Erection of a Gasometer*, 18.

54. "Explosion of a Gas Pipe in Broome-Street, Ex-Alderman Clayton Dangerously Injured," *NYT*, December 25, 1852; "Fire in Nassau-Street. Gas Explosion — Several Persons Injured," *New York Daily Tribune*, May 22, 1854; "Gas Explosion at a Paper Warehouse," *New York Daily Tribune*, August 21, 1854.

55. Burr & Brother, *Solar Gas*, 6–10.

56. Baldwin, *In the Watches of the Night*, 14–103.

57. Foster, *New York by Gas-Light*, 53.

58. "The Overflow," *New Orleans Daily Picayune*, May 12, 1849; "The Gas Works," *New Orleans Daily Picayune*, May 14, 1849; "The Gas Works," *New Orleans Daily Picayune*, May 15, 1849; "The Overflow," *New Orleans Daily Picayune*, May 16, 1849; "The New Orleans Gas Works," *New Orleans Daily Picayune*, June 6, 1849.

59. Boston City Council, Special Committee on Gas Inspection, *Report of the Evidence*, 182–85.

CHAPTER FOUR

1. "Desultory Thoughts on Swine, Written at Cincinnati," *Western Farmer and Gardener* 2 (February 1841): 105.

2. Cist, *Sketches and Statistics of Cincinnati in 1851*, 283–85.

3. Olmsted, *Journey through Texas*, 9; Cist, *Sketches and Statistics of Cincinnati in 1851*, 287.

4. Schrepfer and Scranton, *Industrializing Organisms*; McShane and Tarr, *Horse in the City*; Greene, *Horses at Work*.

5. For further discussion along these lines, see Cronon, *Nature's Metropolis*; Mitchell, *Rule of Experts*; Peck, "Nature of Labor"; Johnson, introduction to *Chattel Principle*, and "Pedestal and the Veil," 299–308; McNeur, *Taming Manhattan*.

6. Cronon, *Nature's Metropolis*, 229; Ross, *Workers on the Edge*, 137–39.

7. My concept of the "pigpen archipelago" is inspired by Aleksandr Solzhenitsyn's *The Gulag Archipelago* (1973) to invoke the essentially carceral nature of this islanded geography of life, labor, and death, and by Thomas Andrews's "workscape," Walter

Johnson's "carceral landscape," and Stephanie Camp's "geography of containment." See Andrews, *Killing for Coal*, 87–196; Johnson, *River of Dark Dreams*, 209–43; and Camp, *Closer to Freedom*, 12–34.

8. Hunter, *Studies in the Economic History of the Ohio Valley*; Stoll, *Larding the Lean Earth*; Pawley, "Accounting with the Fields."

9. Jones, *History of Agriculture in Ohio to 1880*, 123; Robinson, *Facts for Farmers*, 27–28.

10. Robinson, *Facts for Farmers*, 29–30; "Management of Swine—Cooking Food," *Prairie Farmer* 3 (January 1843): 13.

11. Burnett, "Hog Raising and Hog Driving," 94–95.

12. "The Hog Trade of Cincinnati," *Harper's Weekly*, February 4, 1860, 73; Jones, *History of Agriculture in Ohio to 1880*; Stoll, *Larding the Lean Earth*; Pawley, "Accounting with the Fields." For more on hogs, manure, and hogs as manure manufacturers, see "A Hamilton County Farm," *Western Farmer and Gardener* 2 (November 1840): 35; The American Farmer, "Making Pork—Cooking Food for Animals," *Western Farmer and Gardener* 2 (December 1840): 61; Robinson, *Facts for Farmers*, 23–24, 26–27; and "Hog Raising at the South," *Southern Cultivator* 19 (March 1861): 73–75.

13. "The Hog Trade of Cincinnati," *Harper's Weekly*, February 4, 1860, 73; Capper, *Port and Trade of London*, 204–5; Kohl, *Foreign Library*, 106–11, 142, and *Russia*, 500–503; De Tegoborksi, *Commentaries on the Productive Forces of Russia*, 2:86–102, 153–55, 180, 211–15, 341–44, 396, 424.

14. Burnett, "Hog Raising and Hog Driving," 95; Genovese, "Livestock in the Slave Economy of the Old South"; Hilliard, "Pork in the Ante-Bellum South"; McDonald and McWhiney, "Antebellum Southern Herdsman"; McWhiney and McDonald, "Celtic Origins of Southern Herding Practices"; McWhiney, *Cracker Culture*; Cuff, "Weighty Issue Revisited"; McCall, "Never Quite Settled." See also Hahn, *Roots of Southern Populism*; McCurry, *Masters of Small Worlds*; Nelson, "Livestock, Boundaries, and Public Space in Spartanburg"; and Andrews, "Beasts of the Southern Wild."

15. Robinson, *Facts for Farmers*, 62.

16. The Farmer's Register, "Wire Grass Destroyed by Hogs," *Western Farmer and Gardener* 2 (July 1841): 228.

17. "Cooking for Swine," *Prairie Farmer* 10 (October 1849): 305; Robinson, *Facts for Farmers*, 23–24, 26–27. Looked at from a more Marxian perspective, feeding in the field also allowed farmers to create and measure the value of land as an expression of socially average hog labor. According to Solon Robinson's *Facts for Farmers* (26), an average "thrifty pig" would gain 100 lbs. on grass in six months, or "$12 for the acre of pasture"; improving breed or plant would produce even more value. For more on soil and manure in antebellum agriculture, see Stoll, *Larding the Lean Earth*.

18. Davenport Gazette, "Cost of Hogs at Large," *Prairie Farmer* 11 (April 1851): 181.

19. Burnett, "Hog Raising and Hog Driving," 95; B. F. W. Stubbing, "Fattening Hogs, &c.," *Prairie Farmer* 12 (October 1852): 472; Davenport Gazette, "Cost of Hogs at Large," *Prairie Farmer* 11 (April 1851): 181; W. A. J. Russell, "Stone Coal for Hogs. Locust Trees," *Prairie Farmer* 6 (September 1846): 281.

20. B. F. W. Stubbing, "Fattening Hogs, &c.," *Prairie Farmer* 12 (October 1852): 472; Davenport Gazette, "Cost of Hogs at Large," *Prairie Farmer* 11 (April 1851): 181; R. Cheney, "Unruly Animals, Poor Fences, and Neighbors," *Prairie Farmer* 7 (April 1847): 115.

21. Jones, *History of Agriculture in Ohio to 1880*, 124–27.

22. Quoted in Jones, *History of Agriculture in Ohio to 1880*, 125–26.

23. "A Hamilton County Farm," *Western Farmer and Gardener* 2 (November 1840): 34; "Cooking for Swine," *Prairie Farmer* 10 (October 1849): 305. For more on Thomas Affleck and his work to modernize slavery, see Rosenthal, "Slavery's Scientific Management" and *Accounting for Slavery*, 86–95.

24. "Management of Swine—Cooking Food," *Prairie Farmer* 3 (January 1843): 13; Robinson, *Facts for Farmers*, 24. For studies on more recent histories of industrial livestock farming and slaughter, see Pollan, *Omnivore's Dilemma*; Pachirat, *Every Twelve Seconds*; Rosenberg, "Race Suicide among the Hogs"; and Specht, *Red Meat Republic*.

25. Jones, *History of Agriculture in Ohio to 1880*, 125–26; Burnett, "Hog Raising and Hog Driving," 95–96.

26. Robinson, *Facts for Farmers*, 28.

27. Jones, *History of Agriculture in Ohio to 1880*, 127; Green, "Hog Drover's Visit," 136.

28. Cincinnati Gazette, "The Cincinnati Slave Case," *Anti-Slavery Bugle*, February 9, 1856. For more on the Ohio River as a borderland region of crossings of slavery, freedom, and white supremacy, see Salafia, *Slavery's Borderland*.

29. Cincinnati Gazette, "The Pork Trade of Cincinnati," *Farmer's Magazine* 8 (October 1843): 264; Jones, *History of Agriculture in Ohio to 1880*, 128–29; Cist, *Sketches and Statistics of Cincinnati in 1851*, 279; *Mason v. Cowan's Administrator*, 1 B. Mon. 7 (Ky. Ct. App. Oct. 14, 1840); "The Culture of the Hog," *Gallipolis Journal*, January 12, 1854.

30. Jones, *History of Agriculture in Ohio to 1880*, 129; Green, "Hog Drover's Visit," 136.

31. Burnett, "Hog Raising and Hog Driving," 90, 92.

32. Green, "Hog Drover's Visit," 137; Cincinnati Gazette, "The Pork Trade of Cincinnati," *Farmer's Magazine* 8 (October 1843): 264; Illinois, State Legislature, Drovers, Ill., Rev. Stat. 1845.

33. Burnett, "Hog Raising and Hog Driving," 88–89, 99–100, 102–3.

34. Louisville Journal, "Hog Story," *New Albany Daily Ledger* (Ind.), September 22, 1854, 3.

35. Burnett, "Hog Raising and Hog Driving," 91.

36. Olmsted, *Journey through Texas*, 11–12.

37. Walsh, "Pork Packing as a Leading Edge of Midwestern Industry." See also Walsh, "Spatial Evolution of the Mid-Western Pork Industry" and "From Pork Merchant to Meat Packer."

38. "The Hog Trade of Cincinnati," *Harper's Weekly*, February 4, 1860, 73; Green, "Hog Drover's Visit," 137–38; Cincinnati Gazette, "The Hog and Its Product," *Gallipolis Journal*, January 12, 1854.

39. "The Hog Trade of Cincinnati," *Harper's Weekly*, February 4, 1860, 73–74.

40. Green, "Hog Drover's Visit," 138–39. Bill Jenkins's story, written as a narrated interview by a Cincinnati local named Invisible Green, Esq., in the 1856 collection of stories *Green Peas, Picked from the Patch of Invisible Green Esq.*, was published in and for a Cincinnati audience, making it at least evidence of the kinds of stories told and expected by Cincinnati readers.

41. Cincinnati Gazette, "A Cincinnati Slaughter House," *Easton Gazette* (Md.), April 22, 1843; Arms, "From Disassembly to Assembly," 198.

42. Cist, *Sketches and Statistics of Cincinnati in 1851*, 287–88.

43. Cist, *Sketches and Statistics of Cincinnati in 1851*, 280; Ross, *Workers on the Edge*, 144; Le Guin, "Day before the Revolution," 300. See also Ross, *Workers on the Edge*, 123, 141–92.

44. Salafia, *Slavery's Borderland*, 108–36; Cincinnati Gazette, "The Hog and Its Product,"

Gallipolis Journal, January 12, 1854; Cist, *Sketches and Statistics of Cincinnati in 1851*, 280. For statistics on the hog industry and its change over time, see especially Cist, *Sketches and Statistics of Cincinnati in 1851*; Cincinnati Chamber of Commerce, *Annual Report*, 58–62; and Walsh, "Pork Packing as a Leading Edge of Midwestern Industry."

45. Cist, *Sketches and Statistics of Cincinnati in 1851*, 280; "The Hog Trade of Cincinnati," *Harper's Weekly*, February 4, 1860, 74; Cincinnati Gazette, "The Hog and Its Product," *Gallipolis Journal*, January 12, 1854.

46. Cist, *Sketches and Statistics of Cincinnati in 1851*, 228–29.

47. "Pork-Packing," *Harper's Weekly*, September 6, 1873, 778; Cincinnati Gazette, "The Hog and Its Product," *Gallipolis Journal*, January 12, 1854; "The Hog Trade of Cincinnati," *Harper's Weekly*, February 4, 1860, 72; Cist, *Sketches and Statistics of Cincinnati in 1851*, 279.

48. "The Hog Trade of Cincinnati," *Harper's Weekly*, February 4, 1860, 74; Cincinnati Gazette, "The Hog and Its Product," *Gallipolis Journal*, January 12, 1854.

49. Cincinnati Gazette, "The Hog and Its Product," *Gallipolis Journal*, January 12, 1854; Cist, *Sketches and Statistics of Cincinnati in 1851*, 280.

50. Cist, *Sketches and Statistics of Cincinnati in 1851*, 280; Cincinnati Gazette, "The Hog and Its Product," *Gallipolis Journal*, January 12, 1854; Cincinnati Gazette, "A Cincinnati Slaughter House," *Easton Gazette* (Md.), April 22, 1843.

51. Cist, *Sketches and Statistics of Cincinnati in 1851*, 280; Cincinnati Gazette, "The Hog and Its Product," *Gallipolis Journal*, January 12, 1854.

52. Cist, *Sketches and Statistics of Cincinnati in 1851*, 281; "The Hog Trade of Cincinnati," *Harper's Weekly*, February 4, 1860, 74.

53. Cist, *Sketches and Statistics of Cincinnati in 1859*, 264; Cincinnati Gazette, "The Hog and Its Product," *Gallipolis Journal*, January 12, 1854.

54. Cincinnati Gazette, "The Hog and Its Product," *Gallipolis Journal*, January 12, 1854; Professor Faraday, "The Chemical History of a Candle: Lecture I," *Scientific American*, February 9, 1861, 82.

55. Cist, *Sketches and Statistics of Cincinnati in 1851*, 283–85; Cincinnati Chamber of Commerce, *Annual Report*, 60, 44; Cist, *Sketches and Statistics of Cincinnati in 1859*, 266; "Debate on the Tax Bill," *Newark Advocate* (Ohio), April 25, 1862.

56. Cist, *Sketches and Statistics of Cincinnati in 1859*, 266.

57. "Meeting of the Operative Tallow Chandlers," *NYT*, August 29, 1853. See also Virginia Penny, "Candles," in *Employments of Women*, 391–92; "Fire at Cincinnati!!," *Daily Ohio Statesman*, January 15, 1851; and "Fire," *Cleveland Herald*, August 8, 1851.

58. Schivelbusch, *Disenchanted Night*, 157–69.

59. Davis, Gallman, and Gleiter, *In Pursuit of Leviathan*, 356–57; Capper, *Port and Trade of London*, 186–87, 303, 317, 336, 386–88.

60. Capper, *Port and Trade of London*, 204–5; Kohl, *Foreign Library*, 106–11, 142, and *Russia*, 500–503; De Tegoborksi, *Commentaries on the Productive Forces of Russia*, 2:86–102, 153–55, 180, 211–15, 341–44, 396, 424; Circular of John E. Tibbs, "The Russian Tallow Speculation," *Freedom's Journal and Daily Commercial Advertiser* (Dublin), September 3, 1859; Daily News, "The Tallow Trade," *Freedom's Journal and Daily Commercial Advertiser*, September 22, 1859; "The Tallow Speculation," *Daily News* (London), October 25, 1859; "The Tallow Speculation," *Daily News* (London), December 14, 1859; "The Tallow Speculation," *Daily News* (London), December 20, 1859; "The Tallow Speculation," *Daily News* (London), December 26, 1859; "Money Market and City News," *Morning Post* (London), August 20, 1861. For some English-language studies of Russian settler

colonialism, see Moon, *Russian Peasantry*; Khodarkovsky, *Russia's Steppe Frontier*; Sunderland, *Taming the Wild Field*; and Breyfogle, Schrader, and Sunderland, *Peopling the Russian Periphery*.

61. Capper, *Port and Trade of London*, 336; Lovejoy, "Slavery and 'Legitimate Trade' on the West African Coast."

62. "A Visit to the Mammoth Cave of Kentucky," *Fraser's Magazine* 42 (October 1850): 387.

CHAPTER FIVE

1. Richard Toye interview (Dixon, Son, & Evans, Newton Heath, Manchester), 85; Michael Johnstone interview (Dixon, Son, & Evans), 85–86; Martha Cheetham interview (Dixon, Son, & Evans), 86; John Stafford interview (Chamberlayne's, Manchester), 87, all in White, "Lucifer Match Manufacture." All interviews and factory inspections in this chapter are from White, "Lucifer Match Manufacture," unless otherwise indicated.

2. Richardson and Watts, *Complete Practical Treatise*, 182; Penny, *Employments of Women*, 458.

3. "One of the Evils of Match-Making," *Household Words*, May 1, 1852, 153; Emilia Block interview (Marshall's, Aberdeen), 100; Patrick Morgan interview (Malcomson's, Belfast), 103; Richard Chidleigh interview (Bryant & May's, Bow, London), 59; Tardieu, "Étude hygiénique et médico-légale," 23; Dupasquier, "Mémoire relatif," 348.

4. George Bailey interview (Newcastle infirmary), 83; John Bell interview (Thomas Todd's, Bethnal Green), 49, Henry Cooke interview (Nottingham), 95; Pollock, Brown, and Rubin, "'Phossy Jaw' and 'Bis-phossy Jaw'"; Nixon, *Slow Violence and the Environmentalism of the Poor*; Nash, *Inescapable Ecologies*.

5. Wallingford (Conn.) Circular, "The Way Matches Are Made," *Scientific American*, September 10, 1864, 167.

6. Wallingford Circular, "Utilization of Bones," *Scientific American*, June 12, 1869, 373; Monfalcon and de Polinière, *Hygiène de la Ville de Lyon*, 250.

7. Marx, *Capital*, 356, 342.

8. "Life in a Saladero," *Household Words*, January 25, 1851, 417–18; Latham, *States of the River Plate*, 7; Seymour, *Pioneering in the Pampas*, 118; "Life in an Estancia," *Household Words*, November 16, 1850, 191; Sluyter, *Black Ranching Frontiers*, 140–68; Mitchell, *Horse Nations*, 231–301.

9. Mansfield, *Paraguay, Brazil, and the Plate*, 160, 165; Darwin, *Voyage of the Beagle*, 127.

10. Sluyter, *Black Ranching Frontiers*, 169–210; Grandin, *Empire of Necessity*, 97–105.

11. "Life in a Saladero," *Household Words*, January 25, 1851, 418; Brown, *Socioeconomic History of Argentina*, 111; Sluyter, *Black Ranching Frontiers*, 183.

12. "Life in a Saladero," *Household Words*, January 25, 1851, 418; Brown, *Socioeconomic History of Argentina*, 111–12; Grandin, *Empire of Necessity*, 101; Bell, "Early Industrialization in the South Atlantic"; Borucki, Chagas, and Stalla, *Esclavitud y Trabajo*, 21–22.

13. Brown, *Socioeconomic History of Argentina*, 114–16, and *Brief History of Argentina*, 136; Whigham, "Cattle Raising in the Argentine Northeast"; Salvatore, "Autocratic State and Labor Control in the Argentine Pampas"; Garavaglia and Gelman, "Rural History of the Río de la Plata."

14. "Life in a Saladero," *Household Words*, January 25, 1851, 418–19.

15. I am drawing here particularly on the work of generations of black radical thinkers.

Such literature is vast, but a few notable works include Du Bois, *Black Reconstruction in America*; Césaire, *Discourse on Colonialism*; Robinson, *Black Marxism*; Kelley, *Freedom Dreams*; Fields and Fields, *Racecraft*; and Movement for Black Lives, "Vision for Black Lives." I am also inspired by the Japanese writer Haruki Murakami's words in a speech delivered in Jerusalem in 2009, when he said, "Between a high, solid wall and an egg that breaks against it, I will always stand on the side of the egg." Thank you to Christopher Heaney for pointing me toward Murakami. See also Johnson, "On Agency."

16. "Life in a Saladero," *Household Words*, January 25, 1851, 418–19; Latham, *States of the River Plate*, 8; Sluyter, *Black Ranching Frontiers*, 183.

17. Montoya, *Historia de los Saladeros Argentinos*, 85; Capper, *Port and Trade of London*, 186–88, 201–4, 317, 488; Wallingford Circular, "Utilization of Bones," *Scientific American*, June 12, 1869, 373.

18. Threlfall, *Story of 100 Years of Phosphorus Making*, 51; U.S. Department of State, "Sovereignty of Guano Islands in the Caribbean Sea," 108; "A New Manure," *Bristol Mercury* (Bristol, England), July 24, 1858; Wm. Pickford & Co., "Sombrero Guano," *North American and United States Gazette* (Philadelphia), March 14, 1860.

19. "News from the Guano Islands. Our Sombrero Correspondence," *New York Herald*, July 29, 1860.

20. Petersburg Express, "Insurrection at the Sombrero Island. Outrages by Negroes— A Baltimore Vessel Attacked," *North American and United States Gazette*, August 11, 1860.

21. Skaggs, *Great Guano Rush*, 159.

22. "The Guano Diggings," *Household Words*, September 25, 1852, 44 (though this article is unattributed in the original publication, Dickens Journals Online [www.djo.org.uk] identifies the author as Samuel Rinder); "News from the Guano Islands. Our Sombrero Correspondence," *New York Herald*, July 29, 1860.

23. Rinder's account of Peruvian guano mining provides far and away the most detailed description of the material practices of the work, and while no comparable accounts of the Sombrero workings have survived, the inescapable ecology of the labor meant that trimming Sombrero guano would have been unavoidably similar. See "The Guano Diggings," *Household Words*, September 25, 1852, 46, and Skaggs, *Great Guano Rush*, 160.

24. Skaggs, *Great Guano Rush*, 159–69.

25. Cushman, *Guano and the Opening of the Pacific World*, 55; New Orleans Picayune, "Horrors of the Guano Trade," *Barre Patriot* (Mass.), May 26, 1854.

26. Skaggs, *Great Guano Rush*, 160–61; New Orleans Picayune, "Horrors of the Guano Trade," *Barre Patriot* (Mass.), May 26, 1854.

27. Petersburg Express, "Insurrection at the Sombrero Island. Outrages by Negroes— A Baltimore Vessel Attacked," *North American and United States Gazette*, August 11, 1860.

28. Norfolk Argus, "From Sombrero—Arrival of the Ringleader of the Mutiny," *Public Ledger*, August 27, 1860; "General News," *The Press* (Philadelphia); August 28, 1860; "Not Killed," *Maine Farmer*, August 30, 1860; U.S. Department of State, "Sovereignty of Guano Islands in the Caribbean Sea," 110.

29. Dupasquier, "Mémoire relatif," 345–46, 350–52; Glénard, *Sur la fabrication du phosphore*, 8–10, 33; Richardson and Watts, *Complete Practical Treatise*, 117.

30. Dupasquier, "Mémoire relatif," 347, 352; Glénard, *Sur la fabrication du phosphore*, 10; Richardson and Watts, *Complete Practical Treatise*, 117.

31. Monfalcon and Polinière, *Hygiène de la Ville de Lyon*, 250–53; White, *All Round the Wrekin*, 238; Threlfall, *Story of 100 Years of Phosphorus Making*, 34–35, 40–42, 73.

32. Threlfall, *Story of 100 Years of Phosphorus Making*, 13–15, 56, 58, 34, 51–52, 74–76; Dupasquier, "Mémoire relatif," 345; Glénard, *Sur la fabrication du phosphore*, 8; Richardson and Watts, *Complete Practical Treatise*, 111, 112, 123, 125.

33. N. Martindale's factory inspection (Liverpool), 88; interviews from N. Martindale's, Liverpool: Thomas Fisher, 89–90; John Fen, 89; Patrick Lovan, 89; Richard Jones, 89; William Flenn, 90; Edward Murphy, 90; John Fannan, 90; Daniel Dunn, 90; John McKay, 90–91; John Somerton, 91; Isabella Sumner, 91. See also Bristowe, "Occupations Which Have to Do with Phosphorus," 172.

34. N. Martindale's factory inspection (Liverpool), 88; Thomas Fisher interview (N. Martindale's, Liverpool), 89–90; John Fannan interview (N. Martindale's, Liverpool), 90; Richardson and Watts, *Complete Practical Treatise*, 182; "Wood and How to Cut It," *Household Words*, February 19, 1853, 544.

35. "Wood and How to Cut It," *Household Words*, February 19, 1853, 541, 544; Tomlinson, *Cyclopaedia of Useful Arts*, 244; John Fannan interview (N. Martindale's, Liverpool), 90; Roderick McLure interview (Marshal's, Aberdeen), 100; Thomas Moore interview (Marshal's, Aberdeen), 100.

36. John Fannan interview (N. Martindale's, Liverpool), 90; White, "Lucifer Match Manufacture," 42–45. For more on sulfuring, see Henry Walker's factory inspection (Southwark, London), 53; George Gardner interview (Bryant & May's, Bow, London), 59; John Wilson interview (John Jex Long's, Glasgow), 99; Hugh White interview (John Jex Long's, Glasgow), 99; Bristowe, "Occupations Which Have to Do with Phosphorus," 172; Children's Employment Commission, *Second Report of the Commissioners*, 252; and "Illustrations of Cheapness: The Lucifer Match," *Household Words*, April 13, 1850, 56.

37. White, "Lucifer Match Manufacture," 43.

38. Penny, *Employments of Women*, 459.

39. Patrick Lovan interview (N. Martindale's, Liverpool), 89; Richard Jones interview (N. Martindale's, Liverpool), 89; John McKay interview (N. Martindale's, Liverpool), 90–91; John Somerton interview (N. Martindale's, Liverpool), 91; Schmidt, "'Restless Movements Characteristic of Childhood.'"

40. Children's Employment Commission, *Second Report of the Commissioners*, 253–55.

41. William Flenn interview (N. Martindale's, Liverpool), 90.

42. Margery Nicholl interview (Marshall's, Aberdeen), 100; John McKay interview (N. Martindale's, Liverpool), 90–91; White, "Lucifer Match Manufacture," 43.

43. N. Martindale's factory inspection (Liverpool), 88; John Fen interview (N. Martindale's, Liverpool), 89; "Illustrations of Cheapness. The Lucifer Match," *Household Words*, April 13, 1850, 56; Richard Toye interview (Dixon, Son, & Evans, Newton Heath, Manchester), 85; Thomas Harrison interview (Thomas Todd's, Bethnal Green, London) 49; William Wright interview (Lewis Waite's, Bethnal Green, London), 51; James Bennett interview (George Evance's, Somer's Town), 57.

44. Interviews from N. Martindale's, Liverpool: John Fen, 89; Patrick Lovan, 89; Richard Jones, 89; Edward Murphy, 90; Daniel Dunn, 90; John McKay, 90–91; John Somerton, 91.

45. N. Martindale's factory inspection (Liverpool), 88.

46. Isabella Sumner interview (N. Martindale's, Liverpool), 91.

47. Wallingford Circular (Conn.), "The Way Matches Are Made," *Scientific American*, September 10, 1864, 167.

48. "Lucifer-Box Making," *All the Year Round*, April 6, 1867, 352. I have drawn here on

the work of feminist historians of labor. See especially Blackmar, *Manhattan for Rent*; Boydston, *Home and Work*; Morgan, *Laboring Women*; Glymph, *Out of the House of Bondage*; and Rockman, *Scraping By*. For good studies of the legal history of child and dependent labor before the Civil War, see Schmidt, "'Restless Movements Characteristic of Childhood,'" and Winsberg, "Making Work."

49. "Lucifer-Box Making," *All the Year Round*, April 6, 1867, 353–55.

50. Children's Employment Commission, *Second Report of the Commissioners*, 254–55.

51. For "infrapolitics," see Scott, *Domination and the Arts of Resistance*. For some examples of children's mobile practices in chasing and promoting wages and health, see Christopher Bateman interview (Thomas Todd's, Bethnal Green), 50; Henry Walker's factory inspection (Southwark, London), 54; George Rowland interview (Henry Walker's, London), 54; Richard Chidleigh interview (Bryant & May's, Bow, London), 59; John Osborne interview (London Hospital), 60–61; John Thorn interview (Richard Bell & Co.'s, London), 67; Richard Raftry interview (Peele's, Newcastle), 82.

52. William Köhler interview (Birmingham), 91; Arthur Albright, evidence submitted, in White, "Lucifer Match Manufacture," 76.

53. William Hillier interview (Hillier's, Birmingham), 95; David Bermingham interview (David Bermingham's, Birmingham), 91–92.

54. "Illustrations of Cheapness. The Lucifer Match," *Household Words*, April 13, 1850, 54–55.

55. "Statistics of Camphene: Explosions, &c., for the Year," *NYT*, July 23, 1852. For coal miners, matches could, like safety lamps, at once make the mines safer and extend them into more dangerous undergrounds. Miners carrying matches no longer had to worry about absolute darkness should their lamps go out. In a pinch, the glow-in-the-dark match heads could even have served as fireless markers. But matches could also spontaneously ignite in pockets or if stepped on. And the possibility of extinguishing lamps, walking through firedamp in total blackness, and then relighting lamps in a safer section of the mine surely led mine operators and miners to take greater risks underground. See Pohs, *Miner's Flame Light Book*, 557.

56. C. L. B., "Walks among the New-York Poor: Vagrant Boys," *NYT*, January 21, 1853; "The Wants Of The Poor: The Iron Moulder," *NYT*, January 8, 1855.

57. Mayhew, *London Labour and the London Poor*, 482–83.

58. "The Match Question," *Household Words*, June 9, 1860, 209–12. For examples of the ubiquitous ecology of matches, see "How to Get a Better House," *NYT*, June 30, 1856; "Semi-Annual Report of the Fire Marshal," *NYT*, February 25, 1858; and House of Commons, *Report from the Select Committee on Fire Protection*, 471.

59. "Mischief in Matches," *NYT*, October 19, 1859; *Deseret News*, August 15, 1860; "Ordinances," *New-Bedford Mercury*, December 26, 1834; "Lucifer Matches," *Salem Gazette*, September 28, 1838; "Look Out," *Connecticut Courant*, April 18, 1840; "Fire on the Railroad by Friction Matches," *Southern Patriot*, October 24, 1842; "The Fire Marshall's Report," *New York Herald*, February 22, 1860; "Match Safes," *Chicago Press and Tribune*, February 12, 1859; "How to Get a Better House," *NYT*, June 30, 1856.

60. "A Chapter of Fires," *NYT*, July 5, 1852; "Children Playing with Matches," *NYT*, August 16, 1854; "Destruction of a Carpenter's Shop," *NYT*, July 21, 1856; "In Eighth-Avenue," *NYT*, August 16, 1856; "Children Burnt to Death," *NYT*, November 12, 1856; "News of the Day," *NYT*, September 10, 1857; "Illinois Matters," *Chicago Daily Tribune*, January 30, 1858; "City Items," *NYT*, March 26, 1858; "Fire in Varick-Street," *NYT*, June 29,

1859; "Fire in Downing-Street," *NYT*, July 30, 1859; "Great Fire at Galva," *Chicago Tribune*, September 3, 1861; Philadelphia Ledger, "A Curious Fire and Its Origin," *NYT*, January 18, 1862; "Fire in West Thirty-Third-Street," *NYT*, August 21, 1862; "Fire in Broadway," *NYT*, June 10, 1864; "Fire," *NYT*, June 15, 1855; "In Fourth-Avenue," *NYT*, May 26, 1856; "Fire at Bergen Hill," *NYT*, June 7, 1858; "New York Crystal Palace Destroyed by Fire — Other Fires in New York," *Chicago Press and Tribune*, October 6, 1858; "Fire in West Twenty-First-Street," *NYT*, July 11, 1859; "Fire in Orchard-Street," *NYT*, August 19, 1859; "In a Stable," *NYT*, August 3, 1860; "Fire in a Stable — Three Horses Burned," *NYT*, October 29, 1860.

61. "Fire Marshal's Investigations — Committals for Arson," *NYT*, May 7, 1855.

62. Providence Journal, "Miscellaneous," *NYT*, January 2, 1860; Providence Journal, "More Matches in Cotton," *Chicago Press and Tribune*, January 20, 1860.

63. Brown, *Slave Life in Georgia*, 179; Johnson, *River of Dark Dreams*, 174; Providence Journal, "More Matches in Cotton," *Chicago Press and Tribune*, January 20, 1860.

64. "More Trouble in Virginia. A Negro Insurrectionist Sentenced to be Hanged," *Chicago Press and Tribune*, January 10, 1860; Northup, *Twelve Years a Slave*, 272. Court records show access to matches by enslaved people to be considered unremarkable. See, for instance, *State v. Sandy*, an arson case, and *Couch v. Jones*, a case involving railroad work, blasting, and matches.

65. For more on nineteenth-century sacrifice zones, landscapes of energy, and the relations between local and global processes, see Jones, *Routes of Power*, and Pratt, Melosi, and Brosnan, *Energy Capitals*.

CHAPTER SIX

1. Kurlansky, *Salt*, 249–61, 310, 315; Williamson and Daum, *American Petroleum Industry*, 14–17, 77–81. For more on southern railroads and political economy, see Nelson, *Iron Confederacies* and "Who Put Their Capitalism in My Slavery?"; Adams, *Old Dominion, Industrial Commonwealth*; and Majewski, *Modernizing a Slave Economy*.

2. Black, *Petrolia*; Philadelphia Ledger, August 1864, quoted in Sabin, "'Dive into Nature's Great *Grab-bag*,'" 487.

3. Lewis, *Black Coal Miners in America*, 7.

4. Washington, *Up from Slavery*, 24–26.

5. Laing, "Early Development of the Coal Industry," 146.

6. Stealey, "Slavery and the Western Virginia Salt Industry," 116, 119.

7. Lewis, *Black Coal Miners in America*, 9; Stealey, "Slavery and the Western Virginia Salt Industry," 118.

8. Quoted in Stealey, "Slavery and the Western Virginia Salt Industry," 109; Starobin, *Industrial Slavery in the Old South*, 95–109; Zaborney, *Slaves for Hire*, 131–35.

9. Stealey, "Slavery and the Western Virginia Salt Industry," 124–25.

10. Stealey, "Slavery and the Western Virginia Salt Industry," 123–27, 107, 118–20; Lewis, *Black Coal Miners in America*, 8–9.

11. Laing, "Early Development of the Coal Industry"; Stealey, "Slavery and the Western Virginia Salt Industry," 115–17, 107–8; Lewis, *Black Coal Miners in America*, 7–8; Zaborney, *Slaves for Hire*, 122–23.

12. Stealey, "Slavery and the Western Virginia Salt Industry," 129–31; Thomas Pollard to Henry F. Thompson, January 11, 1855, in Frantel, *Chesterfield County Virginia Uncovered*, 175–76. See also Zaborney, *Slaves for Hire*, 128–30.

13. Stealey, "Slavery and the Western Virginia Salt Industry," 111; Washington, *Up from Slavery*, 43, 38; Maury, *Resources of the Coal Field of the Upper Kanawha*, 19.

14. Winifrede Mining and Manufacturing Company, Ledger Book, #92, 1850–58, KCCA; Lewis, *Black Coal Miners in America*, 9; Conley, "Early Coal Development in the Kanawha Valley," 209.

15. "Internal Improvements," *KVS*, August 18, 1857; Maury, *Resources of the Coal Field of the Upper Kanawha*, 6.

16. Adams, *Old Dominion, Industrial Commonwealth*, 84–102.

17. Coal River and Kanawha Mining and Manufacturing Company, Pvt. Acct. Book, #56, 1851–1858, Minutes, March 23, 1854, KCCA.

18. Coal River and Kanawha Mining and Manufacturing Company, Pvt. Acct. Book, Minutes, March 27, 1854, KCCA.

19. Rice, "Coal Mining in the Kanawha Valley to 1861"; Coal River and Kanawha Mining and Manufacturing Company, Pvt. Acct. Book, Minutes, May 6, 1856, KCCA; Boston City Council, Special Committee on Gas Inspection, *Report of the Evidence*.

20. From the Charlottesville Advocate, "Editorial Correspondence," *KVS*, December 23, 1856; "Cannel Coal for Boston," *KVS*, March 29, 1859. Cumberland coal, mined in western Maryland, was an important antebellum fuel for industry and steam engines but was not useful for manufacturing gas. For a useful summary of the geography and chemistry of American gas coals, see Taylor, *Statistics of Coal*, 603–10, and Clegg, *Practical Treatise on the Manufacture and Distribution of Coal-Gas*, 70–73.

21. Coal River and Kanawha Mining and Manufacturing Company, Pvt. Acct. Book, Minutes, August 15, 1855, KCCA; "Cannel Coal Oil," *KVS*, July 28, 1857.

22. "Cannel Coal Oil," *KVS*, July 28, 1857.

23. Williamson and Daum, *American Petroleum Industry*, 51–52; "Cannel Coal Oil," *KVS*, July 28, 1857; Coal River and Kanawha Mining and Manufacturing Company, Pvt. Acct. Book, Minutes, May 6, 1856, KCCA; "Coal-Oil and Coal-Oil Lamps," *Daily Evening Bulletin* (San Francisco), October 11, 1859.

24. "Coal-Oil and Coal-Oil Lamps," *Daily Evening Bulletin* (San Francisco), October 11, 1859. Dietz was almost certainly not the first to invent a lamp capable of burning coal oil or kerosene. In Europe, at least, such lamps were around by 1853, sparking interest in the petroleum fields of Austrian Galicia. See Frank, *Oil Empire*, 56. For an excellent survey of the history of the coal-oil industry, see Williamson and Daum, *American Petroleum Industry*, 43–60.

25. "Coal-Oil and Coal-Oil Lamps," *Daily Evening Bulletin* (San Francisco), October 11, 1859; "Cannel Coal Oil," *KVS*, February 16, 1858.

26. "Coal Oil Manufacture," *KVS*, February 27, 1860; Williamson and Daum, *American Petroleum Industry*, 55–59.

27. "Coal Oil Manufacture," *KVS*, February 27, 1860; "Cannel Coal Oil Factories," *KVS*, April 9, 1861.

28. Correspondence of the Springfield Republican, Gallipolis, Ohio, December 5, 1859, "The Coal Oil Fields of the Ohio Valley," *New Advocate* (Ohio), December 23, 1859.

29. "Internal Improvements," *KVS*, August 27, 1857.

30. An Old Subscriber, *KVS*, March 3, 1857.

31. McConnelsville Enquirer, "Oil from Coal—Wonderful Results," *Newark Advocate* (Ohio), April 8, 1857.

32. Rice, "Coal Mining in the Kanawha Valley to 1861," 415; "Will the Oil Works Bring Money into the Country?," *KVS*, October 10, 1859.

33. "Coal and Salt Works," *KVS*, September 28, 1858.

34. For a comprehensive analysis of the early history of oil in Pennsylvania, the most indispensable account remains Williamson and Daum, *American Petroleum Industry*. See also Tarbell, *History of the Standard Oil Company*; Yergin, *The Prize*; Black, *Petrolia*; and Jones, *Routes of Power*.

35. "Great Kanawha Oil Prospects," *KVS*, March 19, 1861.

36. Williamson and Daum, *American Petroleum Industry*, 44–48, 51–54.

37. Williamson and Daum, *American Petroleum Industry*, 81; Stealey, *Antebellum Kanawha Salt Business and Western Markets*, 49–51, 125, 134–36; Bruns, "Antebellum Industrialization of the Kanawha Valley in the Virginia Backcountry," 57–63, 66–70, 90–91, 121, 128.

38. "Great Kanawha Oil Prospects," *KVS*, March 19, 1861.

39. Shattuck, "Coal-Oil in West Virginia," 526–27.

40. "Virginia News," *Alexandria Gazette*, March 22, 1861; "Virginia Legislature," *Richmond Whig*, March 19, 1861; "Oil! Oil!! Oil!!! A Clevelander's Account of the Virginia Oil Wells," *Plain Dealer*, March 8, 1861; from the Wheeling Intelligencer, March 8, "The Oil Excitement in Virginia—Letting of Claims," *New York Herald*, March 11, 1861.

41. Williamson and Daum, *American Petroleum Industry*, 123, 164–68, 185.

42. Jones, "Energy Landscapes," 165–66.

43. "Oil! Oil!! Oil!!! A Clevelander's Account of the Virginia Oil Wells," *Plain Dealer*, March 8, 1861; S. D. Collins for the Cincinnati Commercial, "More Oil," *Newark Advocate* (Ohio), April 5, 1861; Wheeling Intelligencer, "The Oil Excitement in Virginia," *New York Herald*, March 11, 1861.

44. "Oil! Oil!! Oil!!! A Clevelander's Account of the Virginia Oil Wells," *Plain Dealer*, March 8, 1861.

45. Collins, "More Oil," *Newark Advocate*, April 5, 1861.

46. Shattuck, "Coal-Oil in West Virginia," 528–29; Adams, *Old Dominion, Industrial Commonwealth*, 215.

47. "Tariff on Coal," *KVS*, March 19, 1861.

48. "What Shall Western Virginia Do?," *KVS*, April 9, 1861.

49. Majewski, *Modernizing a Slave Economy*, 7.

50. Du Bois, *Black Reconstruction in America*; Lewis, *Coal, Iron, and Slaves*; Robinson, *Black Marxism*; Robinson, *Bitter Fruits of Bondage*; Hahn, "Did We Miss the Greatest Slave Rebellion in Modern History?," in *Political Worlds of Slavery and Freedom*, 55–114; Manning, *Troubled Refuge*. For arguments that southern nationalists were antimodern, anti-industrialist, and antigovernment, see Genovese, *World the Slaveholders Made*, xxvi, 3–20, 95–102, and *Slaveholders' Dilemma*; McPherson, *Drawn with the Sword*; Egnal, *Divergent Paths*, 52–68, 87–101; Ransom, *Confederate States of America*, 188–97; Einhorn, *American Taxation, American Slavery*; McCardell, *Idea of a Southern Nation*; and Ford, *Origins of Southern Radicalism*.

51. Dolin, *Leviathan*, 309–34; Davis, Gallman, and Gleiter, *In Pursuit of Leviathan*, 363.

52. Outland, *Tapping the Pines*, 123–25; U.S. Congress, *Alcohol in the Manufactures and Arts* and *Reports of a Commission*, 161–62; Williamson and Daum, *American Petroleum Industry*, 322.

53. Benjamin Grist to James R. Grist, Fish River, July 23, August 25, October 9, 1861; "Negros in 1862 sent up March 29. Hired to Shelby Iron Works at 125$ a year," all in box 4, JRGBR.

54. For the significance of the geography of turpentine to the military history of the Civil War, see U.S. War Department, *War of the Rebellion*, ser. 1, 14:559; 29(1):909–10; 35(2):302; 47(1):137, 198, 322–23, 381, 410, 413, 588, 622, 690, 698, 1021; 47(2):521; and 49(1):143, 178–79, 216–19, 232–33, 698, 1049.

55. U.S. Army, *Report of the Quartermaster General*, 601; Major-General J. D. Cox to Major-General Schofield, Hdqrs. Third Division, Twenty-Third Army Corps, Brunswick River Ferry, February 21, 1865, in U.S. War Department, *War of the Rebellion*, ser. 1, 47(2):521; Chauncy W. Curtis, "The Burnside Expedition to Roanoke" (ca. 1900), North Carolina Digital Collections (digital.ncdcr.gov).

56. "The Mechanics Strike in Baltimore," *NYT*, February 17, 1853; "A Strike among the Employes of the Gas Companies. A Large Police Force Called Out—A Serious Riot Apprehended," *NYT*, November 13, 1862; "The Late Strike at the Gas Works," *NYT*, November 17, 1862; "The Late Strike at the Manhattan Gas Works—An Action for False Imprisonment," *NYT*, December 17, 1862; "Strike at the Philadelphia Gas Works—The City in Darkness," *NYT*, July 17, 1868; "The Strike in the Philadelphia Gas Works—The City in Darkness," *NYT*, July 18, 1868; "The Gas Strike in Philadelphia—The Strikers Victorious," *NYT*, July 19, 1868.

57. Baldwin, *In the Watches of the Night*, 17–19.

58. "Pay, Rations and Clothing of the Army," *Albany Journal*, April 27, 1861; "A Word with the Volunteers," *Salem Register*, August 29, 1861; William Allen Clark to George W. Clark, Camp near Boiling Spring, Rutherford Co., Tenn., May 31, 1863, in Clark, Civil War correspondence, 219; "Soldiering in Winter Quarters. The Realities of Camp Life. Everyday Experiences of Volunteers," *Albany Evening Journal*, January 25, 1862; Chauncey H. Cooke to Sister, Columbus, KY, Headquarters 25th Wisconsin Volunteers Infantry, May 3, 1863, in Cooke, "Letters of a Badger Boy in Blue," 337. For studies of how correspondence and camp life during the Civil War helped create a national culture, see Faust, *This Republic of Suffering*, and Manning, *What This Cruel War Was Over*.

59. "A Word with the Volunteers," *Salem Register*, August 29, 1861; Taylor, Diary, October 7, 1862, 227; "The Extra Army Ration and What Becomes of It," *Milwaukee Daily Sentinel*, August 22, 1862; "Military Affairs," *Hartford Daily Courant*, May 6, 1861.

60. Cincinnati Chamber of Commerce, *Annual Report*, 60, 44; Chicago Board of Trade, *Annual Statement*, 46; Cist, *Sketches and Statistics of Cincinnati in 1859*, 266; "Debate on the Tax Bill," *Newark Advocate* (Ohio), April 25, 1862.

61. "News of the Day," *NYT*, July 10, 1861; "Precautions to Be Observed in the Transportation of Troops by Sea," *New York Herald*, October 22, 1861; "How Can Slavery Be Turned against the Rebellion," *NYT*, December 5, 1861.

62. Fleming, "Ohio Columbus Barber," 18–19; Bounds, "Wilmington Match Companies," 10–11.

63. Yergin, *The Prize*, 30; "Petroleum Oil as Valuable as Cotton," *Public Ledger*, February 3, 1862.

64. "Destruction of a Kerosene Refinery," *New York Commercial Advertiser*, August 27, 1861. For the greater–New York fires at kerosene refineries, see "Fire in Stanton Street," *New York Herald*, May 15, 1861; "Fire in Stanton Street," *New York Commercial Advertiser*, July 16, 1861; "Destruction of a Kerosene Refinery," *New York Commercial Advertiser*, August 27,

1861; "Destruction of a Kerosene Factory," *New York Commercial Advertiser*, September 5, 1861; "Another Kerosene Factory Burned," *Philadelphia Inquirer*, September 27, 1861; "Brooklyn. Fire," *New York Commercial Advertiser*, September 27, 1861; "Destructive Fire at Red Hook.—A Kerosene Oil Factory Destroyed.—Loss $30,000," *New York Commercial Advertiser*, September 28, 1861; "Fires," *New York Daily Tribune*, September 28, 1861; "Destructive Fire in Jersey City," *New York Commercial Advertiser*, October 16, 1861; "Fire at Hunter's Point," *New York Commercial Advertiser*, October 17, 1861; "A Kerosene Oil Factory Destroyed by Fire," *New York Daily Tribune*, November 6, 1861; "Arrest for Manufacturing Kerosene Oil without a License," *New York Commercial Advertiser*, December 9, 1861; "Another Explosion in a Kerosene Oil Factory. Destruction of the Newton Creek Works—Loss, $120,000," *Albany Evening Journal*, December 28, 1861; "Brooklyn. Another Explosion in a Kerosene Oil Factory," *New York Evening Post*, January 15, 1862; "Fire in a Kerosene Factory," *New York Daily Herald*, January 16, 1862; "The Recent Explosion in a Kerosene Oil Factory in Brooklyn," *New York Tribune*, April 12, 1862; "The Late Disastrous Explosion in a Kerosene Oil Factory," *New York Tribune*, April 15, 1862; and "Visit of the Brooklyn Common Council Committee to the Kerosene Works," *New York Tribune*, June 13, 1862. For fires among distributors and storage depots, see "Fire and Loss of Life," *Boston Post*, September 9, 1861; "Fire—Narrow Escape of Two Men," *Philadelphia Inquirer*, September 11, 1861; "Destructive Fire in Jersey City," *New York Commercial Advertiser*, September 25, 1861; "Uninsurable Merchandise," *New York Commercial Advertiser*, December 24, 1861; *New York Herald*, December 25, 1861; "The Scene of Monday Night's Fire," *North American and United States Gazette*, May 14, 1862; "Disastrous Fire in Williamsburgh," *New York Tribune*, May 30, 1862; "Extensive Fires," *Boston Daily Advertiser*, October 23, 1862; "The Pirate Semmes Still on the Coast," *Boston Post*, December 8, 1862; and *Alexandria Gazette*, January 22, 1863.

65. "The Petroleum Region—the Rock Oil Business—The Extent and Sources of Supply," *Scientific American*, February 22, 1862, 122; Enri, *Coal Oil and Petroleum*, 22, 89.

66. "Will Kerosene Explode," *Daily Evening Bulletin* (San Francisco), March 11, 1861; Niles Republican (March 9, 1861), "Kerosene Explosion," *Kalamazoo Gazette*, March 15, 1861; "Fatal Result from the Explosion of a Kerosene Lamp," *New York Tribune*, March 29, 1861; "Kerosene Explosions," *Salem Register*, April 1, 1861; *New York Daily Tribune*, April 4, 1861, 5; "Kerosene Oil Explosive," *New York Daily Tribune*, April 6, 1861; "Explosive Kerosene," *Springfield Weekly Republican* (Mass.), April 13, 1861; "How to Test Kerosene," *Newport Mercury*, April 20, 1861; "Kerosene—Danger of Explosion," *New York Tribune*, July 21, 1861; "Explosive Qualities of Kerosene Oil," *North American and United States Gazette*, September 9, 1861; "Fire on a Canal Boat," *New York Commercial Advertiser*, September 19, 1861; Alex. W. Blackburn, Fire Marshal, "Local Intelligence," *Philadelphia Inquirer*, January 29, 1862; Examiner, "Coal Oils," *Public Ledger*, February 18, 1862; "The Dangers of Our Artificial Lights," *Scientific American*, May 4, 1867, 285; "The Sale of Explosive Illuminating Fluids," *Scientific American*, August 20, 1870, 119; "Cleveland Non-Explosive Lamp Company's Works—Their Importance, Extent and Benefit to the Public," *Cleveland Leader*, February 2, 1872, in "Lamp Fillers: Notes and Queries, Quotes and News: Protection to Life and Property," *Rushlight* 66 (June 2000): 14–17; "Insurance Companies and Under Test Oils," *Oil, Paint and Drug Reporter* 20 (November 1881): 901–2; "Petroleum for Light," *Scientific American*, September 15, 1883, 160; Williamson and Daum, *American Petroleum Industry*, 523–24; Nolan and Jones, "Cleveland Non-Explosive Lamp Company."

67. "Sewing and Starving," *New York Ledger*, December 12, 1863; "The Sewing Women

of New York. How Northern Philanthropy Is Supported and Miscegenation Encouraged," *Philadelphia Daily Age*, March 23, 1864; U.S. Army, *Report of the Quartermaster General*, 15, 13.

68. Beckert, *Monied Metropolis*, 111–44. For more on the work and temporal politics of sewing women during the Civil War, see Penny, *Employments of Women*, 296–98, 345, 348, 351; "The Sewing Women," *New York Tribune*, December 2, 1863; "The Sewing Women and Their Employers," *New York Evening Post*, March 22, 1864; "Working Women," *Boston Post*, March 24, 1864; and "Tremendous and Enthusiastic Meeting of Working Women," *Philadelphia Daily Age*, January 20, 1865.

69. Beecher and Stowe, *American Woman's Home*, 363–64.

70. Boydston, *Home and Work*, 113; Diary of Lydia Maria Child, 1864, as quoted in Boydston, *Home and Work*, 85; Brox, *Brilliant*, 86–87.

71. Boydston, *Home and Work*, 91, 86; Brox, *Brilliant*, 86; Adams, *Home Fires*, 93–104.

72. Penny, *Employments of Women*, 310; Stansell, *City of Women*, 113–14.

EPILOGUE

1. Receipt, "Trip to New York (Electric Light)," November 28, 1882, box 3, folder 18, MS Thr 432, BBTCR; "The 'Bijou' Features of the Opening Night of the New Theatre," *Boston Daily Advertiser*, December 12, 1882. The engine room was located in a basement room rented from the printers Cashman, Keating & Co., 550 feet from the Bijou. See receipt for $5,625, January 7, 1883, for "rent of basement . . ." Bought of Cashman, Keating & Co., in Orders Book, 1882–1884, box 4, item 24, BBTCR.

2. "The Edison Light," *Boston Herald*, December 18, 1882.

3. Re—inquest of Jerry Toomey, deceased, June 1, 1887, Misc. 264, Coroner's Inquest Reports, Butte, Mont. For a more thorough treatment of the politics, hazards, and work of Montana copper mining, see Zallen, "'Dead Work.'"

4. Curtis, *Gambling on Ore*. For an excellent study of the hazards of the Butte underground, see Shovers, "Miners, Managers, and Machines" and "Perils of Working in the Butte Underground." For additional studies of western mining and its dangers, see Wyman, *Hard Rock Epic*; Peck, "Manly Gambles"; Punke, *Fire and Brimstone*; and Andrews, *Killing for Coal*.

5. Jerry Toomey inquest, Coroner's Inquest Reports, Butte, Mont.

6. "The Opening of the Bijou Theatre," *Boston Evening Transcript*, Tuesday, December 12, 1882; "A Talk with Edison," *Boston Herald*, Thursday, December 14, 1882.

7. "The Opening of the Bijou Theatre," *Boston Evening Transcript*, Tuesday, December 12, 1882; Schivelbusch, *Disenchanted Night*, 50–51; "The Edison Light," *Boston Herald*, December 18, 1882.

8. Sicilia, "Selling Power," 131–32; Orders Book, 1882–1884, BBTCR.

9. Invisible were those who had installed the system, the Menlo Park factory hands churning out light bulbs and dynamos, the Connecticut copper wire mills, the migrant Irishmen laboring in the copper mines of Michigan and Montana, and the globally proliferating subterranean worlds of coal and iron mining. So too were the employees of J. J. McNutt Builder and Manufactures who razed the old Gaiety Theatre, leaving only the walls, and built the fireproofed Bijou in its place. No mention was made of either Patrick or Jerry O'Connor, who guarded the Bijou every day for at least a year, Patrick as night watchman and Jerry as day doorkeeper. Nor did the weekly labor of Joe McElroy,

the gasman/electrician, receive public notice. And these were only a few of the more permanent members of the backstage crew. See receipts for building expenses, 1882–1883, box 4, folder 28, and box 5, folder 29, and Salaries, box 3, folders 18 and 19, BBTCR.

10. Painter, *Standing at Armageddon*. See also White, *Railroaded* and *Republic for Which It Stands*, and Hahn, *Nation without Borders*.

11. For more on the practice and culture of cultivating *disbelief*, see Warren, *Buffalo Bill's America*.

12. From 1883 to 1884, large work crews labored to make Anaconda into a smelter town, with enormous reduction works, including a massive concentrator and smelter. The San Francisco–based syndicate, in partnership with Comstock veteran and Anaconda superintendent Marcus Daly, poured millions of dollars into developing the mine, at least $4 million into the plant alone, and that just to get it operational. See Malone, *Battle for Butte*, 24–31. See also Wyman, *Hard Rock Epic*, 6, and Punke, *Fire and Brimstone*, 22. For some additional studies of Butte and its mining industry, see Lingenfelter, *Hard Rock Miners*; Shovers, "Miners, Managers, and Machines" and "Perils of Working in the Butte Underground"; Calvert, *The Gibraltar*; Emmons, *Butte Irish*; Quivik, "Smoke and Tailings"; and Mercier, *Anaconda*. For an excellent new history of copper smelting and the environmental history of open-pit copper mining (which follows the period of mining explored here), see LeCain, *Mass Destruction*.

13. Curtis, *Gambling on Ore*, 140.

14. Patten and Patten interview; Shovers, "Miners, Managers, and Machines," 22–23. Herb Mickelson remembered how smart (and stubborn) underground mules were. They "could tell how many cars you hooked them up to" and would refuse to be overworked, but he noted that "many mules became blind after years underground" (Mickelson interview).

15. Shovers, "Miners, Managers, and Machines," 46–48; Claude T. Rice, "Prevention of Accidents in Metal Mines," *Engineering and Mining Journal*, February 6, 1909, 302. For more on the politics and structures of risk in mining, see Adams, "Dark as a Dungeon."

16. "All other supplies of whatsoever character embracing candles, powder, and such miscellaneous articles, as may be found necessary, will be furnished to said Receiver at the collar of said Berkeley Shaft, at the actual cost of same, to this Company." See Contract between Butte & Boston Consolidated Mining Company and Snohomish and Tramway Mines, March 24, 1900, box 368, folder 16, ACMCR; General Journal A (Dec. 1896– June 1899), 4–81, MF 426, ACMCR; "Inventory—Machinery and Supplies—Anaconda Mine—July 1st, 1895," box 58, folder 6, ACMCR; Shovers, "Miners, Managers, and Machines," 19; Wyman, *Hard Rock Epic*, 13; and Pohs, *Miner's Flame Light Book*, 125–29, 147–97.

Paraffin candles made from refined petroleum were cheap enough but were too soft and melted too easily. By the time the Butte underground was being delved, miners' candle manufacturers in Cincinnati (such as Procter & Gamble), Pittsburgh, St. Louis, and Chicago had expanded their operations to encompass the fatty waste not only from pork packing but of cattle and beef as well. From 1900 to 1901, W. & H. Walker of Pittsburgh sent dozens of letters requesting a contract with the Butte & Boston Consolidated Mining Company. Excited at the prospect of gaining such a lucrative client, the representatives of W. & H. Walker wrote, "We are sending you by express, prepaid, a two pound sample of our Stearic Acid, 14oz 6's, Mining Candles, which we quote in carload quantities, packed 40 sets per box, at $4.45 per box, freight paid to Butte. We trust you will give the candles we are sending you a thorough test, as we are confident you will find them perfectly satisfactory in every way, and equal in quality to any candles you have ever used." W. & H. Walker repeatedly asserted the superiority of its candles, claiming proof in "exhaustive" tests "with all leading brands in hot, damp and draughty mines, and the fact that our customers are ordering repeatedly in spite of sharp competition. These tests have invariably shown ours to be the hardest, to burn the longest and to give the strongest and most brilliant light" (W. & H. Walker to Butte & Boston Consolidated Mining Co., July 19, 1901, box 368, folder 16, ACMCR).

17. Shovers, "Miners, Managers, and Machines," 45–53, 85–89; Wyman, *Hard Rock Epic*, 67–68. Miners also fought to change and adapt to the conditions of their work through unions, fraternal ethnic orders, and legislative campaigns. In Comstock in 1867 and in Butte by 1900, an eight-hour day was won. The Butte Miners' Union paid injured and sick miners' families $10 a week and $90 for funeral expenses. The Ancient Order of Hibernians, meanwhile, paid $8 a week to sick and injured miners for up to thirteen weeks, for a monthly membership fee of 50 cents. Ethnic societies for Germans, Finns, Croats, Italians, and Austrians also offered health benefits.

18. Emmons, *Butte Irish*, 153–54; Shovers, "Miners, Managers, and Machines," 36–37.

19. Shovers, "Miners, Managers, and Machines," 76.

20. Mercier, *Anaconda*, 14; Calvert, *The Gibraltar*; Malone, *Battle for Butte*, 76–77. For more on "whiteness" and the politics of western industrial labor, see Peck, *Reinventing Free Labor*, 160–236, and Jameson, *All That Glitters*, 140–60. Workers finally achieved a closed shop only after threatening to lynch the superintendent of the Bluebird Mine, who was the last holdout. Considering the widespread use of lynching as an instrument of white terror in the post-Reconstruction South, the symbolism intended by threatening the Bluebird superintendent seemed clear. Like countless other nineteenth-century movements, copper miners' struggles were refracted and articulated through the powerful politics of whiteness, manhood, and nationalism.

21. Malone, *Battle for Butte*, 39.

22. Lingenfelter, *Hard Rock Miners*, 196–228. The Western Federation of Miners' constitution included the following objects related to safety: "First: To secure an earning fully compatible with the dangers of our employment. . . . Third: To procure the introduction and use of any and all suitable, efficient appliances for the preservation of life, health and limbs of all employees, and thereby preserve to society the lives of large numbers of wealth producers annually. . . . Fourth: To labor for the enactment of suitable mining laws, with a sufficient number of inspectors, who shall be practical miners, for the proper enforcement of such laws."

23. Nye, *Electrifying America*, 33–36, 47.

24. Nye, *Electrifying America*, 37–38; Graff, "Dream City, Plaster City," 704.

25. Silkenat, "Workers in the White City," 268.

26. R. A. F. Penrose Jr., "Notes on the State Exhibits in the Mines and Mining Building at the World's Columbian Exposition, Chicago," *Journal of Geology* 1 (July–August 1893): 464–67.

27. Graff, "Dream City, Plaster City," 708–9. For new takes on the role of corruption, failure, and showmanship in late-nineteenth-century American capitalism, see Warren, *Buffalo Bill's America*, and White, *Railroaded*.

28. "The Cycle of a Day: The Western Electric Company Give an Exhibition in Their Electric Scenic Theater That Surpasses Anything in This Line Heretofore Attempted,"

Chicago Daily Tribune, August 27, 1893; Barrett, *Electricity at the Columbian Exposition*, 13–15. The forces of consolidation and monopolization transforming copper mining were equally on display in the Electricity Building. By 1893, General Electric and Westinghouse had established a virtual duopoly over electrical systems, and they dominated the fair's electrical exhibits. See Nye, *Electrifying America*, 33–38, 40–41, 170.

29. Silkenat, "Workers in the White City," 294–95.

30. "A Woman Fatally Burned: An Incident of a Baxter Street Tenement House Fire," *NYT*, November 22, 1891; Young interview. The 1888 match strike led by the girls and women working at the London Quaker firm of Bryant & May led to increasingly organized and successful unions of women match makers across the English industry. Over the next few decades, however, Ivar Kreuger consolidated power first in the Swedish and then in other European and American match industries, becoming known as the Match King. During the 1920s he was considered one of the wealthiest men in the world. By the early 1930s, when he became a supporter of Adolf Hitler, Kreuger owned 250 match factories in over forty countries, controlling three-fourths of the world's match trade. See Emsley, *13th Element*, 65–130.

31. Baldwin, *In the Watches of the Night*, 158–59; Covello and D'Agostino, *Heart Is the Teacher*, 46.

32. For more on the twentieth-century environmental history of copper and electrification, see LeCain, *Mass Destruction*.

Bibliography

MANUSCRIPTS AND SPECIAL COLLECTIONS

Massachusetts

    Houghton Library, Harvard University, Cambridge

        Boston Bijou Theatre Company Records, 1882–1927,

        Harvard Theatre Collection

        Daniel B. Fearing Logbook Collection

            Richard W. Hixson. "Journal of the Voyage of

            the 'Maria,'" 1832–1834. F 6870.53.20

    Nantucket Historical Association, Nantucket

        Christopher Mitchell & Co. Papers

        Ships' Log Collection

Montana

        Anaconda Copper Mining Company Records, 1895–1964,

          Montana Historical Society Archives, Helena

        Coroner's Inquest Reports, Office of the Clerk of the

        Court, Butte–Silver Bow Courthouse, Butte

North Carolina

        David M. Rubenstein Rare Book & Manuscript Library, Duke University, Durham

        James Redding Grist Business Records

        Jeremiah T. Jones Papers

            "Account Book," 1841–1843

          M. Jones, "Account Book, 1856–1859"

West Virginia

        Kanawha County Court Archives, West Virginia & Regional History

        Collection, West Virginia University Downtown Library, Morgantown

ORAL HISTORIES, INTERVIEWS, AND SONGS

Holden, Charles W. Interview. Library of Congress, Manuscript Division, WPA Federal
    Writers' Project Collection. http://hdl.loc.gov/loc.mss/wpalh3.34060718.

Mickelson, Herbert H. Interview by Laurie Mercier, January 12, 1982 (OH 225). Montana
    Historical Society Archives, Helena.

Patten, James, and Phyllis McLeod Patten. Interview by Laurie Mercier, February 9, 1983
    (OH 460). Montana Historical Society Archives, Helena.

Tour Guide. Underground Mine Tour, May 18, 2012. The World Museum of Mining, Butte, Montana.

Travis, Merle. "Dark as a Dungeon." Recorded August 8, 1946, Hollywood, California.

Young, Truman. Interview by Jeremy Zallen and Ken Krantz, October 16, 2010. Digital audio recording, Rutland, Vermont.

ONLINE ARCHIVES AND DATABASES

19th Century U.S. Newspapers. infotrac.galegroup.com.ezp-prod1.hul.harvard.edu.

"American Offshore Whaling Voyages: A Database." National Maritime Digital Library. nmdl.org/aowv/whsearch.cfm.

America's Historical Newspapers. infoweb.newsbank.com.

Digital Southern Historical Collection. dc.lib.unc.edu/cdm/archivalhome/collection /ead.

Google Books. books.google.com.

Hathi Trust. www.hathitrust.org.

Hitchcock, Tim, Robert Shoemaker, Clive Emsley, Sharon Howard, Jamie McLaughlin, et al. *The Old Bailey Proceedings Online, 1674–1913.* www.oldbaileyonline.org, version 7.1. April 2013.

Internet Archive. archive.org.

"Martha Ballard's Diary Online." *Do History.* dohistory.org/diary/index.html.

"North American Immigrant Letters, Diaries, and Oral Histories" database. solomon .imld.alexanderstreet.com.

North Carolina Digital Collections. digital.ncdcr.gov.

"On the Water." Online exhibition of the National Museum of American History. amhistory.si.edu/onthewater.

ProQuest Historical Newspapers. search.proquest.com.

State Archives of North Carolina's Photostream. www.flickr.com/photos/north-carolina -state-archives.

Westlaw. westlaw.com.

PUBLISHED MANUSCRIPT COLLECTIONS

Baltimore Life Insurance Company Records. In *Chesterfield County Virginia Uncovered: The Records of Death and Slave Insurance Records for the Coal Mining Industry, 1810–1895,* edited by Nancy C. Frantel, appendix 6, 175–80. Westminster, Md.: Heritage Books, 2008.

Berlin, Ira, and Barbara J. Fields, Thavolia Glymph, Joseph P. Reidy, and Leslie S. Rowland, eds. *Freedom: A Documentary History of Emancipation, 1861–1867.* Ser. 1, vol. 1, *The Destruction of Slavery.* New York: Cambridge University Press, 1985.

Clark, William Allen. Civil War correspondence. In "'Please Send Stamps': The Civil War Letters of William Allen Clark," pt. 2, by Margaret Black Tatum. *Indiana Magazine of History* 91, no. 2 (June 1995): 197–225.

Cooke, Chauncey H. "Letters of a Badger Boy in Blue: Into the Southland." *Wisconsin Magazine of History* 4, no. 3 (March 1921): 322–44.

Taylor, Isaac Lyman. Diary. In "Campaigning with the First Minnesota: A Civil War Diary," by Hazel C. Wolf. *Minnesota History* 25, no. 3 (September 1944): 224–57.

Thistlewood, Thomas. Diary. Reproduced in *In Miserable Slavery: Thomas Thistlewood in Jamaica, 1750–86*, edited by Douglas C. Hall. Kingston: University of the West Indies Press, 1999.

PERIODICALS

Advocate of Moral Reform and Family Guardian (New York)
All the Year Round (London)
American Farmer (Baltimore)
Arizona Mining Journal (Tucson)
Cincinnati Miscellany (Cincinnati)
DeBow's Review (New Orleans)
Farmer's Magazine (London)
Farmer's Register (Petersburg, Va.)
Frank Leslie's Illustrated Newspaper (New York)
Fraser's Magazine (London)
Harleian Miscellany (London)
Harper's New Monthly Magazine (New York)

Harper's Weekly (New York)
Household Words (London)
Journal of Geology (Chicago)
Louisiana Planter and Sugar Manufacturer (New Orleans)
New Monthly Magazine (London)
New York Daily Graphic (New York)
Oil, Paint and Drug Reporter (New York)
Prairie Farmer (Chicago)
Scientific American (New York)
Southern Cultivator (Augusta, Ga.)
Vanity Fair (New York)
Western Farmer and Gardener (Cincinnati)

NEWSPAPERS

Albany (N.Y.) Evening Journal
Alexandria (Va.) Gazette
Anti-Slavery Bugle (New Lisbon, Ohio)
Baltimore Gazette and Daily Advertiser
Barre (Mass.) Patriot
Boston Courier
Boston Daily Advertiser
Boston Daily Atlas
Boston Daily Globe
Boston Evening Transcript
Boston Herald
Boston Post
Boston Recorder
Bristol (U.K.) Mercury
Charlottesville (Va.) Advocate
Chicago Tribune
Cincinnati Commercial
Cincinnati Gazette
Cleveland Daily Herald
Cleveland Leader
Cleveland Plain Dealer
Connecticut Courant (Hartford)
Daily National Intelligencer (Washington, D.C.)

Daily Ohio Statesman (Columbus)
Davenport (Ohio) Gazette
Deseret News (Salt Lake City, Utah)
Fayetteville (N.C.) Observer
Gallipolis (Ohio) Journal
Goldsboro (N.C.) Patriot
Hartford Daily Courant
Indiana Democrat (Indianapolis)
Kalamazoo (Mich.) Gazette
Kanawha Valley Star (Charleston, Va. [W.Va.])
Lawrenceburg (Ind.) Register
London Daily News
Louisville Journal
Maine Farmer (Augusta, Me.)
McConnelsville (Ky.) Enquirer
Milwaukee Daily Sentinel
New Albany (Ind.) Daily Ledger
Newark (Ohio) Advocate
New-Bedford Mercury
Newburyport (Mass.) Herald
New Jersey State Gazette (Trenton)
New Orleans Daily Picayune
Newport (R.I.) Mercury

New York Commercial Advertiser
New York Daily Herald
New York Daily Times
New York Daily Tribune
New York Evening Post
New York Ledger
New York Spectator
New York Spirit of the Times
Niles (Mich.) Republican
North American and United States
Gazette (Philadelphia)
Philadelphia Daily Age
Philadelphia Examiner
Philadelphia Inquirer
Philadelphia Public Ledger
Pittsburgh Gazette
Plattsburgh (N.Y.) Republican
The Press (Philadelphia)
Public Ledger (Philadelphia)
Raleigh Register

Richmond Compiler
Richmond Daily Dispatch
Richmond Enquirer
Richmond Whig
Salem (Mass.) Gazette
Salem (Mass.) Register
San Francisco Daily Evening Bulletin
Southern Patriot (Charleston, S.C.)
Springfield (Mass.) Weekly Republican
Tarboro (N.C.) Southerner
Wabash Courier (Terre Haute, Ind.)
Washington (D.C.) Evening Star
Washington (N.C.) Dispatch
Whalemen's Shipping List, and Merchant's
Transcript (New Bedford, Mass.)
Wheeling (Va.) Intelligencer
Wilmington (N.C.) Chronicle
Wilmington (N.C.) Herald
Wilmington (N.C.) Journal

REPORTS AND PUBLISHED PRIMARY SOURCES

Barrett, J. P. *Electricity at the Columbian Exposition: Including an Account of the Exhibits.* Chicago: R. R. Donnelley & Sons, 1894.

Beecher, Catherine E. *A Treatise on Domestic Economy, for the Use of Young Ladies at Home, and at School.* Boston: Thomas H. Webb & Co., 1843.

Beecher, Catherine E., and Harriet Beecher Stowe. *The American Woman's Home.* New York: J. B. Ford and Co., 1869.

Boston City Council. *Report of a Committee of the Consumers of Gas, of the City of Boston, February, 1844.* Boston: John H. Eastburn, 1844.

———. *Report on the Erection of a Gasometer in Mason Street.* City Document, no. 39, July 19, 1852.

———. Special Committee on Gas Inspection. *Report of the Evidence and Other Matter Presented before a Joint Committee of the City Council of Boston upon the Subject of Gas.* Boston: Geo. C. Rand & Avery, 1867.

Boston Gas Light Company. *Trial. Boston Gas Light Company versus William Gault, containing the Arguments of Counsel, and the Charge of the Judge.* Boston: Eastburn's Press, 1848.

———. *Report of the Hearings before the Board of Mayor and Aldermen, upon the Remonstrances against the South Gasometer, and the Extension of the Works of the Gas Company at the North End.* Boston: Wright & Hasty, 1852.

Bristowe, Dr. "The Occupations Which Have to Do with Phosphorus." In *Public Health: Fifth Report of the Medical Officer of the Privy Council, with Appendix, 1862,* 162–205. Ordered by the House of Commons to be printed, April 14, 1863.

Brown, John. *Slave Life in Georgia: A Narrative of the Life, Sufferings, and Escape of John Brown, a Fugitive Slave, Now in England.* London: W. M. Watts, 1855.

Browne, John Ross. *Etchings of a Whaling Cruise: With Notes of a Sojourn on the Island of Zanzibar.* New York: Harper & Brothers, 1846.

Burr & Brother. *Solar Gas, made by a neat, simple, and economical apparatus, patented by James Crutchett, of Washington City, and which rapidly produces a cheap light of unequalled brilliancy, for lighting cities, blocks of buildings, churches, hotels, theatres, public halls, restaurants, steam-boats, mills, factories, and private residences.* New York: Jared W. Bell, 1848.

Capper, Charles. *The Port and Trade of London, Historical, Statistical, Local, and General.* London: Smith, Elder & Co., 1862.

Carey, Mathew. *Miscellaneous Essays.* Philadelphia: Carey & Hart, 1830.

Chicago Board of Trade. *Annual Statement.* 1864.

Children's Employment Commission. *Second Report of the Commissioners: Trades and Manufacturers: Presented to both Houses of Parliament by Command of her Majesty.* London: William Clowes and Sons, 1843.

Cincinnati Chamber of Commerce. *Annual Report of the Cincinnati Chamber of Commerce and Merchant's Exchange, for the Commercial Year Ending August 31, 1866.* Cincinnati: Gazette Steam Printing House, 1866.

Cist, Charles. *Sketches and Statistics of Cincinnati in 1851.* Cincinnati: Wm. H. Moore & Co., 1851.

———. *Sketches and Statistics of Cincinnati in 1859.* Cincinnati: n.p., 1859.

Clark, James. "History of an Aneurism of the Crural Artery, with Singular Circumstances." *Medical and Philosophical Commentaries* (Edinburgh, Scotland) 13 (1788): 326–43.

Clark v. Manufacturers' Ins. Co., Circuit Court, D, Massachusetts, May Term, 1847 (accessed through Westlaw, May 1, 2011): 2 Woodb. & M. 472, 5 F.Cas. 889, No. 2829.

Clegg, Samuel. *A Practical Treatise on the Manufacture and Distribution of Coal-Gas, its Introduction and Progressive Improvement.* 3rd ed. London: John Weale, 1859.

Couch v. Jones, 49 N.C. 402 (1857).

Covello, Leonard, and Guido D'Agostino. *The Heart Is the Teacher.* New York: McGraw-Hill, 1958.

Darwin, Charles. *The Voyage of the Beagle.* Vol. 29. Edited by Charles W. Elliot. New York: P. F. Collier & Son, 1909.

Davis, William M. *Nimrod of the Sea; or, The American Whaleman.* New York: Harper & Brothers, 1874.

de Beer, E. S. "The Early History of London Street-Lighting." *History* 25 (March 1941): 323–24.

de Crèvecoeur, J. Hector St. John. *Letters from an American Farmer.* 1782. Reprint, New York: Fox, Duffield & Co., 1904.

de Tegoborksi, M. L. *Commentaries on the Productive Forces of Russia.* Vol. 2. London: Longman, Brown, Green, and Longmans, 1856.

Dietz, Fred. *A Leaf from the Past, Dietz Then and Now.* New York: R. E. Dietz Co., 1914.

Dod, Santiago. "Stray Glimpses of the Cuban Sugar Industry," pt. 8. *Louisiana Planter and Sugar Manufacturer* 29 (August 9, 1902): 91–93.

Douglass, Frederick. *Narrative of the Life of Frederick Douglass, an American Slave.* 1845. Reprinted with an introduction by Kwame Anthony Appiah and notes and biographical notes by Joy Viveros in *Narrative of the Life of Frederick Douglass, an American Slave & Incidents in the Life of a Slave Girl.* New York: Modern Library, 2000.

Dupasquier, Alph. "Mémoire relatif aux effets des émanations phosphorées sur les ouvriers employés dans las fabriques de phosphore et les ateliers ou l'on prépare les allumettes chimiques." *Annales D'Hygiène Publique et de Médecine Légale*, 36 (July 1846): 342–58.

Enri, Henry. *Coal Oil and Petroleum: Their Origin, History, Geology, and Chemistry, with a View of Their Importance in Their Bearing upon National Industry*. Philadelphia: Henry Carey Baird, 1865.

Foster, George G. *New York by Gas-Light: With Here and There a Streak of Sunshine*. New York: Dewitt & Davenport, Tribune Buildings, 1850.

Freedley, Edwin T. *Philadelphia and Its Manufactures*. Philadelphia: Edward Young, 1858.

Gardner, Will. *Three Bricks and Three Brothers: The Story of the Nantucket Whale-Oil Merchant Joseph Starbuck*. Cambridge, Mass.: Riverside Press, 1945.

Glénard, A. *Sur la fabrication du phosphore et des allumettes phosphorées a Lyon*. Rapport au Conseil D'Hygiène Publique et de Salubrité. Lyon: imprimerie d'Aimé Vingtrinier, 1856.

Green, Invisible, Esq. "The Hog Drover's Visit; or, Bill Jenkins' First Impressions of the Queen City." In *Green Peas, Picked from the Patch of Invisible Green Esq.*, 136–39. Cincinnati: Moore, Wilstach, Keys & Overend, 1856.

Herrick, Rufus Frost. *Denatured or Industrial Alcohol*. New York: John Wiley & Sons, 1907.

Hinckley, Mary. *The Camphene Lamp; or, Touch Not, Taste Not, Handle Not*. Lowell: J. P. Walker, 1852.

House of Commons (Great Britain). *Abridgement of the Minutes of the Evidence, Taken before a Select Committee of the Whole House, to Whom it was Referred to Consider of the Slave-Trade*. No. 3. 1790.

———. *An Abstract of the Evidence Delivered before a Select Committee of the House of Commons, in the Years 1790 and 1791; on the Part of the Petitioners for the Abolition of the Slave Trade*. Published by the Society in Newcastle for Promoting the Abolition of the Slave-trade, 1791.

———. *Protectors of Slaves Reports*. June 12, 1829.

———. *Report from the Select Committee on the Extinction of Slavery throughout the British Dominions*. August 11, 1832.

———. *Class B., Correspondence with British Ministers and Agents in Foreign Countries, and with Foreign Ministers in England, relating to the Slave Trade, From April 1, 1857, to March 31, 1858, Presented to both Houses of Parliament by Command of Her Majesty*. London: Harrison and Sons, 1858.

———. *Report from the Select Committee on Fire Protection*. July 25, 1867.

Howe, Henry. *Historical Collections of Virginia*. Charleston, S.C.: Babcock & Co., 1845.

Hutchinson, Thomas. *The History of the Province of Massachusets-Bay*. Vol. 2. London: J. Smith, 1768.

Illinois. State Legislature. Drovers, Ill., Rev. Stat. 1845, Chap. 35 (passed February 27, 1845).

Jacobs, Harriet Ann. *Incidents in the Life of a Slave Girl: Written by Herself*. Originally published under the name Linda Brent in Boston, 1861. Reprinted with an introduction by Kwame Anthony Appiah and notes and biographical notes by Joy Viveros in *Narrative of the Life of Frederick Douglass, an American Slave & Incidents in the Life of a Slave Girl*. New York: Modern Library, 2000.

Kohl, J. G. *Foreign Library, No. 1: Russia and the Russians, in 1842*. Philadelphia: Carey and Hart, 1843.

———. *Russia: St. Petersburg, Moscow, Kharkoff, Riga, Odessa, the German Provinces on the Baltic, the Steppes, the Crimea, and the Interior of the Empire*. London: Chapman and Hall, 1844.

Latham, Wilfred. *The States of the River Plate: Their Industries and Commerce*. London: Longmans, Green, and Co., 1866.

Long, Edward. *The History of Jamaica*. Vol. 1. London: T. Lowndes, 1774.

——. *The History of Jamaica*. Vol. 2. London: T. Lowndes, 1774.

Mansfield, C. B. *Paraguay, Brazil, and the Plate: Letters Written in 1852–1853*. Cambridge: Macmillan, 1856.

Marx, Karl. *Capital: A Critique of Political Economy*. Vol. 1. 1867. Translated by Ben Fowkes and introduced by Ernest Mandel. New York: Penguin, 1976.

——. *Early Writings*. London: Penguin, 1992.

Mason v. Cowan's Administrator, 1 B. Mon. 7 (Ky. Ct. App. Oct. 14, 1840).

Maury, M. F., Jr. *The Resources of the Coal Field of the Upper Kanawha*. Baltimore: Sherwood & Co., 1873.

Maxwell v. Eason, Supreme Court of Alabama, July 1828 (accessed through Westlaw, May 1, 2011): 1 Stew. 514, 1828 WL 500 (Ala.).

Mayhew, Henry. *London Labour and the London Poor: The Condition and Earnings of Those That Will Work, Cannot Work, and Will Not Work*. Vol. 1, *London Street-Folk*. London: Charles Griffin and Co., 1864.

Melville, Herman. *Moby-Dick; or, The Whale*. 1851. New York: Signet Classic, 1961.

Meriam, E. "Deaths and Injuries from the Use of Camphene." In *The New-York Almanac and Weather Book for the Year 1857*, 114–22. New York: Mason Brothers, 1857.

Mid-Lothian Coal Mining Company. *Charter, scheme, and conditions of subscriptions of the Mid Lothian Coal Mining Company*. Richmond: S. Sheperd, 1835. In *Chesterfield County Virginia Uncovered: The Records of Death and Slave Insurance Records for the Coal Mining Industry, 1810–1895*, edited by Nancy C. Frantel, appendix 2, 123–42. Westminster, Md.: Heritage Books, 2008.

Monfalcon, J. B., and A. P. I. de Polinière. *Hygiène de la Ville de Lyon, ou Opinions et Rapports du Conseil de Salubrité du Département du Rhône*. Paris: J. B. Baillière, Librairie de l'Académie Royale de Médecine, 1845.

The New Orleans Gas Light and Banking Company v. George R. Botts, 9 La. 305 (1844).

Northup, Solomon. *Twelve Years a Slave: Narrative of Solomon Northup, a Citizen of New-York, Kidnapped in Washington City in 1841, and Rescued in 1853, from a Cotton Plantation Near the Red River in Louisiana*. Auburn: Derby and Miller, 1853.

Olmsted, Frederick Law. *A Journey in the Seaboard Slave States, with Remarks on Their Economy*. New York: Dix & Edwards, 1856.

——. *A Journey through Texas*. New York: Dix, Edwards & Co., 1857.

Orpen, Adelea Elizabeth Richards. *Memories of Old Emigrant Days in Kansas, 1861–1865*. London: William Blackwood and Sons, 1926.

Parnell, Edward Andrew, ed. *Applied Chemistry: In Manufactures, Arts, and Domestic Economy*. New York: D. Appleton & Co., 1844.

Parton, James. *General Butler in New Orleans. History of the Administration of the Department of the Gulf in the Year 1862: With an Account of the Capture of New Orleans, and a Sketch of the Previous Career of the General, Civil and Military*. New York: Mason Brothers, 1864.

Penny, Virginia. *The Employments of Women: A Cyclopaedia of Woman's Work*. Boston: Walker, Wise, & Co., 1863.

Perry, G. W. *A Treatise on Turpentine Farming*. New Bern, N.C.: Muse & Davies, 1859.

Philadelphia Common Council. *Journal of the Common Council of the Consolidated City of*

Philadelphia, beginning November 1, 1855, ending May 8, 1856. Vol. 4. Philadelphia: Wm. H Sickels, 1856.

Platt, O. H., of Connecticut, U.S. Senator. "Invention and Advancement." Address delivered to Congress, April 1891. Published by the Executive Committee in *Proceedings and Addresses: Celebration of the Beginning of the Second Century of the American Patent System at Washington City, D.C., April 8, 9, 10, 1891,* 57–76. Washington: Gedney & Roberts Co., 1892.

Ramsay, James. *An Essay on the Treatment and Conversion of African Slaves in the British Sugar Colonies.* London: James Phillips, 1784.

Richardson, Thomas, and Henry Watts. *Complete Practical Treatise on Acids, Alkalies, and Salts; Their Manufacture and Application.* 2nd ed., vol. 2. London: H. Baillière, 1867.

Ripley, George, and Charles A. Dana, eds. *The New American Cyclopaedia: A Popular Dictionary of General Knowledge.* Vol. 3, *Beam-Browning.* New York: D. Appleton and Co, 1859.

Robinson, Solon, ed. *Facts for Farmers.* New York: Johnson & Ward, 1865.

Roughley, Thomas. *The Jamaica Planter's Guide; or, A System for Planting and Managing a Sugar Estate, or Other Plantations in the Island, and throughout the British West Indies in General.* London: Longman, Hurst, Rees, Orme, and Brown, 1823.

Roy, Andrew. *The Coal Mines.* Cleveland: Robison, Savage & Co., 1876.

Seymour, Richard Arthur. *Pioneering in the Pampas; or, the First Four Years of a Settler's Experience in the La Plata Camps.* 2nd ed. London: Longmans, Green, and Co., 1870.

Shattuck, C. H. "Coal-Oil in West Virginia." In U.S. Dept. of Agriculture, *Report of the Commission of Agriculture for the Year 1863,* 525–29. Washington, D.C.: Government Printing Office, 1863.

Simonin, Louis. *Underground Life; or, Mines and Miners.* Translated and edited by H. W. Bristow. London: Chapman & Hall, 1869.

Six Hundred Dollars a Year. A Wife's Effort at Low Living, Under High Prices. Boston: Ticknor and Fields, 1867.

Society for the Reformation of Juvenile Delinquents, in the City and State of New York. *Documents Relative to the House of Refuge, Instituted by the Society for the Reformation of Juvenile Delinquents in the City of New-York, in 1828.* New York: M. Day, 1832.

———. Annual Reports of the Managers (1839–1851). New York: Mahlon Day.

Starbuck, Alexander. *History of the American Whale Fishery from Its Earliest Inception to the Year 1876.* Waltham, Mass.: Alexander Starbuck, 1878.

State v. Sandy, 25 N.C. 570 (1843).

Stephen, James. *The Slavery of the British West India Colonies.* Vol. 2. London: Saunders and Benning, 1830.

Tardieu, Ambroise. "Étude hygiénique et médico-légale sur la fabrication et l'emploi des allumettes chimiques." *Annales D'Hygiène Publique et de Médecine Légale,* ser. 2, vol. 6 (July 1856): 5–54.

Taylor, R. C. *Statistics of Coal: Including Mineral Bituminous Substances Employed in Arts and Manufactures; with their Geographical, Geological, and Commercial Distribution, and Amount of Production and Consumption on the American Continent.* 2nd ed. Philadelphia: J. W. Moore, 1855.

Thompson, John. *The Life of John Thompson, a Fugitive Slave; Containing His History of 25 Years in Bondage, and His Providential Escape: Written by Himself.* Worcester, Mass.: John Thompson, 1856.

Tomlinson, Charles, ed. *Cyclopaedia of Useful Arts, Mechanical and Chemical, Manufactures, Mining, and Engineering.* Vol. 2, *Hammer to Zirconium.* London: George Virtue & Co., 1854.

U.S. Army. *Report of the Quartermaster General of the United States Army to the Secretary of War for the Year Ending June 30, 1865.* Washington, D.C.: Government Printing Office, 1865.

U.S. Congress. "Reduction, since 1860, in the Production and Consumption of Distilled Spirits in the United States." In *Reports of a Commission Appointed for a Revision of the Revenue System of the United States, 1865–'66,* 2:161–67. Washington, D.C.: Government Printing Office, 1866.

———. *Reports of a Commission Appointed for a Revision of the Revenue System of the United States, 1865–'66.* Washington, D.C.: Government Printing Office, 1866.

———. *Alcohol in the Manufactures and Arts.* S. Rep. No. 411–55 (1897).

U.S. Department of State. "The Sovereignty of Guano Islands in the Caribbean Sea." Vol. 2 of "The Sovereignty of Islands Claimed under the Guano Act and of the Northwest Hawaiian Islands Midway and Wake." Unpublished typescript ms., 3 vols., 1932–1933 (continuously paginated).

U.S. War Department. *The War of the Rebellion: A Compilation of the Official Records of the Union and Confederate Armies.* Ser. 1, vols. 14, 29, 35, 47, 49. Washington, D.C.: Government Printing Office, 1885–97.

Washington, Booker T. *Up from Slavery: An Autobiography.* New York: Doubleday, 1907.

White, John Edward. "The Lucifer Match Manufacture." In *Children's Employment Commission (1862), First Report of the Commissioners, With Appendix, Presented to both Houses of Parliament by Command of Her Majesty,* xlviii–lvi, 41–104. London: George Edward Eyre and William Spottiswoode, 1863.

White, Walter. *All Round the Wrekin.* London: Chapman and Hall, 1860.

Wilson, John. *New York Industrial Exhibition: Special Report, presented to the House of Commons by Command of Her Majesty, in pursuance of their address of February 6th, 1854.* London: Harrison and Sons, 1854.

SECONDARY SOURCES

Adams, Branden C. "Dark as a Dungeon: Work Inside Nineteenth-Century Coal Mines." Senior thesis, Harvard University, 2011.

Adams, Sean Patrick. *Old Dominion, Industrial Commonwealth: Coal, Politics, and Economy in Antebellum America.* Baltimore: Johns Hopkins University Press, 2004.

———. *Home Fires: How Americans Kept Warm in the Nineteenth Century.* Baltimore: Johns Hopkins University Press, 2014.

Anderson, Jennifer L. *Mahogany: The Costs of Luxury in Early America.* Cambridge, Mass.: Harvard University Press, 2012.

Andrews, Thomas G. *Killing for Coal: America's Deadliest Labor War.* Cambridge, Mass.: Harvard University Press, 2008.

———. "Beasts of the Southern Wild: Slaveholders, Slaves, and Other Animals in Charles Ball's *Slavery in the United States.*" In *Animals, Bodies, Places, Politics,* edited by Marguerite S. Shaffer and Phoebe S. K. Young, 21–47. Philadelphia: University of Pennsylvania Press, 2015.

Arms, Richard G. "From Disassembly to Assembly: Cincinnati: The Birthplace of Mass-

Production." *Bulletin of the Historical and Philosophical Society of Cincinnati* 17 (1959): 198–200.

Bailey, Ronald. "The Slave(ry) Trade and the Development of Capitalism in the United States: The Textile Industry in New England." *Social Science History* 14 (Autumn 1990): 373–414.

———. "The Other Side of Slavery: Black Labor, Cotton, and Textile Industrialization in Great Britain and the United States." *Agricultural History* 68 (Spring 1994): 35–50.

Baldwin, Peter C. *In the Watches of the Night: Life in the Nocturnal City, 1820–1930*. Chicago: University of Chicago Press, 2012.

Baptist, Edward E. "Toxic Debt, Liar Loans, Collateralized and Securitized Human Beings, and the Panic of 1837." In *Capitalism Takes Command: The Social Transformation of Nineteenth-Century America*, edited by Michael Zakim and Gary J. Kornblith, 69–92, 304–8. Chicago: University of Chicago Press, 2012.

Beattie, J. M. *Policing and Punishment in London, 1660–1750*. New York: Oxford University Press, 2001.

Beckert, Sven. *The Monied Metropolis: New York City and the Consolidation of the American Bourgeoisie, 1850–1896*. New York: Cambridge University Press, 2001.

———. *Empire of Cotton: A Global History*. New York: Vintage, 2014.

Beckert, Sven, and Seth Rockman, eds. *Slavery's Capitalism: A New History of American Economic Development*. Philadelphia: University of Pennsylvania Press, 2016.

Bell, Stephen. "Early Industrialization in the South Atlantic: Political Influences on the *charqueadas* of Rio Grande do Sul before 1860." *Journal of Historical Geography* 19, no. 4 (1993): 399–411.

Black, Brian. *Petrolia: The Landscape of America's First Oil Boom*. Baltimore: Johns Hopkins University Press, 2000.

Blackbourn, David. *The Conquest of Nature: Water, Landscape, and the Making of Modern Germany*. New York: Norton, 2006.

Blackmar, Elizabeth. *Manhattan for Rent, 1785–1850*. Ithaca, N.Y.: Cornell University Press, 1989.

Bolster, W. Jeffrey. "'To Feel like a Man': Black Seamen in the Northern States, 1800–1860." *Journal of American History* 76 (March 1990): 1173–99.

———. *Black Jacks: African American Seamen in the Age of Sail*. Cambridge, Mass.: Harvard University Press, 1997.

Borucki, Alex, Karla Chagas, and Natalia Stalla. *Esclavitud y Trabajo: Un Estudio sobre los Afrodescendientes en la Frontera Uruguaya (1835–1855)*. Montevideo, Uruguay: Pulmón Ediciones, 2004.

Bounds, Harvey. "Wilmington Match Companies." *Delaware History* 10, no. 1 (April 1962): 3–32.

Bowers, Brian. *Lengthening the Day: A History of Lighting Technology*. New York: Oxford University Press, 1998.

Boydston, Jeanne. *Home and Work: Housework, Wages, and the Ideology of Labor in the Early Republic*. New York: Oxford University Press, 1990.

Breyfogle, Nicholas B., Abby Schrader, and Willard Sunderland, eds. *Peopling the Russian Periphery: Borderland Colonization in Eurasian History*. New York: Routledge, 2007.

Brown, Jonathan C. *A Socioeconomic History of Argentina, 1776–1860*. New York: Cambridge University Press, 1979.

———. *A Brief History of Argentina*. 2nd ed. New York: Facts on File, 2010.

Brown, Vincent. "Eating the Dead: Consumption and Regeneration in the History of Sugar." *Food and Foodways* 16, no. 2 (April 2008): 117–26.

———. *The Reaper's Garden: Death and Power in the World of Atlantic Slavery*. Cambridge, Mass.: Harvard University Press, 2008.

Brox, Jane. *Brilliant: The Evolution of Artificial Light*. Boston: Mariner Books, 2010.

Bruce, Kathleen. *Virginia Iron Manufacture in the Slave Era*. New York: Century Co., 1931.

Bruns, Thomas. "'. . . the whole river is a bustle some about their Children, Brothers and Husbands and the rest of us about our salt.': The Antebellum Industrialization of the Kanawha Valley in the Virginia Backcountry." Master's thesis, Western Carolina University, 2013.

Burkett, Paul, and John Bellamy Foster. "Metabolism, Energy, and Entropy in Marx's Critique of Political Economy: Beyond the Podolinsky Myth." *Theory and Society* 35 (February 2006): 109–56.

Burnett, Edmund Cody. "Hog Raising and Hog Driving in the Region of the French Broad River." *Agricultural History* 20, no. 2 (April 1946): 86–103.

Burnett, Graham. *The Sounding of the Whale: Science and Cetaceans in the Twentieth Century*. Chicago: University of Chicago Press, 2012.

Calvert, Jerry W. *The Gibraltar: Socialism and Labor in Butte, Montana, 1895–1920*. Helena: Montana Historical Society Press, 1988.

Camp, Stephanie M. H. *Closer to Freedom: Enslaved Women and Everyday Resistance in the Plantation South*. Chapel Hill: University of North Carolina Press, 2004.

Cecelski, David S. *The Waterman's Song: Slavery and Freedom in Maritime North Carolina*. Chapel Hill: University of North Carolina Press, 2001.

Césaire, Aimé. *Discourse on Colonialism*. Translated by Joan Pinkham. New York: Monthly Review Press, 1972.

Chakrabarty, Dipesh. *Provincializing Europe: Political Thought and Historical Difference*. Princeton, N.J.: Princeton University Press, 2000.

Conley, Phil. "Early Coal Development in the Kanawha Valley." *West Virginia History* 8 (January 1947): 206–15.

Cowan, Brian. *The Social Life of Coffee: The Emergence of the British Coffeehouse*. New Haven: Yale University Press, 2005.

Crary, Jonathan. "Techniques of the Observer." *October* 45 (Summer 1988): 3–35.

Cronon, William. *Changes in the Land: Indians, Colonists, and the Ecology of New England*. New York: Hill and Wang, 1983, 2003.

———. "Revisiting the Vanishing Frontier: The Legacy of Frederick Jackson Turner." *Western History Quarterly* 18 (April 1987): 157–76.

———. *Nature's Metropolis: Chicago and the Great West*. New York: Norton, 1991.

Crosby, Alfred W. *Children of the Sun: A History of Humanity's Unappeasable Appetite for Energy*. New York: Norton, 2006.

Cuff, Timothy. "A Weighty Issue Revisited: New Evidence on Commercial Swine Weights and Pork Production in Mid-Nineteenth Century America." *Agricultural History* 66 (Autumn 1992): 55–74.

Curtis, Kent A. *Gambling on Ore: The Nature of Metal Mining in the United States, 1860–1910*. Boulder: University Press of Colorado, 2013.

Cushman, Gregory T. *Guano and the Opening of the Pacific World: A Global Ecological History*. New York: Cambridge University Press, 2013.

Davis, Lance E., Robert E. Gallman, and Karin Gleiter. *In Pursuit of Leviathan: Technology,*

Institutions, Productivity, and Profits in American Whaling, 1816–1906. Chicago: University of Chicago Press, 1997.

Davis, Mike. *Late Victorian Holocausts: El Niño Famines and the Making of the Third World*. New York: Verso, 2001.

de Boer, Tycho. "The Corporate Forest: Capitalism and Environmental Change in Southeastern North Carolina's Longleaf Pine Belt, 1790–1940." Ph.D. diss., Vanderbilt University, 2002.

———. *Nature, Business, and Community in North Carolina's Green Swamp*. Gainesville: University Press of Florida, 2008.

de la Peña, Carolyn Thomas. *The Body Electric: How Strange Machines Built the Modern American*. New York: New York University Press, 2003.

Dening, Greg. *Mr. Bligh's Bad Language: Passion, Power, and Theatre on the Bounty*. New York: Cambridge University Press, 1992.

Dillon, Maureen. *Artificial Sunshine: A Social History of Domestic Lighting*. London: National Trust, 2002.

Dolin, Eric Jay. *Leviathan: The History of Whaling in America*. New York: Norton, 2007.

Du Bois, W. E. B. *Black Reconstruction in America, 1860–1880*. 1935, 1962. Reprinted with an introduction by David Levering Lewis, New York: Free Press, 1992.

Edgerton, David. *The Shock of the Old: Technology and Global History since 1900*. New York: Oxford University Press, 2006.

Egnal, Marc. *Divergent Paths: How Culture and Institutions Have Shaped North American Growth*. New York: Oxford University Press, 1996.

Einhorn, Robin L. *American Taxation, American Slavery*. Chicago: University of Chicago Press, 2006.

Ekirch, A. Roger. *At Day's Close: Night in Times Past*. New York: Norton, 2005.

Ellis, Richard. *Men and Whales*. New York: Knopf, 1991.

———. *Monsters of the Sea*. New York: Knopf, 1995.

Emmons, David M. *The Butte Irish: Class and Ethnicity in an American Mining Town, 1875–1925*. Urbana: University of Illinois Press, 1989.

Emsley, John. *The 13th Element: The Sordid Tale of Murder, Fire, and Phosphorus*. New York: John Wiley & Sons, 2000.

Essig, Mark. *Edison and the Electric Chair: A Story of Light and Death*. New York: Walker & Co., 2003.

Faust, Drew Gilpin. *This Republic of Suffering: Death and the American Civil War*. New York: Knopf, 2008.

Fields, Karen E., and Barbara J. Fields. *Racecraft: The Soul of Inequality in American Life*. New York: Verso, 2012.

Finn, Janet L. *Tracing the Veins: Of Copper, Culture, and Community from Butte to Chuquicamata*. Berkeley: University of California Press, 1998.

Fischer, John Ryan. *Cattle Colonialism: An Environmental History of the Conquest of California and Hawai'i*. Chapel Hill: University of North Carolina Press, 2015.

Fleming, William Franklin. "Ohio Columbus Barber, 1841–1920, an American Industrialist." Ph.D. thesis, Case Western Reserve University, 1977.

Follett, Richard. *The Sugar Masters: Planters and Slaves in Louisiana's Cane World, 1820–1860*. Baton Rouge: Louisiana State University Press, 2005.

Ford, Lacy K., Jr. *The Origins of Southern Radicalism: The South Carolina Upcountry, 1800–1860*. New York: Oxford University Press, 1988.

Foster, John Bellamy. *Marx's Ecology: Materialism and Nature.* New York: Monthly Review Press, 2000.

Foster, Mark. "The Hadwen & Barney Candle Factory: How It Worked." *Historic Nantucket* 59 (Fall 2009): 4–9.

Frank, Alison Fleig. *Oil Empire: Visions of Prosperity in Austrian Galicia.* Cambridge, Mass.: Harvard University Press, 2005.

Frantel, Nancy C. *Chesterfield County Virginia Uncovered: The Records of Death and Slave Insurance Records for the Coal Mining Industry, 1810–1895.* Westminster, Md.: Heritage Books, 2008.

Freeberg, Ernest. *The Age of Edison: Electric Light and the Invention of Modern America.* New York: Penguin, 2013.

Garavaglia, Juan Carlos, and Jorge D. Gelman. "Rural History of the Río de la Plata, 1600–1850: Results of a Historiographical Renaissance." *Latin American Research Review* 30, no. 3 (1995): 75–105.

Genovese, Eugene D. "Livestock in the Slave Economy of the Old South—A Revised View." *Agricultural History* 36 (July 1962): 143–49.

———. *The World the Slaveholders Made: Two Essays in Interpretation.* Middletown, Conn.: Wesleyan University Press, 1988.

———. *The Slaveholders' Dilemma: Freedom and Progress in Southern Conservative Thought, 1820–1860.* Columbia: University of South Carolina Press, 1992.

Glymph, Thavolia. *Out of the House of Bondage: The Transformation of the Plantation Household.* New York: Cambridge University Press, 2008.

Graff, Rebecca S. "Dream City, Plaster City: Worlds' Fairs and the Gilding of American Material Culture." *International Journal of Historical Archaeology* 16 (December 2012): 696–716.

Grandin, Greg. *The Empire of Necessity: Slavery, Freedom, and Deception in the New World.* New York: Metropolitan Books, 2014.

Greene, Ann Norton. "War Horses: Equine Technology in the American Civil War." In *Industrializing Organisms: Introducing Evolutionary History,* edited by Susan R. Schrepfer and Philip Scranton, 143–65. New York: Routledge, 2004.

———. *Horses at Work: Harnessing Power in Industrial America.* Cambridge, Mass.: Harvard University Press, 2008.

Gutman, Herbert G. *Work, Culture, and Society in Industrializing America.* New York: Vintage, 1976.

Hahn, Steven. *The Roots of Southern Populism: Yeoman Farmers and the Transformation of the Georgia Upcountry, 1850–1890.* 1983. Updated ed., New York: Oxford University Press, 2006.

———. *The Political Worlds of Slavery and Freedom.* Cambridge, Mass.: Harvard University Press, 2009.

———. *A Nation without Borders: The United States and Its World in an Age of Civil Wars, 1830–1910.* New York: Viking, 2016.

Hall, Douglas C., ed. *In Miserable Slavery: Thomas Thistlewood in Jamaica, 1750–86.* Kingston: University of the West Indies Press, 1999.

Harvey, David. *Justice, Nature, and the Geography of Difference.* Cambridge, Mass: Blackwell, 1996.

Hilliard, Sam B. "Pork in the Ante-Bellum South: The Geography of Self-Sufficiency." *Annals of the Association of American Geographers* 59 (September 1969): 461–80.

Hobsbawm, E. J. *Industry and Empire: The Birth of the Industrial Revolution*. 1968. Reprint, New York: New Press, 1999.

Howard, Mark. "Coopers and Casks in the Whaling Trade, 1800–1850." *Mariner's Mirror* 82 (November 1996): 436–50.

Huang, Nian-Sheng. "Franklin's Father Josiah: Life of a Colonial Boston Tallow Chandler, 1657–1745." *Transactions of the American Philosophical Society* 90, no. 3 (2000): 1–155.

Hughes, Thomas P. *Networks of Power: Electrification in Western Society, 1880–1930*. Baltimore: Johns Hopkins University Press, 1993.

———. *Human-Built World: How to Think about Technology and Culture*. Chicago: University of Chicago Press, 2005.

Hunter, Louis C. *Studies in the Economic History of the Ohio Valley; Seasonal Aspects of Industry and Commerce before the Age of Big Business: The Beginnings of Industrial Combination*. Northampton, Mass.: Department of History of Smith College, 1934.

Hymer, Stephen. "Robinson Crusoe and the Secret of Primitive Accumulation." *Monthly Review*, September 1971. Reprinted September 2011.

Jackson, Jeremy B. "When Ecological Pyramids Were Upside Down." In *Whales, Whaling, and Ocean Ecosystems*, edited by J. A. Estes, D. P. Demaster, D. F. Doak, T. M. Williams, and R. L. Brownell, 27–37. Berkeley: University of California Press, 2006.

Jakle, John A. *City Lights: Illuminating the American Night*. Baltimore: Johns Hopkins University Press, 2001.

James, C. L. R. *Mariners, Renegades, and Castaways: The Story of Herman Melville and the World We Live In*. 1953, 1978. Complete text with an introduction by Donald E. Pease, Hanover, N.H.: Dartmouth College Press, 2001.

Jameson, Elizabeth. *All That Glitters: Class, Conflict, and Community in Cripple Creek*. Urbana: University of Illinois Press, 1998.

Johnson, Walter. *Soul by Soul: Life inside the Antebellum Slave Market*. Cambridge, Mass.: Harvard University Press, 1999.

———. "On Agency." *Journal of Social History* 37 (Autumn 2003): 113–24.

———. "The Pedestal and the Veil: Rethinking the Capitalism/Slavery Question." *Journal of the Early Republic* 24 (Summer 2004): 299–308.

———. *The Chattel Principle: Internal Slave Trades in the Americas*. New Haven: Yale University Press, 2004.

———. *River of Dark Dreams: Slavery and Empire in the Cotton Kingdom*. Cambridge, Mass.: Harvard University Press, 2013.

Jones, Christopher F. "Energy Landscapes: Coal Canals, Oil Pipelines, Electricity Transmission Wires in the Mid-Atlantic, 1820–1930." Ph.D. diss., University of Pennsylvania, 2009.

———. *Routes of Power: Energy and Modern America*. Cambridge, Mass.: Harvard University Press, 2014.

Jones, Robert Leslie. *History of Agriculture in Ohio to 1880*. Kent, Ohio: Kent State University Press, 1983.

Jonnes, Jill. *Empires of Light: Edison, Tesla, Westinghouse, and the Race to Electrify the World*. New York: Random House, 2003.

Joyce, Patrick. *The Rule of Freedom: Liberalism and the Modern City*. New York: Verso, 2003.

Kaye, Anthony E. "The Second Slavery: Modernity in the Nineteenth-Century South and the Atlantic World." *Journal of Southern History* 75 (August 2009): 627–50.

Kelley, Robin D. G. *Freedom Dreams: The Black Radical Imagination*. Boston: Beacon, 2002.

Kern, Stephen. *The Culture of Time and Space, 1880–1918*. Cambridge, Mass.: Harvard University Press, 1983.

Khodarkovsky, Michael. *Russia's Steppe Frontier: The Making of a Colonial Empire, 1500–1800*. Bloomington: Indiana University Press, 2002.

Kopytoff, Igor. "The Cultural Biography of Things." In *The Social Life of Things: Commodities in Cultural Perspective*, edited by Arjun Appadurai, 64–91. New York: Cambridge University Press, 1986.

Koslofsky, Craig. *Evening's Empire: A History of the Night in Early Modern Europe*. New York: Cambridge University Press, 2011.

Kugler, Richard C. *The Whale Oil Trade, 1750–1775*. Old Dartmouth Historical Sketch 79. New Bedford: Colonial Society of Massachusetts, 1980.

Kurlansky, Mark. *Salt: A World History*. New York: Penguin, 2002.

Laing, James T. "The Early Development of the Coal Industry in the Western Counties of Virginia." *West Virginia History* 27 (January 1966): 144–55.

Langston, Nancy. *Toxic Bodies: Hormone Disruptors and the Legacy of DES*. New Haven: Yale University Press, 2010.

LeCain, Timothy J. *Mass Destruction: The Men and Giant Mines That Wired America and Scarred the Planet*. New Brunswick, N.J.: Rutgers University Press, 2009.

Lefebvre, Henri. *The Production of Space*. Translated by Donald Nicholson-Smith. Cambridge, Mass.: Blackwell, 1991.

Le Guin, Ursula K. "The Day before the Revolution." In *The Wind's Twelve Quarters: Stories*, 285–303. New York: Harper & Row, 1975.

Lewis, Ronald L. *Coal, Iron, and Slaves: Industrial Slavery in Maryland and Virginia, 1715–1865*. Contributions in Labor History, Number 6. Westport, Conn.: Greenwood Press, 1979.

———. *Black Coal Miners in America: Race, Class, and Community Conflict, 1780–1980*. Lexington: University Press of Kentucky, 1987.

Linebaugh, Peter. *The London Hanged: Crime and Civil Society in the Eighteenth Century*. 2nd ed. London: Verso, 2003.

Linebaugh, Peter, and Marcus Rediker. *The Many-Headed Hydra: Sailors, Slaves, Commoners, and the Hidden History of the Revolutionary Atlantic*. Boston: Beacon, 2000.

Lingenfelter, Richard E. *The Hard Rock Miners: A History of the Mining Labor Movement in the American West, 1863–1893*. Berkeley: University of California Press, 1974.

Lipman, Andrew. *The Saltwater Frontier: Indians and the Contest for the American Coast*. New Haven: Yale University Press, 2015.

Long, Priscilla. *Where the Sun Never Shines: A History of America's Bloody Coal Industry*. New York: Paragon House, 1989.

Lovejoy, Paul E. "Slavery and 'Legitimate Trade' on the West African Coast." In *Transformations in Slavery: A History of Slavery in Africa*, 160–84. New York: Cambridge University Press, 2012.

Luckiesh, Matthew. *Torch of Civilization: The Story of Man's Conquest of Darkness*. New York: G. P. Putnam's Sons, 1940.

Majewski, John. *Modernizing a Slave Economy: The Economic Vision of the Confederate Nation*. Chapel Hill: University of North Carolina Press, 2009.

Malone, Michael P. *The Battle for Butte: Mining and Politics on the Northern Frontier, 1864–1906*. Helena: Montana Historical Society Press, 1981.

Manning, Chandra. *What This Cruel War Was Over: Soldiers, Slavery, and the Civil War*. New York: Knopf, 2007.

———. *Troubled Refuge: Struggling for Freedom in the Civil War*. New York: Vintage, 2016.

Manuel, Jeffrey T. "The Original Bridge Fuel: Camphene Lighting in Antebellum America and the Origins of the Oil and Gas Energy Regime." Paper presented at Society for the History of Technology annual meeting, Philadelphia, October 2017.

Massachusetts Historical Commission. "Mann's Cotton Mill Double Worker Housing." MHC Inventory Form B, nos. 97, 98, 99, 100. Town of Sharon, July 2008. http://mhc-macris.net.

McCall, Ronald J. "Never Quite Settled: Southern Plain Folk on the Move." Master's thesis, East Tennessee State University, 2013.

McCardell, James. *The Idea of a Southern Nation: Southern Nationalists and Southern Nationalism, 1830–1860*. New York: Norton, 1979.

McCurry, Stephanie. *Masters of Small Worlds: Yeoman Households, Gender Relations, and the Political Culture of the Antebellum South Carolina Low Country*. New York: Oxford University Press, 1995.

———. *Confederate Reckoning: Power and Politics in the Civil War South*. Cambridge, Mass.: Harvard University Press, 2010.

McDonald, Forrest, and Grady McWhiney. "The Antebellum Southern Herdsman: A Reinterpretation." *Journal of Southern History* 41 (May 1975): 147–66.

McNeil, J. R. *Something New under the Sun: An Environmental History of the Twentieth-Century World*. New York: Norton, 2000.

McNeur, Catherine. *Taming Manhattan: Environmental Battles in the Antebellum City*. Cambridge, Mass.: Harvard University Press, 2014.

McPherson, James M. *Drawn with the Sword*. New York: Oxford University Press, 1996.

McShane, Clay, and Joel A. Tarr. *The Horse in the City: Living Machines in the Nineteenth Century*. Baltimore: Johns Hopkins University Press, 2007.

McWhiney, Grady. *Cracker Culture: Celtic Ways in the Old South*. Tuscaloosa: University of Alabama Press, 1988.

McWhiney, Grady, and Forrest McDonald. "Celtic Origins of Southern Herding Practices." *Journal of Southern History* 51 (May 1985): 165–82.

Meinig, D. W. *The Shaping of America: A Geographical Perspective on 500 Years of History*. Vol. 2, *Continental America, 1800–1867*. New Haven: Yale University Press, 1993.

Melbin, Murray. *Night as Frontier: Colonizing the World after Dark*. New York: Free Press, 1987.

Melosi, Martin. *The Sanitary City: Environmental Services in Urban America from Colonial Times to the Present*. Abridged ed. Pittsburgh: University of Pittsburgh Press, 2008.

Mercier, Laurie. *Anaconda: Labor, Community, and Culture in Montana's Smelter City*. Urbana: University of Illinois Press, 2001.

Mintz, Sidney W. *Sweetness and Power: The Place of Sugar in Modern History*. New York: Penguin, 1985.

Mitchell, Peter. *Horse Nations: The Worldwide Impact of the Horse on Indigenous Societies Post-1492*. New York: Oxford University Press, 2015.

Mitchell, Timothy. *Rule of Experts: Egypt, Techno-Politics, Modernity*. Berkeley: University of California Press, 2002.

Moment, David. "The Business of Whaling in America in the 1850's." *Business History Review* 31 (October 1, 1957): 261–91.

Montoya, Alfredo J. *Historia de los Saladeros Argentinos*. Buenos Aires: Editorial El Coloquio, 1956.

Moon, David. *The Russian Peasantry, 1600–1930: The World the Peasants Made*. New York: Longman, 1999.

Moore, Jason W. "'The Modern World-System' as Environmental History? Ecology and the Rise of Capitalism." *Theory and Society* 32 (June 2003): 307–77.

Morgan, Jennifer L. *Laboring Women: Reproduction and Gender in New World Slavery*. Philadelphia: University of Pennsylvania Press, 2004.

Morse, Kathryn. *The Nature of Gold: An Environmental History of the Klondike Gold Rush*. Seattle: University of Washington Press, 2003.

The Movement for Black Lives. "A Vision for Black Lives: Policy Demands for Black Power, Freedom & Justice." policy.m4bl.org.

Mumford, Lewis. *Technics and Civilization*. 1934, 1963. Reprinted with a new foreword by Langdon Winner, Chicago: University of Chicago Press, 2010.

Murphy, Sharon A. "Securing Human Property: Slavery, Life Insurance, and Industrialization in the Upper South." *Journal of the Early Republic* 25 (Winter 2005): 615–52.

———. *Investing in Life: Insurance in Antebellum America*. Baltimore: Johns Hopkins University Press, 2010.

Nantucket Historical Association. "An Island in Time . . . , '1846: The Great Fire.'" *Historic Nantucket* 49 (Winter 2000): 12–38.

Nash, Linda. *Inescapable Ecologies: A History of Environment, Disease, and Knowledge*. Berkeley: University of California Press, 2006.

Nead, Lynda. *Victorian Babylon: People, Streets, and Images in Nineteenth-Century London*. New Haven: Yale University Press, 2000.

Nelson, Scott Reynolds. *Iron Confederacies: Southern Railways, Klan Violence, and Reconstruction*. Chapel Hill: University of North Carolina Press, 1999.

———. "Livestock, Boundaries, and Public Space in Spartanburg: African American Men, Elite White Women, and the Spectacle of Conjugal Relations." In *Sex, Love, Race: Crossing Boundaries in North American History*, edited by Martha Hodes, 313–27. New York: New York University Press, 1999.

———. *Steel Drivin' Man: John Henry, the Untold Story of an American Legend*. New York: Oxford University Press, 2006.

———. "Who Put Their Capitalism in My Slavery?" *Journal of the Civil War Era* 5 (June 2015): 289–310.

Nixon, Rob. *Slow Violence and the Environmentalism of the Poor*. Cambridge, Mass.: Harvard University Press, 2011.

Noble, Dennis L. *Lighthouses and Keepers: The U.S. Lighthouse Service and Its Legacy*. Annapolis: Naval Institute Press, 1997.

Nolan, Marianne, and Norman Jones. "The Cleveland Non-Explosive Lamp Company: Perkins and House's Metallic Safety Lamps." *Rushlight* 65 (December 1999): 2–17.

Norling, Lisa. *Captain Ahab Had a Wife: New England Women and the Whalefishery, 1720–1870*. Chapel Hill: University of North Carolina Press, 2000.

Novak, William J. *The People's Welfare: Law and Regulation in Nineteenth-Century America*. Chapel Hill: University of North Carolina Press, 1996.

Nye, David E. *Electrifying America: Social Meanings of a New Technology, 1880–1940.* Cambridge, Mass.: MIT Press, 1990.

———. *American Technological Sublime.* Cambridge, Mass.: MIT Press, 1996.

———. *American Illuminations: Urban Lighting, 1800–1920.* Cambridge, Mass.: MIT Press, 2018.

O'Dea, William T. *The Social History of Lighting.* New York: Macmillan, 1958.

Otter, Chris. *The Victorian Eye: A Political History of Light and Vision in Britain, 1800–1910.* Chicago: University of Chicago Press, 2008.

Outland, Robert B., III. *Tapping the Pines: The Naval Stores Industry in the American South.* Baton Rouge: Louisiana State University Press, 2004.

Pachirat, Timothy. *Every Twelve Seconds: Industrialized Slaughter and the Politics of Sight.* New Haven: Yale University Press, 2011.

Painter, Nell Irvin. *Standing at Armageddon: A Grassroots History of the Progressive Era.* New York: Norton, 1987, 2008.

Pawley, Emily. "Accounting with the Fields: Chemistry and Value in Nutriment in American Agricultural Improvement, 1835–1860." *Science as Culture* 19, no. 4 (2010): 461–82.

Peck, Gunther. "Manly Gambles: The Politics of Risk on the Comstock Lode, 1860–1880." *Journal of Social History* 26 (Summer 1993): 701–23.

———. *Reinventing Free Labor: Padrones and Immigrant Workers in the North American West, 1880–1930.* New York: Cambridge University Press, 2000.

———. "The Nature of Labor: Fault Lines and Common Ground in Environmental and Labor History." *Environmental History* 11 (April 2006): 212–38.

Perry, Percival. "The Naval-Stores Industry in the Old South, 1790–1860." *Journal of Southern History* 34, no. 4 (November 1968): 509–26.

Pohs, Henry A. *The Miner's Flame Light Book: The Story of Man's Development of Underground Light.* Denver: Flame Publishing, 1995.

Pollan, Michael. *The Omnivore's Dilemma: A Natural History of Four Meals.* New York: Penguin, 2006.

Pollock, Richard A., Ted W. Brown Jr., and David M. Rubin. "'Phossy Jaw' and 'Bis-phossy Jaw' of the 19th and the 21st Centuries: The Diuturnity of John Walker and the Friction Match." *Craniomaxillofacial Trauma & Reconstruction* 8, no. 3 (September 2015): 262–70.

Pratt, Joseph A., Martin Melosi, and Kathleen Brosnan, eds. *Energy Capitals: Local Impact, Global Influence.* Pittsburgh: University of Pittsburgh Press, 2014.

Punke, Michael. *Fire and Brimstone: The North Butte Mining Disaster of 1917.* New York: Hyperion, 2006.

Pyne, Stephen. *Vestal Fire: An Environmental History, Told through Fire, of Europe and Europe's Encounter with the World.* Seattle: University of Washington Press, 1997.

Quivik, Frederic L. "Smoke and Tailings: An Environmental History of Copper Smelting Technologies in Montana, 1880–1930." Ph.D. diss., University of Pennsylvania, 1998.

Ralph, Michael. "The Price of Life: From Slavery to Corporate Life Insurance." *Dissent* (Spring 2017).

Ransom, Roger L. *The Confederate States of America: What Might Have Been.* New York: Norton, 2005.

Rediker, Marcus. *Between the Devil and the Deep Blue Sea: Merchant Seamen, Pirates, and the Anglo-American Maritime World, 1700–1750.* New York: Cambridge University Press, 1987.

———. *Villains of All Nations: Atlantic Pirates in the Golden Age*. Boston: Beacon, 2004.

———. *The Slave Ship: A Human History*. New York: Viking, 2007.

Reilly, Kevin S. "Slavers in Disguise: American Whaling and the African Slave Trade, 1845–1862." *American Neptune* 53 (Summer 1993): 177–89.

Reséndez, Andrés. *The Other Slavery: The Untold Story of Indian Enslavement in America*. New York: Houghton Mifflin Harcourt, 2016.

Rice, Otis K. "Coal Mining in the Kanawha Valley to 1861: A View of Industrialization in the Old South." *Journal of Southern History* 31 (November 1965): 393–416.

Richards, John F. *The Unending Frontier: An Environmental History of the Early Modern World*. Berkeley: University of California Press, 2006.

Richardson, Ralph A., and General Motors Corporation. *Optics and Wheels: A Story of Lighting from the Primitive Torch to the Sealed Beam Headlamp*. Detroit: General Motors Corporation, 1940.

Robins, F. W. *The Story of the Lamp (and the Candle)*. New York: Oxford University Press, 1939.

Robinson, Armstead L. *Bitter Fruits of Bondage: The Demise of Slavery and the Collapse of the Confederacy, 1861–1865*. Charlottesville: University of Virginia Press, 2005.

Robinson, Cedric J. *Black Marxism: The Making of the Black Radical Tradition*. 1983. Reprinted with a foreword by Robin D. G. Kelley and a new preface by the author, Chapel Hill: University of North Carolina Press, 2000.

Rockman, Seth. *Scraping By: Wage Labor, Slavery, and Survival in Early Baltimore*. Baltimore: Johns Hopkins University Press, 2009.

———. "What Makes the History of Capitalism Newsworthy?" *Journal of the Early Republic* 34 (Fall 2014): 439–66.

Rogers, Nicholas. *Mayhem: Post-War Crime and Violence in Britain, 1748–53*. New Haven: Yale University Press, 2012.

Rohe, Randall. "Man and the Land: Mining's Impact in the Far West." *Arizona and the West* 28, no. 4 (Winter 1986): 299–338.

Rood, Daniel B. *The Reinvention of Atlantic Slavery: Technology, Labor, Race, and Capitalism in the Greater Caribbean*. New York: Oxford University Press, 2017.

Rosenberg, Charles E. *The Cholera Years: The United States in 1832, 1849, and 1866*. 2nd ed. Chicago: University of Chicago Press, 1987.

Rosenberg, Gabriel N. "A Race Suicide among the Hogs: The Biopolitics of Pork in the United States, 1865–1930." *American Quarterly* 68 (March 2016): 49–73.

Rosenthal, Caitlin. "Slavery's Scientific Management: Masters and Managers." In *Slavery's Capitalism: A New History of American Economic Development*, edited by Sven Beckert and Seth Rockman, 62–86. Philadelphia: University of Pennsylvania Press, 2016.

———. *Accounting for Slavery: Masters and Management*. Cambridge, Mass.: Harvard University Press, 2018.

Ross, Steven J. *Workers on the Edge: Work, Leisure, and Politics in Industrializing Cincinnati, 1788–1890*. New York: Columbia University Press, 1985.

Ryder, Karen K. "'Permanent Property': Slave Life Insurance in the Antebellum United States." Ph.D. diss., University of Delaware, 2012.

Sabin, Paul. "'A Dive into Nature's Great *Grab-bag*': Nature, Gender, and Capitalism in the Early Pennsylvania Oil Industry." *Pennsylvania History* 66 (Autumn 1999): 472–505.

Salafia, Matthew. *Slavery's Borderland: Freedom and Bondage along the Ohio River*. Philadelphia: University of Pennsylvania Press, 2013.

Salvatore, Ricardo D. "Autocratic State and Labor Control in the Argentine Pampas: Buenos Aires, 1829–1852." *Peasant Studies* 18 (Summer 1991): 251–74.

Schivelbusch, Wolfgang. *Disenchanted Night: The Industrialization of Light in the Nineteenth Century.* Translated by Angela Davies. Berkeley: University of California Press, 1995.

Schmidt, James D. "'Restless Movements Characteristic of Childhood': The Legal Construction of Child Labor in Nineteenth-Century Massachusetts." *Law and History Review* 23 (Summer 2005): 315–50.

Schrepfer, Susan R., and Philip Scranton, eds. *Industrializing Organisms: Introducing Evolutionary History.* New York: Routledge, 2004.

Scott, James C. *Domination and the Arts of Resistance: Hidden Transcripts.* New Haven: Yale University Press, 1990.

———. *Seeing Like a State: How Certain Schemes to Improve the Human Condition Have Failed.* New Haven: Yale University Press, 1998.

———. *The Art of Not Being Governed: An Anarchist History of Upland Southeast Asia.* New Haven: Yale University Press, 2009.

Scott, Julius Sherrard, III. "The Common Wind: Currents of Afro-American Communication in the Era of the Haitian Revolution." Ph.D. diss., Duke University, 1986.

Shoemaker, Nancy. *Native American Whalemen and the World: Indigenous Encounters and the Contingency of Race.* Chapel Hill: University of North Carolina Press, 2015.

Shovers, Brian Lee. "Miners, Managers, and Machines: Industrial Accidents and Occupational Disease in the Butte Underground, 1880–1920." Master's thesis, Montana State University, Bozeman, 1987.

———. "The Perils of Working in the Butte Underground: Industrial Fatalities in the Copper Mines, 1880–1920." *Montana: The Magazine of Western History* 37 (Spring 1987): 26–39.

Sicilia, David. "Selling Power: Marketing and Monopoly at Boston Edison, 1886–1929." Ph.D. diss., Brandeis University, 1991.

Silkenat, David. "Workers in the White City: Working Class Culture at the World's Columbian Exposition of 1893." *Journal of the Illinois State Historical Society* 104 (Winter 2011): 266–300.

Sinclair, Harold. *The Port of New Orleans.* Garden City, N.Y.: Doubleday, Doran, 1942.

Skaggs, Jimmy M. *The Great Guano Rush: Entrepreneurs and American Overseas Expansion.* New York: St. Martin's Press, 1994.

Sluyter, Andrew. *Black Ranching Frontiers: African Cattle Herders of the Atlantic World, 1500–1900.* New Haven: Yale University Press, 2012.

Smallwood, Stephanie E. *Saltwater Slavery: A Middle Passage from Africa to American Diaspora.* Cambridge, Mass.: Harvard University Press, 2007.

Smith, Fitz-Henry. *The Story of Boston Light: With Some Account of the Beacons in Boston Harbor.* Boston: privately printed, 1911.

Smith, Jordan. "The Invention of Rum." Ph.D. diss., Georgetown University, 2018.

Smith, Merritt Roe, and Leo Marx. *Does Technology Drive History? The Dilemma of Technological Determinism.* Cambridge, Mass.: MIT Press, 1994.

Sohn, Daniel. "The Other Side of the Story." *Rushlight* 72 (June 2006): 6–12.

Specht, Joshua. *Red Meat Republic: A Hoof-to-Table History of How Beef Changed America.* Princeton: Princeton University Press, 2019.

Stackpole, Edouard A. "Nantucket Whale Oil and Street Lighting." *Historic Nantucket* 32 (April 1985): 24–28.

Stansell, Christine. *City of Women: Sex and Class in New York, 1789–1860.* New York: Knopf, 1986.

Starling, Robert B. "The Plank Road Movement in North Carolina," pt. 1. *North Carolina Historical Review* 16 (January 1939): 1–22.

———. "The Plank Road Movement in North Carolina," pt. 2. *North Carolina Historical Review* 16 (April 1939): 147–73.

Starobin, Robert S. *Industrial Slavery in the Old South.* New York: Oxford University Press, 1970.

Stealey, John Edmund, III. "Slavery and the Western Virginia Salt Industry." *Journal of Negro History* 59 (April 1974): 105–31.

———. *The Antebellum Kanawha Salt Business and Western Markets.* Lexington: University Press of Kentucky, 1993.

Stoll, Steven. *Larding the Lean Earth: Soil and Society in Nineteenth-Century America.* New York: Hill and Wang, 2002.

Strasser, Susan. *Never Done: A History of American Housework.* New York: Pantheon Books, 1982.

Sunderland, Willard. *Taming the Wild Field: Colonization and Empire on the Russian Steppe.* Ithaca, N.Y.: Cornell University Press, 2004.

Tandeter, Enrique. "Forced and Free Labour in Late Colonial Potosí." *Past & Present* 93, no. 1 (November 1981): 98–136.

Tarbell, Ida M. *The History of the Standard Oil Company.* Vol. 1. New York: McClure, Phillips & Co., 1904.

Tatum, Margaret Black. "'Please Send Stamps': The Civil War Letters of William Allen Clark," pt. 2. *Indiana Magazine of History* 91, no. 2 (June 1995): 197–225.

Thompson, E. P. *The Making of the English Working Class.* New York: Vintage, 1963.

Threlfall, Richard E. *The Story of 100 Years of Phosphorus Making, 1851–1951.* Oldbury, England: Albright & Wilson, 1951.

Tomory, Leslie. *Progressive Enlightenment: The Origins of the Gaslight Industry, 1780–1820.* Cambridge, Mass.: MIT Press, 2012.

Tsing, Anna Lowenhaupt. *Friction: An Ethnography of Global Connection.* Princeton, N.J.: Princeton University Press, 2005.

Tutino, John. *Making a New World: Founding Capitalism in the Bajío and Spanish North America.* Durham, N.C.: Duke University Press, 2011.

Ulrich, Laurel Thatcher. *A Midwife's Tale: The Life of Martha Ballard, Based on Her Diary, 1785–1812.* New York: Vintage, 1990.

Vollmers, Gloria. "Industrial Slavery in the United States: The North Carolina Turpentine Industry, 1849–1861." *Accounting, Business & Financial History* 13, no. 3 (November 2003): 369–92.

Walker, Brett. *Toxic Archipelago: A History of Industrial Disease in Japan.* Seattle: University of Washington Press, 2010.

Walsh, Margaret. "Pork Packing as a Leading Edge of Midwestern Industry, 1835–1875." *Agricultural History* 51, no. 4 (October 1977): 702–17.

———. "The Spatial Evolution of the Mid-Western Pork Industry, 1835–75." *Journal of Historical Geography* 4, no. 1 (1978): 1–22.

————. "From Pork Merchant to Meat Packer: The Midwestern Meat Industry in the Mid Nineteenth Century." *Agricultural History* 56, no. 1 (January 1982): 127–37.

————. *The Rise of the Midwestern Meat Packing Industry.* Lexington: University of Kentucky Press, 1982.

Warner, Jessica. *Craze: Gin and Debauchery in an Age of Reason.* New York: Random House, 2002.

Warren, Louis S. *Buffalo Bill's America: William Cody and the Wild West Show.* New York: Vintage, 2005.

Way, Peter. *Common Labour: Workers and the Digging of North American Canals, 1780–1860.* New York: Cambridge University Press, 1993.

Weaver, Jace. *The Red Atlantic: American Indigenes and the Making of the Modern World, 1000–1927.* Chapel Hill: University of North Carolina Press, 2014.

West, Elliott. *The Contested Plains: Indians, Goldseekers, and the Rush to Colorado.* Lawrence: University Press of Kansas, 1998.

Whigham, Thomas. "Cattle Raising in the Argentine Northeast: Corrientes, c. 1750–1870." *Journal of Latin American Studies* 20 (November 1988): 313–35.

White, Richard. *"It's Your Misfortune and None of My Own": A New History of the American West.* Norman: University of Oklahoma Press, 1991.

————. "'Are You an Environmentalist or Do You Work for a Living?': Work and Nature." In *Uncommon Ground: Rethinking the Human Place in Nature,* edited by William Cronon, 171–85. New York: Norton, 1995.

————. *The Organic Machine: The Remaking of the Columbia River.* New York: Hill and Wang, 1995.

————. "What is Spatial History?" *The Spatial History Project,* February 1, 2010. http://www.stanford.edu/group/spatialhistory/cgi-bin/site/pub.php?id=29. Accessed April 23, 2014.

————. *Railroaded: The Transcontinentals and the Making of Modern America.* New York: Norton, 2011.

————. *The Republic for Which It Stands: The United States during Reconstruction and the Gilded Age, 1865–1896.* New York: Oxford University Press, 2017.

Whitehead, Hal. "Estimates of the Current Global Population Size and Historical Trajectory for Sperm Whales." *Marine Ecology Progress Series* 242 (October 25, 2002): 295–304.

Williams, Eric. *Capitalism and Slavery.* 1944. Reprinted with a new introduction by Colin A. Palmer. Chapel Hill: University of North Carolina Press, 1994.

Williams, Michael. *Americans and Their Forests: A Historical Geography.* New York: Cambridge University Press, 1989.

Williams, Raymond. *Marxism and Literature.* New York: Oxford University Press, 1977.

Williamson, Harold F., and Arnold R. Daum. *The American Petroleum Industry: The Age of Illumination, 1859–1899.* Evanston, Ill.: Northwestern University Press, 1959.

Winsberg, Sarah. "Making Work: Lawyers and the Boundaries of Labor, 1780–1860." Forthcoming dissertation, University of Pennsylvania.

Wyman, Mark. *Hard Rock Epic: Western Miners and the Industrial Revolution, 1860–1910.* Berkeley: University of California Press, 1979.

Yellin, Jean Fagan. *Harriet Jacobs: A Life.* New York: Basic Civitas Books, 2004.

Yergin, Daniel. *The Prize: The Epic Quest for Oil, Money, and Power.* New York: Simon & Schuster, 1991.

Zaborney, John J. *Slaves for Hire: Renting Enslaved Laborers in Antebellum Virginia*. Baton Rouge: Louisiana State University Press, 2012.

Zakim, Michael. *Ready-Made Democracy: A History of Men's Dress in the American Republic, 1760–1860*. Chicago: University of Chicago Press, 2003.

Zakim, Michael, and Gary J. Kornblith, eds. *Capitalism Takes Command: The Social Transformation of Nineteenth-Century America*. Chicago: University of Chicago Press, 2012.

Zallen, Jeremy. "'Dead Work,' Electric Futures, and the Hidden History of the Gilded Age." *Montana: The Magazine of Western History* 66 (Summer 2016): 39–65.

Index

Baldwin County, Ala., 100

Ball, Robert, 30

Ballard, Martha, 1–5, 8, 31, 262, 277n1, 278n2

Baltimore, Md., 29, 97, 98, 109, 122, 180, 182, 185, 208, 237

Baltimore and Ohio Railroad, 238, 246

Baltimore Life Insurance Company, 110–11, 113, 222–23

banking, 119–20, 292n36

Banks, J., 68

Barnes, Henry, 119

Beecher, Catherine, 252

Belfast, U.K., 169

Ben (enslaved man killed in Lewis Ruffner's coal mine), 222

Berryville, Va., 212

Big Creek, 152

Billey (enslaved collier, ran away in Blue Ridge Mountains), 217

Birdsell (captain of the *Warren*), 185

Birmingham, U.K., 169, 171, 173, 183, 204

black Americans: black lives, 26, 103–4; in coal mining, 11, 98–113, 215–26; in coal oil and petroleum industries, 233–36, 238; enslaved, 2, 4, 5, 7, 9, 11, 16, 18, 20, 26–29, 32, 55, 97–98, 113, 117–21, 133–35, 140, 143–44, 152, 155, 173–75, 211, 241, 250, 255, 270; free, 149, 155, 158, 209, 247, 250, 282n40; freedom struggles of, 10, 34–35, 55, 67, 79, 81–82, 106, 119–20, 149, 173, 176–77, 181, 210–12, 218, 220–21, 249–50, 254, 260, 264, 282n40, 297n15; on guano islands, 172, 180–83, 185; in hog industries, 149–50, 152, 155, 158; in saladeros, 171–72, 175–76, 207; in salt industry, 217–18, 236; in turpentine industry, 6, 11, 59–61, 65–89, 92, 244–46, 269; in the whale fishery, 34–35

Black Heath Pits, 222; explosion of 1839, 105–8

black radical thinkers, 297n15

Blake, John, 126

Block, Emilia, 169

Blue Ridge Mountains, 217–18

bodies and health, 17, 35, 39–40, 76, 108, 118, 177, 188–89, 207, 223, 259; reshaped by labor and light, 63–64, 77, 82, 88–89, 103, 111, 119, 142–43, 148, 169–70, 172, 184, 197–200, 203–4, 211–13, 261, 269, 308n17

bones, 2, 9, 36, 45, 101, 154, 162–63, 170; from battlefields, 171; bone ash, 179–80, 189, 212; from cattle, 171, 174–75, 177–79, 212; global trade of, 171, 179, 207; from hospitals, 171; use of in phosphorus manufacture, 11, 135, 166, 171–73, 187–89

Boone County, Ky., 149

Borden, Spencer, 256

borrowing fire, 2

Boston, Absalom, 34

Boston, Mass., 13–14, 18–19, 29, 34, 38, 39, 50, 52–53, 89, 179, 230–31, 233; Bijou Theatre, 256–59, 267; complaints about South End gasworks, 127; gas explosion at Hanover Street, 129–30; gasworks in, 97, 122–30, 134; Washington Street, 126, 259

Boston, Prince, 34

Boston Daily Atlas, 54

Boston Edison, 259

Boston Evening Transcript, 259

Boston Gas Light Company, 122, 125–27

Boston Herald, 257

Boston Light (lighthouse), 13–14, 18–19, 53

Boston-Providence Railroad, 50

Botts, George, 118–19

Breckenridge Company (coal oil manufacturer), 230, 233

Bresee, O. F., 112

British debt peons, 99

British Empire, 180, 292n16; mobilization and demobilization of seamen, 22, 25; organizing maritime movements, 19; Royal Navy, 19, 22, 66; suppression of piracy, 19, 23; trade within, 11, 18, 25–26, 29, 32, 55, 143, 165–66, 179

Broadman, Tom, 159

Brock, Eliza, 46

Brooklyn, N.Y., 60, 91, 99, 226, 250; value of camphene industry in, 90

Brown, Annie, 169

Brown, David, 21–22

Brown, John (escaped slave), 211

Brown, John (radical abolitionist), 211, 249

Brown, Vincent, 26

coal, 3, 6–7, 11–12; anthracite, 98, 231; best kinds of for coal oil, 231; best kinds of for gasworks, 98–99, 215, 225–27; bituminous, 94, 98, 116, 225, 227, 231, 290n4; British, 99, 121–22, 126, 226–27; cannel, 121, 220, 225–33, 241; coke, 97, 114–16, 123; Cumberland, 227, 301n20; dealers of, 124; geography of coal trade, 124; Kanawha, 215, 224–32; Nova Scotia, 126; Pittsburgh, 98–99, 114, 121, 225, 241, 290n4; Richmond, 98–99, 124, 225, 290n4; tariffs and prices of, 98, 121, 124, 225, 232, 241

Coal and Oil Company of Braxton County, 237

coal gaslight, 3, 7; adjustable flames for, 126; boosters for, 92, 97; as bourgeois light, 64, 98; burners for, 54, 64, 126, 131; carving out middle-class night spaces, 132; in clothing houses, 64; coin-operated gaslights, 268–69; and control over the workday, 64, 131; in Cuban sugar plantations, 54; consumer advocates for, 92; exclusion of from drawing rooms, 165; globe lamps, 126, 165; incandescent gas mantles and electric competition, 268–69; inequality of access to, 63–65, 90, 98; price of gas for, 97, 121–22; pushing working class toward camphene, 131; social justice limits of cheap gas, 122; in streetlighting, 54; in tailoring shops, 126; in theaters, 259; use of in bourgeois households, 98, 164–65, 227

coal gas systems, 6, 11, 97–99, 113–34, 215, 225–27, 229, 231, 241, 246, 270, 286n18, 290n4; in Boston, 97, 124–30; after Civil War, 247, 259, 268–69; during Civil War, 246; class power and uneven distribution, 97–98, 125–29, 131–34, 246–47, 268–69; and class struggle on shop floors, 131–32; coal consumption at gasworks, 98–99, 114, 121, 123–24, 226, 290n4; as collective lamps, 115; descriptions of as self-operating, 115–16; ecological and spatial politics of, 97, 99, 125–34; gas explosions and the risks of gaslight, 129–31; gas manufacturing

process, 97, 115; as monuments to modern age, 98; in New Orleans, 97, 113–21; in New York, 97, 121–22; in Philadelphia, 97, 122–24; private gas systems for factories, 131–32; properties of coal gas, 97; as vehicles for hierarchy, 131

—infrastructure of: condensers and scrubbers, 97; fixtures, 64, 114, 125, 130–31, 227, 246; furnaces, 114–16; gasometers and holders, 64, 97, 113–15, 125, 128, 134; lamp posts, 123, 132, 246; mains, 54, 64, 113, 117, 123, 126, 130, 134; meters, 115, 122–23, 125–27, 268–69; pipes, 64, 97, 113–15, 123, 130; retorts, 64, 97, 116, 121, 123, 126–27, 133

—as institutions: as banks, 119, 292n36; as chartered private corporations, 98–99, 119, 123, 125, 127–28; city governments and, 64, 123–24, 127, 129; government investigations of, 134; and the liberal city, 125, 132; as monopolies, 11, 125, 134, 227, 250; and the "public good," 98, 120, 124, 128, 134; relations between consumer and company, 64, 99, 115, 121–22, 125, 127–28; relations between corporation and the people, 128; role of public officials, 119–20, 123–25, 134; small business and property owners organizing against gas corporations, 126–29; suits against, 126–27

—workers in: clerks, 115, 122–23, 125–26; descriptions of workmen as "salamanders," 115–16; engineers, 97, 117, 122–23; enslaved gasworkers, 97–98, 117–21, 134; firemen, 122, 246; fitters, 123, 126, 131; inspectors, 123, 126, 131; Irish and German immigrants as, 98, 122, 126–27; lamp cleaners, 123; new construction and maintenance, 123; stokers, 97, 116–17, 123; strikes by during Civil War, 246; superintendents, 115, 117–18, 124, 126; wages and salaries of, 123, 126

coal mining (eastern Virginia), 11–12, 94–113, 217–18, 222–23; building shared regional interest in industrial slavery, 107–8; corporate ownership of enslaved

hogs, and spaces of death, 11, 135, 136–39,
143, 154–67; death work, 136, 139, 157–59,
161; "human chopping machine," 137,
139, 158; lard oil and candle factories,
143, 161–67; packinghouses, 136–37, 162;
slaughterhouses, 157–62

hogs, and spaces of life, 11, 137–58;
agricultural reformers, 138, 141, 144–47;
animals and the industrial revolution,
138; differences south and north of Ohio
River, 143–44; ecology and metabolism,
140, 142; farmers and drovers as overseers
of hogwork, 138; hog as producer
and product, 138; hogs as captive and
semicaptive workers, 137; lifework of
hogs, 137, 141, 144, 148, 153, 157, 161; new
rural system of exploitation, 137; pens, 8,
136–39, 142, 140, 144–47, 150–56, 158–59;
pigpen archipelago, 139, 149–50, 153–54,
158, 160, 167, 174, 213, 293n7; private
property and the commons, 138–40,
143–45, 153; relation to slavery, 140,
143–44, 149

—droving: assembling droves, 149–50;
chasing food across the landscape, 153;
convergence in Cincinnati, 154; depot
pens, 154–55; drovers and drivers, 149;
enslaved drivers and geographies of
freedom, 149; ferries, 153–54; final drove
through the city, 155–56; geography of
hog transportation, 150; hog hotels or
stands, 152–53; importance of sound, 151;
"lazy," resistant, and fugitive hogs, 151;
multinational and multiracial labor force
of urban drivers, 155; as post-harvest
activity, 149; railroads, 154; recruiting
helpers, 149; steamboats, 155; the struggle
between the pens, 151, 153–54; terrorist
sows, 153; torturing and disciplining hogs
up to speed, 150–51; trails and roads,
150–54

—fattening: in autumn, 140; and class,
146; at distilleries, 142; feedlots, 146,
153; field fattening, 148; geography, 146;
growers vs. fatteners, 146; importance of
corn in, 140; industrializing the pigpen,
147–48; and the new chemistry of candle

making, 142–43; race to weight, 146;
sonic conditioning with whips and calls,
146–47, 151

—raising: contesting fences, 144–45;
disciplining and training hogs, 144;
feeding strategies for, 142; fertilizing
land, 138, 142; foraging and "running at
large," 142, 145–46; "making the pig-pen
valuable," 144; moveable pigpens, 144,
148

—reproduction: breeding pens, 141;
castration, 142; crossness, 141–42;
culling, 141; disciplining sows, 141;
transforming hogs into property, 142

Honolulu, Hawaii, 46–47
horses, 29, 50, 101, 138, 146, 154, 173–74, 177–
78, 180, 209, 211, 237, 239, 248
household production, 1, 10, 57–58, 64–65,
91, 165, 202–3, 206, 230, 253–54, 269
Household Words, 175–76, 178, 181, 192, 205
humans, 4, 7–10, 20, 33, 36, 39–40, 44, 50, 55,
67, 71, 87, 89, 91–92, 110, 115–16, 120–22,
125, 134–35, 137–41, 146–47, 152–53, 158,
160, 162, 167, 171, 174, 176–77, 196, 211,
233–35, 254, 257–58, 265, 269, 279nn14–15
Hunt, Samuel, 103
Hussey & Murray (turpentine shipping
agents), 89
Hutchinson, Thomas, 25

Idaho, 263–64
Illinois, 121, 215
illumination. *See* light
imperialism, 19, 25, 53, 278n7
Inca empire, 5
Indian Ocean, 13, 30
industrialization, 6–8, 54, 65, 93, 138, 141,
147, 158, 218, 242, 255, 269
industrial revolution, 11–12, 95, 97, 138, 217
industrial slavery, 7, 105, 107–8, 119, 121, 134,
217–19, 221, 241–42, 245, 254, 269–70
industry, 10–11, 14, 22, 25–27, 30, 32–34,
55–56, 57, 60, 62–64, 66–67, 74, 87, 90,
92, 104, 110, 136, 138, 157, 161–62, 164–66,
172, 175, 179, 188–89, 203, 215, 218–19, 228,
230–33, 235–36, 240, 242–45, 248, 254–55,
257, 263, 269–70, 309n30

Massachusetts, 1, 14, 25, 30

masterlessness, 19

matches: campaigns for nontoxic matches, 204, 268; cheapness of, 11, 165, 171, 191, 205–6; global geography of production of, 171; and the industrial family, 201; inescapable ecology of phosphorus, 12, 170, 173, 190, 207, 212–13; kaleidoscopic politics of work, 171–72; Marx's vampire capitalists, 172; match stamp tax during Civil War, 250; match workers' strikes, 309n30; monopoly consolidation of industry, 250; nested hierarchies, 172; as revolutionary technology, 11, 170–71, 205–6; tug-of-war between explosions, toxicity, and profit, 205; ubiquity and accessibility of, 171, 207–10; unstable and uncontrollable power of, 173. *See also* phosphorus

—manufacture of: antebellum labor laws, 196; bells, 196, 199–200; boxing, 199–200; box making, 200–203; boys in, 168–69, 190–93, 195–201; child labor in, 168–73, 177, 190–205; class politics of age and gender, 200–201; collective action to diffuse violence of phosphorus, 198–99; "compo," "flute," or "jaw disease," 12, 170, 189, 197, 204; cutting, 199–200; dipping, 197–98; disaggregated chemistry of flame, 191; discipline, 196–97; drying, 198; evidence from European state investigations, 189; factory owners, 169, 171, 176, 200–201, 204; filling frames, 193–96; food and hunger, 196; gendered spaces, 200–201; Germans in, 193, 202–4; girls in, 168–69, 172, 190–91, 193, 199–201, 203; glowing workers, 168–71, 173, 190, 197–98, 207, 213; incredible work speeds, 193–96; Irish in, 190, 193, 202; match factories, 169–70, 172–73, 190–201; men in, 191–93, 195–97, 200–203; mixing "compo" paste, 198–99; mothers as child labor contractors, 196, 203; outwork in, 201–3; phosphorus poisoning, 12, 170; poverty of workers, 190, 195–96, 202–3; power of adults to contract out children,

196, 203; prison labor, 203–4; quotas and piecework tokens, 191; resistance to red phosphorus, 204; sandpaper, 191, 202; smell, 168–70, 173, 193, 197–98, 200, 213; spatial politics of health and safety, 200–201, 204–5; splinting, 192; spontaneous combustion of workers, 169, 198; sulfuring, 192–93; theft and moral economies, 196–97; use of phosphorus in, 169–70, 189–90, 195, 197–200; use of pine in, 169, 191–92; use of potash in, 169, 191, 195, 197, 204; use of sulfur in, 169, 190–91, 193; vapors and fumes, 190, 198, 200; ventilation, 169, 190, 200; wages and wage theft, 169, 196, 200–203; washing, 169, 198–99, 201; women in, 172, 191, 200–202; workers' struggles, allies, and raising public awareness, 189, 200–201, 203–5; worktimes and breaks, 196, 199–200

—selling of: child match peddlers, 206; elderly match peddlers, 206–7; prices and earnings, 206–7; stores, 206

—use of: accidental combustion, 207–10; as antislavery weapons, 210–11; arson, 208–9, 211; attempts to ban and regulate matches, 208; by children, 207–9; by the enslaved during Civil War, 249–50; by enslaved people, 210–12; helping poor families economize, 206; improvement over flint, steel, and tinder, 205–6; increasing fires, 208; match fires started by mice, 208; by miners, 300n55; by outworkers, 207; playing with matches, 207–9; poisonings, 207; by seamstresses, 207; by smokers, 207; by soldiers during Civil War, 249; in symbiosis with camphene, 206; weapons of the weak and powerless, 208

Maui, 35, 47

Mayhew, Henry, 207

Maysville, Ky., 229

McKay, John, 198–99

Melville, Herman, 6, 8, 13, 15, 17, 35, 44, 53, 271, 282n39; *Moby-Dick*, 6, 13

men: clothing, 57–58, 61, 65, 201, 251–52; democratization of women's

46, 250, 269; use of sound and singing in, 71; violence and the limits of white power in the camps, 76; Wilmington as principle depot for, 68–69; winter work, 70; woodsmen refusing to chip low boxes, 76; woodsmen trying to stay near families, 76. *See also* camphene

Tyson, C. B., 83

Underwriters' Association of the Northwest, 251

Union Company (coal oil manufacturer, Union, Ky.), 230

United Fire Underwriters of America, 251

United Kingdom, 29, 32, 97, 104, 106–7, 122, 168, 171, 188, 190, 203–5, 227, 230–31; bone imports in, 179; House of Commons, 122; palm oil imports in, 166; Parliament, 55, 172, 256, 280n21; tallow imports in, 165–66

United States: Congress, 3, 248; Department of State, 180, 185; Lighthouse Board, 243; Navy, 185; North American West, 62, 67, 142, 145, 255, 257; relations of North and South, 12, 60–61, 65, 99, 121, 132, 143, 241, 255, 269–70; sovereignty of guano islands, 179; Union army, 119–20, 247, 251–52

United States Life Insurance Company, 110–12

University of Virginia, 219

urban spaces, 8, 91, 131

Uruguay, 165, 171, 173, 176, 179

Uruguay River, 173

Valley Insurance Company, 110

vampires, 172

Vermont, 268

violence: chemical, 187, 192, 204; conservation of, 204; ecologies of, 59; fast, 204; gendered violence of camphene, 60, 91, 131; and the history of light, 8; and the limits of white power in turpentine camps, 76; mechanical, 192; of phosphorus, 199, 204; and the pigpen archipelago, 139, 151; resin as response to, 74; slow, 11, 204; of the state, 19, 23;

structures of, 197; temporal violence of sugar making, 26–28; of whales and whaling, 14–15, 41–45

Virginia: eastern, 219, 240; General Assembly, 220, 225, 227, 232, 237; western, 12, 114, 215–16, 218, 220, 224, 228, 230, 232, 234–36, 240–41, 270, 290n4

Virgin Islands, 179, 182

wage labor, 6, 9, 18, 92, 117, 121, 158, 187, 201, 221

Wallingford, Conn., 201

W. & H. Walker (miners' candle manufacturer), 307n16

War of Austrian Succession, 23

War of Spanish Succession, 280n21

Warren (ship), 182, 185–86

Washington, Booker T., 218, 223

Washington, D.C., 247

Washington, N.C., 68

watermen, 22, 79

weapons of the weak, 36, 208

West Africa: "legitimate trade" movement and palm oil, 165–66, 282n37; as market for spermaceti candles, 26, 32; and the slave trade, 26, 32

Western Electric Company, 266

Western Federation of Miners, 263, 265, 267, 308n22

West Indies, 2, 4, 11, 26–28, 55, 171–73, 181, 184, 207, 281n25

Westinghouse Electric Company, 264, 309n28

West Virginia, 215, 217–18, 240, 243

Whalemen's Shipping List, 55

whale oil (non-sperm, refined): as intimate light, 165; lamps for, 27–28, 99, 165; matches and increased economy and convenience of, 165; as portable light, 165; use of as heavyweight industrial lubricant, 11, 58; use of in coal mining, 99; use of in households, 58, 165; use of in streetlighting, 19; use of in sugar production, 27–28. *See also* spermaceti candles; sperm oil

whaling voyages: barrels accumulated per voyage, 49; barrels and coopers, 17, 18,

37, 38, 41, 45; barrels of oil per whale, 17, 49, 50; boat crews, 15, 41, 43; capturing and carrying tortoises, 39, 45; carpenters and coopers on board, 38; children and indentured workers, 35–36; during Civil War, 242–44; colonial American whale fishery, 29; communities and families, 20; cutting in, 15–17; dangers and destruction during war, 29; desertion, mutiny, and arson, 46, 55; difference between head and body oils, 15–16; as fugitive geography, 34–36, 46; fugitive slaves in the fishery, 34–35; gamming, 37, 47; government, discipline, and violence on ships, 45–49, 55; hunting whales, 14–15; importance of Nantucket and New Bedford, 29–30; information and mail networks, 39; the lay system, 46, 55; lengthening voyages and debates about whale populations, 44–45; livestock and produce on whaleships, 39–40, 45; living conditions on whaleships, 36, 39–40; missionaries, brothels, and bars, 47; New England Indians in the fishery, 19, 34–35; nineteenth-century U.S. whale fishery, 30; officers and organization of work, 41; outfitting and provisioning voyages, 36–37; political lives of the weak, 36; practicing and training to catch whales, 38; rats and cockroaches, 40; reaching whaling grounds, 13–14, 30, 32–39; recruitment stories for whalemen, 34–35; repairing and maintaining the ship, 38; representations of, 33–34; resupplying in the Pacific, 37, 39–40; role of Quakers in the fishery, 18, 20, 23, 29, 34; ship agents, 49–51, 54; shore leave, 46–47; trying out crude oil, 16–17, 18, 20, 25, 26, 28, 29, 33, 39, 41, 45, 49, 56, 68; uneven geography of law, 48; vulnerabilities to piracy and naval war, 19, 29; watches, 41; weather and storms, 38, 39; whaleboats, 14–15, 35, 37, 38, 41–48; whaleships as factories and vectors, 41; whales' resistance and violence, 41–45; work and dangers of sailing, 38. See also Nantucket, Mass.

White, John Edward, 172, 190–91, 193, 197–99, 201, 203

whites: fears held by, 185, 211, 220–21, 245–46, 258; on guano islands, 185; in hog trade, 143, 155, 158; middle-class, 260; nativism, 267; poor, 27, 66, 143; racial surplus wage of, 104–5, 233, 236; safety of, 104–5, 112; in slave-worked coal mines, 102–5, 112, 106; in southern gasworks, 117; southern yeomen, 143, 242; white supremacy, 137, 209, 242, 263; white terrorism, 264, 308n20; white women, 4, 60, 140, 269

Whitfield, W. G., 84, 89

wicks, 1–2, 4–5, 13, 27–28, 30–31, 52, 58–59, 61, 95, 99, 115, 164, 211, 231, 253, 277n1

William Thompson (whaleship), 48

Wilmington, Del., 250

Wilmington, N.C., 59–60, 67–69, 79, 84, 89–90, 244, 286n22, 287n24

Wilmington and Manchester Railroad, 286n22, 288n39

Wilmington and Raleigh Railroad, 245

Winifrede Mining and Manufacturing Company, 223–24

women: elderly, 206–7; elite, 7, 29; enslaved, 4, 26, 112, 149, 211, 218, 242, 255; as factory workers, 32, 172, 191, 200–201, 309n30; in labor histories, 8, 279n15; making lights, 1–2, 3, 19, 205, 277n1; middle-class, 98, 252–54; as mothers, 1–2, 9, 168–69, 196, 203, 205, 207, 218; as outworkers, 6, 57–58, 61–65, 90–93, 201–2, 251–52, 253–54, 269; poor, 7, 21, 125; as prisoners, 21, 280n20; as servants, 6–7, 11, 64, 91, 130–31, 165, 209; tending lights, 60, 252–54, 270; white, 4, 61, 140, 269; as wives, 9, 46, 69, 76, 94, 106, 119, 152, 188; workingwomen, 61, 90–91, 93, 253–54

wood, 2, 15, 17–18, 36–37, 45, 51, 61, 66, 69, 72–74, 87, 89, 100–101, 123, 130, 134, 139, 145, 159–63, 171, 175, 177, 179, 191–93, 195, 201–2, 205, 208, 213, 216, 223–24, 239, 254

woodsmen, 6, 61, 65–67, 69–72, 74–75, 77–79, 82, 87–89, 92, 134, 245

Wooldridge, Abraham S., 105, 107–8, 110, 112–13

working class, 18, 58, 62, 92, 105, 112, 124, 131, 164, 172, 201, 252–53, 260, 263, 269–70

Working Women's Protective Union, 251

workscape, 76, 222, 234, 293n7

World's Columbian Exposition: Edison Electric Tower, 266; Electricity Building, 265–66; Electric Scenic Theater, 266; fires at, 267; gilded contradictions of dead work, 265; lighting and infrastructure of, 264–65; as metastasis of Bijou spectacle, 266; Mines and Mining Building, 265; Montana Mining Exhibit, 265; "tramps" at, 267; work and workers at, 265

Young, Truman, 268

Zacatecas, 5